GUIDE TO
STATE-OF-THE-ART
ELECTRON DEVICES

GUIDE TO STATE-OF-THE-ART ELECTRON DEVICES

Edited by

Prof. Dr. Joachim N. Burghartz

Institute for Microelectronics Stuttgart (IMS CHIPS), Germany

IEEE PRESS

A John Wiley & Sons, Ltd., Publication

This edition first published 2013
© 2013, John Wiley & Sons Ltd

Registered office

John Wiley & Sons Ltd, The Atrium, Southern Gate, Chichester, West Sussex, PO19 8SQ, United Kingdom

For details of our global editorial offices, for customer services and for information about how to apply for permission to reuse the copyright material in this book please see our website at www.wiley.com.

Library of Congress Cataloging-in-Publication Data

Guide to state-of-the-art electron devices / edited by Joachim N. Burghartz.
    pages cm
  Papers by members of the IEEE Electron Devices Society.
  Includes bibliographical references and index.
  ISBN 978-1-118-34726-3 (hardback)
1. Electronic apparatus and appliances. I. Burghartz, Joachim N. II. IEEE Electron Devices Society.
  TK7870.G83 2013
  621.3815′28–dc23

                                                    2012040303

A catalogue record for this book is available from the British Library.

Print ISBN: 9781118347263

Typeset in 10/12pt Times by Laserwords Private Limited, Chennai, India.

Contents

Contents

Foreword

No one can ever doubt that electronic devices provide the basis of the way large components of our civilization operate today. This book refreshingly provides both a well-organized history and state of the art of the many technologies involved, how they are connected and most importantly, how this evolution (and frequently revolution) took place. I am overjoyed to see that the IEEE Electron Device Society has taken on this challenging undertaking and succeeded so well. The editor has made excellent choices of authors to cover such a daunting challenge. I have known many of them personally and can attest to that claim.

The initiation of the EDS and its growth has coincided with this revolution and it has been the principal organization in creating conferences and providing publications in which people can present results and interact with others working in their field. The importance of this cannot be overemphasized since the trading of ideas, not to mention competition, between people in a field provides fertilizer for new ideas. Incidentally, it is not too much of an exaggeration to say that, at conferences, as much information has been traded in the hallways as in the technical sessions. The inception of this book is a continuation of that fine tradition of publicizing information.

Although each chapter in this book covers a separate subject, they all start with a historical and tutorial mix before attacking the current state of the art. This is very beneficial to both students and experts in a given field who wish to broaden their horizons. I especially applaud the use of the blue sidebars which explain terms and concepts which require no explanation for those in the field but are enigmas to those less knowledgeable.

George E. Smith

Note: Dr. George E. Smith received the 2009 Nobel Prize in Physics for "the invention of an imaging semiconductor circuit – the CCD sensor". He is also a Celebrated Member of the IEEE Electron Devices Society.

Foreword

This book marks the 35th anniversary of the IEEE Electron Devices Society (EDS), a journey that began with the formation of the IRE Professional Group on Electron Devices 60 years ago. The major technical advancements in the field of Electron Devices are commemorated chronologically through the "video clips" at the bottom of the pages throughout the book. These clips represent snapshots of many pioneers whose aspirations and dedication led to the discovery of numerous device concepts and their implementation into practical use. This historical time line links well to the electronic booklet "50 Years of Electron Devices," which is freely available on the EDS web site.

Although the invention and development of vacuum tube devices for communications and sensing pre-dated the age of transistor, it was the advent of the solid-state "triode" in 1947, followed by the integrated circuits in the late 1950s, that has ceaselessly pushed new frontiers in computers, communications, and many other emerging areas in the several recent decades. The work described in this book has revolution-ized the way we live and the way we think. As the continuous, insatiable demand for higher performing electronics drives the search for new materials and devices, the global electron devices community will again and again respond to new challenges with novel solutions.

This book was compiled from contributions from many volunteer leaders of the Society. It is our sincere wish that its technical content will serve as a bridge between the last sexagenary cycle and the next.

Paul Yu
IEEE EDS President 2012–2013

Preface

Electronics and power electronics have grown to be indispensable technologies supporting our lifestyle, our health and our safety, social interaction and security. None of today's industries would be viable without electronic communication, automation and control. Through electronics people stay connected, get supported in their professional life, and can enjoy leisure entertainment. Transportation and energy supply depend on electronics as well. Electron devices are the foundation of electronics and power electronics. They enable all kinds of electronic signal processing and allow for switching and steering electrical energy.

This book features a concise guide to state-of-the-art electron devices. It is written by 67 specialists who are members of the IEEE Electron Devices Society (EDS). In 21 chapters they share their expert view on a particular group of electron devices or device aspects. The chapters not only illustrate the broad variety of electron device and device aspects but they are also a mirror of the diversity within the EDS. There are contributions from industry, academic and government institutions. Authors come from all five continents – a true world class team which includes the top-level industry manager as well as young engineer, the renowned university professor, and the young academic. The editor therefore tried to keep the apparent differences in style and language intact to reflect the rich diversity in regional background and affiliation.

Most of the authors are members of one of the 14 Technical Area Committees (TACs) in EDS, which assist the Executive Committee (ExCom) and Board of Governors (BoG) of EDS with their expertise in decision making and strategy processes. The current TACs in EDS are listed in Table 1.

The Institute of Electrical and Electronics Engineers (IEEE), the world's largest professional association, has its roots in the American Institute of Electrical Engineers (AIEE; founded 1884) and the Institute of Radio Engineers (IRE; founded 1912), which merged in 1963 to form the IEEE. The IRE already paid attention to the significance of electron devices by establishing an 'Electron Tube and Solid-State Devices Committee' in 1951 shortly after the invention of the transistor. In 1952 the committee's name was changed to 'IRE Professional Group on Electron Devices'. With the merger of AIEE and IRE an 'IEEE Electron Devices Group' was established in 1964, which in 1976 became the 'IEEE Electron Devices Society (EDS)'.

Today, we look back at the 35-year history of the IEEE Electron Devices Society and at its foundation in the IRE 60 years ago: two great reasons to celebrate and provide the members and potential new members of EDS with this concise guide to state-of-the-art electron devices at a very affordable price. This was made possible by substantial sponsorship through EDS and by dedicated volunteer contributions.

Table 1   Technical Area Committees of the IEEE Electron Devices Society

Compact Modeling	CM
Compound Semiconductor Devices and Circuits	CSDC
Device Reliability Physics	DRP
Electronics Materials	EM
Microelectromechanical Systems	MEMS
Nanotechnology	NT
Optoelectronic Devices	OD
Organic Electronics	OE
Photovoltaic Devices	PD
Power Devices and ICs	PDIC
Semiconductor Manufacturing	SM
Technology Computer Aided Design	TCAD
Vacuum Devices VLSI Technology and Circuits	VLSI

The book is organized in three parts, of which Part II (5 chapters) and Part III (11 chapters) are closely aligned with the TACs of EDS (see Figure 1). Part I of the book introduces in five chapters the fundamentals of electron devices. Sidebars are used in all chapters to define important figures-of-merit or definitions for a particular electron device and to offer an easy entry into the topic to the novice.

		CSDC (Chapter 14)	MEMS (Chapter 18)	NT (Chapter 21)	OD (Chapter 17)	OE (Chapter 20)	PD (Chapter 16)	PDIC (Chapter 15)	VD (Chapter 19)	VLSI (Chapter 11-13)
PART II	CM (Chapter 7)	●●	●●	●●	●●	●●	●	●	●	●●●
	DRP (Chapter 9)	●●	●●	●	●●●	●●	●●●	●●●	●●	●●●
	EM (Chapter 6)	●●●	●●	●●●	●●	●●●	●●	●●	●●	●●●
	SM (Chapter 10)	●●	●●	●	●●●	●	●●●	●●●	●●	●●●
	TCAD (Chapter 8)	●●	●●●	●	●●	●	●●	●●●	●	●●●

Figure 1   Relationship of the Technical Area Committees in the IEEE Electron Devices Society and their organization in Parts II and III of the book

A highlight of the book is the comprehensive timeline at the page bottom, featuring the historical milestones of electron devices in chronological order in three eras, after 1976, between 1952 and 1976, and before 1952. These eras mark the time periods of EDS, of electron devices in the IRE prior to EDS, and of the early electron devices, respectively.

Besides the actual contributors many people have helped 'behind the scenes' to turn the idea of this anniversary book into reality. They are duly acknowledged in the *Acknowledgments* section.

Joachim N. Burghartz

Contributors

- **David K. Abe**, Naval Research Laboratory, Washington DC, USA (Chapter 19)
- **Arman Ahnood**, Cambridge University, Cambridge, United Kingdom (Chapter 17)
- **Timothy J. Anderson**, University of Florida, Gainesville, FL, USA (Chapter 16)
- **Carter M. Armstrong**, L-3 Communications Electron Devices Inc., San Carlos, CA, USA (Chapter 19)
- **Kaustav Banerjee**, University of California, Santa Barbara, CA, USA (Chapter 11)
- **Supyrio Bandyopadhyay**, Virginia Commenwealth University, Richmond, VA, USA (Chapter 5)
- **Gennadi Bersuker**, Sematech, Austin, TX, USA (Chapter 9)
- **Richard C. Blish**, RBlish-CS, Saratoga, CA, USA (Chapter 9)
- **Fabrizio Bonani**, Politecnico di Torino, Torino, Italy (Chapter 14)
- **Joachim N. Burghartz**, Institute for Microelectronics Stuttgart (IMS CHIPS), Stuttgart, Germany (Editor; Chapter 4)
- **Marc Cahay**, University of Cincinnati, Cincinnati, OH, USA (Chapter 5)
- **Kristy A. Campbell**, Boise State University, Boise, ID, USA (Chapter 13)
- **Luigi Colombo**, Texas Instruments, Dallas, TX, USA (Chapter 10)
- **John D. Cressler**, Georgia Institute of Technology, Atlanta, GA, USA (Chapter 1)
- **Mohamed N. Darwish**, Maxpower Semiconductor Inc., Campbell, CA, USA (Chapter 15)
- **Alain Diebold**, College of Nanoscale Science and Engineering, University at Albany, Albany, NY, USA (Chapter 10)
- **Robert Doering**, Texas Instruments, Dallas, TX, USA (Chapter 10)
- **David Esseni**, University of Udine, Udine, Italy (Chapter 8)
- **Giovanni Ghione**, Politecnico di Torino, Torino, Italy (Chapter 14)
- **Avik Ghosh**, University of Virginia, Richmond, VA, USA (Chapter 5)
- **Thomas Grant**, Communication & Power Industries LLC, Palo Alto, CA, USA (Chapter 19)
- **Martin A. Green**, University of New South Wales, Sydney, Australia (Chapter 16)
- **Xiaojun Guo**, Shanhai Jiao Tong University, Shanghai, China (Chapter 17)
- **Wilfried Haensch**, IBM Co. Yorktown Heights, NY, USA (Chapter 21)
- **James S. Harris**, Stanford University, Stanford, CA, USA (Chapter 6)
- **James A. Hutchby**, Semiconductor Research Corporation, Research Triangle Park, NC, USA (Chapter 12)
- **Shuji Ikeda**, Tei Solutions Inc., Onogawa Tsukuba Ibaraki, Japan (Chapter 11)
- **Hiroshi Iwai**, Tokyo Institute of Technology, Yokohama, Japan (Chapter 2)
- **Chennupati Jagadish**, Australian National University, Canberra, Australia (Chapter 20)
- **Dong Jin Jung**, Samsung Electronics Co. Ltd, Gyunggi-Do, South Korea (Chapter 3)
- **Christoph Jungemann**, RWTH Aachen, Aachen, Germany (Chapter 8)
- **Erich Kasper**, University of Stuttgart, Stuttgart, Germany (Chapter 14)
- **Lu Kasprzak**, Siemens Healthcare Diagnostics Inc., Newark, DE, USA (Chapter 9)
- **Hanseup Kim**, University of Utah, Salt Lake City, UT, USA (Chapter 18)

- **Kinam Kim**, Samsung Electronics Co. Ltd, Gyunggi-Do, South Korea (Chapter 3)
- **Jackson Lai**, Research in Motion Inc., Waterloo, Ontario, Canada (Chapter 17)
- **Theodore J. Letavic**, IBM Co., Hopewell Junction, NY, USA (Chapter 15)
- **Baruch Levush**, Naval Research Laboratory, Washington DC, USA (Chapter 19)
- **Jürgen Lorenz**, Fraunhofer Institute for Integrated Systems and Device Technology, Erlangen, Germany (Chapter 8)
- **Leda Lunardi**, North Carolina State University, Raleigh, NC, USA (Chapter 20)
- **Colin C. McAndrew**, Freescale Semiconductor Inc., Tempe, AZ, USA (Chapters 4 and 7)
- **William L. Menninger**, L-3 Communications Electron Technologies Inc., Torrance, CA, USA (Chapter 19)
- **Sudha Mokkapati**, Australian National University, Canberra, Australia (Chapter 20)
- **Chandra Mouli**, Micron Technology, Boise, ID, USA (Chapter 13)
- **Laurence W. Nagel**, Omega Enterprises Consulting, Kensington, CA, USA (Chapter 7)
- **Arokia Nathan**, Cambridge University, Cambridge, United Kingdom (Chapter 17)
- **Anthony S. Oates**, TSMC Co. Ltd, Hsinchu, Taiwan (Chapter 9)
- **Mikael Östling**, KTH, Stockholm, Sweden (Chapter 15)
- **Pierpaolo Palestri**, University of Udine, Udine, Italy (Chapter 8)
- **Stephen Parke**, Northwest Nazarene University, Nampa, ID, USA (Chapter 13)
- **Ruediger Quay**, Fraunhofer Institute of Applied Solid-State Physics, Freiburg, Germany (Chapter 14)
- **Ken Rim**, IBM Corporation, Hopewell Junction, NY, USA (Chapter 6)
- **Steven A. Ringel**, Ohio State University, Columbus, OH, USA (Chapter 16)
- **Enrico Sangiorgi**, University of Bologna, Cesena, Italy (Chapter 8)
- **Klaus Schuegraf**, Applied Materials, San Jose, CA, USA (Chapter 10)
- **Luca Selmi**, University of Udine, Udine, Italy (Chapter 8)
- **Rajendra Singh**, Clemson University, Clemson, SC, USA (Chapters 10 and 16)
- **George E. Smith**, Nobel Laureate and EDS Celebrated Member, USA (Foreword)
- **James C. Sturm**, Princeton University, Princeton, NJ, USA (Chapter 6)
- **Simon Min Sze**, National Chiao Tung University, Hsinchu, Taiwan (Chapter 2)
- **Yuan Taur**, University of California at San Diego, San Diego, CA, USA (Chapter 2)
- **Robert J. Walters**, Naval Research Laboratory, Washington DC, USA (Chapter 16)
- **Katsuyoshi Washio**, Tohoku University, Sendai, Japan (Chapter 1)
- **Richard K. Williams**, Advanced Analogic Technologies Inc., Santa Clara, CA, USA (Chapter 15)
- **Hei Wong**, City University of Hong Kong, Hong Kong, China (Chapter 2)
- **Chung-Chih Wu**, National Taiwan University, Taipei, Taiwan (Chapter 6)
- **Darrin J. Young**, University of Utah, Salt Lake City, UT, USA (Chapter 18)
- **Paul Yu**, University of California at Sn Diego, San Diego, CA, USA, and EDS President 2012–2013 (Foreword)
- **Bin Zhao**, Fairchild Semiconductor, Irvine, CA, USA (Chapter 12)

Acknowledgments

The editor wishes to acknowledge the EDS Executive Committee (ExCom) and the EDS Board of Governors (formerly the Advisory Committee (AdCom)) for supporting this anniversary book project and their vote for financial sponsorship, making it possible to offer this book to EDS members at a very low price. **Christopher Jannuzzi** (EDS Executive Director) stepped in with hands-on support and helped at various ends, particularly in working with the **IEEE History Center**, New Brunswick, New Jersey from where we received numerous images used in the historical timeline.

The editor, on behalf of all contributors, considers it a great honor that this book is introduced by a foreword of **Nobel Laureate George E. Smith**. His kind words helped us to complete this ambitious project.

Joachim Deh from IMS CHIPS in Stuttgart helped with editing some of those and other images used in the timeline and in Chapter 4 and with arranging copyright permissions for such images. **Fernando Guarin** (EDS Secretary) carried out a critical review of parts of the book prior to the copyediting by the publisher. **Peter Mitchell, Laura Bell, Liz Wingett, Richard Davies,** and **Genna Manaog** from John Wiley & Sons as well as **Sangeetha Parthasarathy** from Laserwords Pvt Ltd. are acknowledged for arranging a smooth copyediting and publication process; particular thanks goes to **Peter Mitchell** for his patience in negotiating the terms of the publication contracts. Advance acknowledgments go to **Bin Zhao** (EDS Vice President Meetings), **Xing Zhou** (EDS Vice President Regions & Chapters), **Jean Bae** (EDS Executive Office), again **Chris Jannuzzi**, and all organizers of EDS-sponsored meetings and EDS chapter chairs involved in distributing the book within the EDS community.

The authors of Chapter 6 gratefully acknowledge **Yi-Hsiang Huang** of National Taiwan University and **Noah Jafferis** of Princeton for expert assistance with graphics.

The authors of Chapter 7 give credit to **Gennady Gildenblat**, professor at Arizona State University, who provided a great amount of support to that chapter without being a co-author. His credentials in the field of compact modelling are reflected in the historical timeline. The authors of Chapter 8 gratefully acknowledge **Claudio Fiegna,** University of Bologna, for many helpful discussions.

The authors of Chapter 15 would like to thank **Dr. Phil Krein** and **Professor Emeritus Nick Holonyak** of the ECE Department of the University of Illinois at Urbana-Champaign for their assistance in identifying records and references chronicling early developments in power semiconductors.

The authors of Chapter 16 would like to extend a special acknowledgement to **Dr. Tyler Grassman** of The Ohio State University for his key contributions to both the technical content and integration of information throughout the chapter.

The authors of Chapter 19 gratefully acknowledge the valuable contributions from **Heinz Bohlen** (Communications and Power Industries, USA, retired), **Ernst Bosch** (Thales Electron Devices GmbH, Germany), **Gregory Nusinovich** (Institute for Research in Electronics and Applied Physics, University of Maryland, College Park, USA), and **Edward Wright** (Beam-Wave Research Inc., USA).

Introduction: Historic Timeline

Electron devices go back a long time: from the vacuum tube to the inventions of the transistor and the integrated circuit, through 40 years of scaling microelectronics to the exciting possibilities brought about by the current investigations on emerging research devices. The history of electron device applications is also captured here. Looking back in time means learning from the past so that future progress can be made more efficiently. It also means taking pride in the pioneers' achievement and viewing them as role models empowering young electron device engineers.

The historic timeline that runs as a movie strip at the bottom of the pages in chronological order through this entire anniversary book marks key milestones of electron device development and applications in more than 1000 slides. Landmarks of world history and other technical breakthroughs place those milestones into historical perspective. The book can, thus, be read in two ways; chapter-by-chapter, or along the timeline of device history.

Part I

BASIC ELECTRON DEVICES

| 1623 | Wilhelm Schickard invents the calculator | ELECTRON DEVICES SOCIETY | | |

Chapter 1

Bipolar Transistors

John D. Cressler and Katsuyoshi Washio

1.1 Motivation

In terms of its influence on the development of modern technology and hence, global civilization, the invention of the point contact transistor on December 23, 1947 at Bell Labs in New Jersey by Bardeen and Brattain was by any reckoning a watershed moment in human history [1]. The device we know today as a bipolar junction transistor was demonstrated four years later in 1951 by Shockley and co-workers [2] setting the stage for the transistor revolution. Our world has changed profoundly as a result [3].

Interestingly, there are actually seven major families of semiconductor devices (only one of which includes transistors!), 74 basic classes of devices within those seven families, and another 130 derivative types of devices from those 74 basic classes (Figure 1.1) [4]. Here we focus only on three basic devices: (1) the *pn* homojunction junction diode (or *pn* junction or diode), (2) the homojunction bipolar junction transistor (or BJT), and (3) the special variant of the BJT called the silicon-germanium heterojunction bipolar transistor (or SiGe HBT). As we will see, diodes are useful in their own right, but also are the functional building block of all transistors.

Surprisingly, all semiconductor devices can be built from a remarkably small set of materials building blocks (Figure 1.2), including [4]:

- the metal–semiconductor interface (e.g., Pt/Si; a "Schottky barrier")
- the doping transition (e.g., a Si *p*-type to *n*-type doping transition; the *pn* junction)
- the heterojunction (e.g., *n*-AlGaAs/*p*-GaAs)
- the semiconductor/insulator interface (e.g., Si/SiO$_2$)
- the insulator/metal interface (e.g., SiO$_2$/Al).

Guide to State-of-the-Art Electron Devices, First Edition. Edited by Joachim N. Burghartz.
© 2013 John Wiley & Sons, Ltd. Published 2013 by John Wiley & Sons, Ltd.

1650   1675   ELECTRON DEVICES SOCIETY®   1700   1725

The Transistor Food Chain

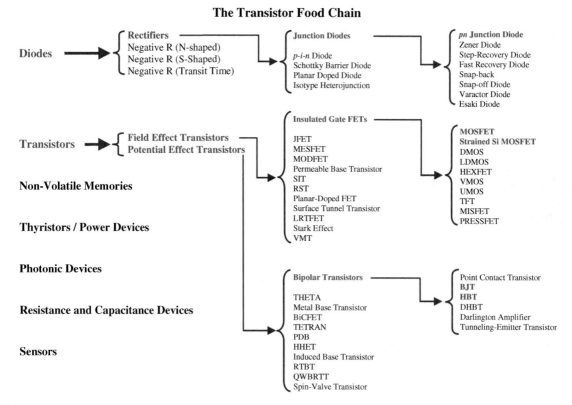

Figure 1.1 The transistor "food chain" showing all major families of semiconductor devices. Reproduced with permission from Cressler, J. D.; *Silicon Earth: Introduction to the Microelectronics and Nanotechnology Revolution*; 2009, Cambridge University Press

Why do we actually need transistors in the first place? Basically, because nature attenuates all electrical signals. By this we mean that the magnitude of all electrical signals (think "1s" and "0s" inside a computer, or an EM radio signal from a cell phone) necessarily decreases as it moves from point A to point B, something we call "loss". When we present an (attenuated) input signal to the transistor, the transistor is capable of creating an output signal of larger magnitude (i.e., "gain"), and hence the transistor serves as a "gain block" to "regenerate" (recover) the attenuated signal in question, an essential concept for electronics. In the electronics world, when the transistor is used as a source of signal gain, we refer to it as an "amplifier." Amplifiers are ubiquitous to all electronic systems.

Naming of the Transistor

The name "transistor" was actually coined by J.R. Pierce of Bell Labs, following an office betting pool which he won. He started with a literal description of what the device actually does electronically, a "transresistance amplifier," which he first shortened to "trans-resistor," and then finally "transistor" [3].

1745   E. von Kleist and P. van Musschenbroek invent the capacitor (Leyden Bottle)   **E**LECTRON **D**EVICES **S**OCIETY

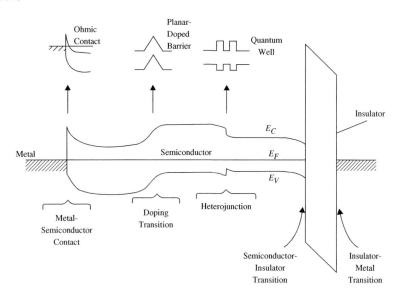

Figure 1.2 The essential building blocks of all semiconductor devices. Reproduced with permission from Cressler, J. D.; *Silicon Earth: Introduction to the Microelectronics and Nanotechnology Revolution*; 2009, Cambridge University Press

Not only can the transistor serve as a wonderful nanoscale sized amplifier, but importantly it can also be used as a tiny "regenerative switch"; meaning, an on/off switch that does NOT have loss associated with it. Why is this so important? Well, imagine that the computational path through a microprocessor requires 1 000 000 binary switches (think light switch on the wall – on/off, on/off) to implement the complex digital binary logic of a given computation. If each of those switches even contributes a tiny amount of loss (which it inevitably will), multiplying that tiny loss by 1 000 000 adds up to unacceptably large system loss. That is, if we push a logical "1" or "0" in, it rapidly will get so small during the computation that it gets lost in the background noise. If, however, we implement our binary switches with gain-enabled transistors, then each switch is effectively regenerative, and we can now propagate the signals through the millions of requisite logic gates without excessive loss, maintaining their magnitude above the background noise level.

In short, the transistor can serve in one of two fundamental capacities: (1) an amplifier or (2) a regenerative switch. Amplifiers and regenerative switches work well only because the transistor has the ability to produce gain. So a logical question becomes, where does transistor gain come from? To answer this, first we need to understand *pn* junctions.

1.2 The *pn* Junction and its Electronic Applications

Virtually all semiconductor devices (both electronic and photonic) rely on *pn* junctions (a.k.a., "diodes", a name which harkens back to a vacuum tube legacy) for their functionality. The simplest embodiment of a *pn* junction is the *pn* "homojunction", meaning that within a single piece of semiconductor (e.g., silicon – Si) we have a transition between p-type doping and n-type doping (e.g., p-Si/n-Si). The opposite would be

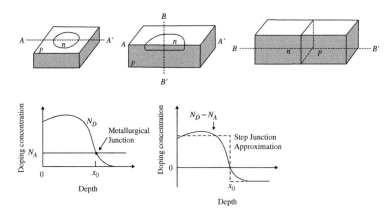

Figure 1.3 Cartoons of a pn junction, showing doping transition from n-type to p-type. Reproduced with permission from Cressler, J. D.; *Silicon Earth: Introduction to the Microelectronics and Nanotechnology Revolution*; 2009, Cambridge University Press

a *pn* heterojunction, in which the p-type doping is within one type of semiconductor (e.g., p-GaAs), and the n-type doping is within another type of semiconductor (e.g., n-AlGaAs).

As shown in Figure 1.3, to build a *pn* junction we might, for instance, ion implant and then diffuse *n*-type doping into a *p*-type wafer. The important thing is the resultant "doping profile" as one moves through the junction ($N_D(x) - N_A(x)$, which is just the net doping concentration). At some point in the doping transition, $N_D = N_A$, and we thus have a transition between net n-type and net p-type doping. This point is called the "metallurgical junction" (x_0 in Figure 1.3) and all of the important electrical action of the junction is centered here. To make the physics easier, two simplifications are typically made: (1) Let us assume a "step junction" approximation to the real *pn* junction doping profile, which is just what it says, an abrupt change (a step) in doping occurring at the metallurgical junction (Figure 1.3). (2) Let us assume that all of the dopant impurities are ionized (one donor atom equals one electron, etc., an excellent approximation for common dopants in silicon at 300 K).

So, how does a *pn* junction actually work? The operation of ALL semiconductor devices is best understood at an intuitive level by considering the energy band diagram, which plots electron and hole energy as a function of position as we move physically through a device. An n-type semiconductor is electron rich (i.e., majority carriers), and hole poor (i.e., minority carriers). Conversely, a p-type semiconductor is hole-rich and electron-poor. If we imagine bringing an n-type and p-type semiconductor into "intimate electrical contact" where they can freely exchange electrons and/or holes from *n* to *p* and *p* to *n*, the final equilibrium band diagram shown in Figure 1.4 will result. Note, that under equilibrium conditions, there is no NET current flow across the junction.

We might logically wonder what actually happened inside the junction to establish this equilibrium condition. When brought into contact, the *n*-type side of the junction is electron rich, while the *p*-type side is electron poor. That is, there is a large driving force for electrons to diffuse from the *n* region to the *p* region. Recall, that there are in fact two ways to move charge in a semiconductor: (1) drift, whose driving force is the electric field (voltage/length), and (2) diffusion, whose driving force is the carrier density gradient (change in carrier density per unit distance). The latter process is what is operative here.

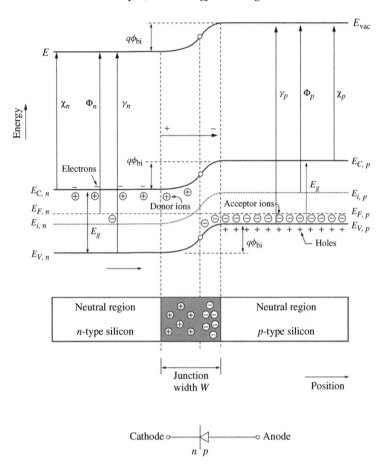

Figure 1.4 Energy band diagram of a pn junction at equilibrium. Reproduced with permission from Cressler, J. D.; *Silicon Earth: Introduction to the Microelectronics and Nanotechnology Revolution;* 2009, Cambridge University Press

Once in electrical contact an electron moves from the *n*-side to the *p*-side, leaving behind a positively charged donor impurity (N_D^+). Note, that far away from the junction, for each charged donor impurity there is a matching donated electron, hence the semiconductor is charge neutral. Once the electron leaves the *n*-side, however, there is no balancing charge, and a region of "space charge" results. The same thing happens on the *p*-side. Hole moves from *p* to *n*, leaving behind an uncompensated acceptor impurity (N_A^-) behind. This resultant charge "dipole" produces an electric field, pointing from $+$ to $-$ (to the right in this case). How does that induced field affect the diffusion-initiated side-to-side transfer of charge just described? It opposes the diffusive motion of both electron and holes via Coulomb's law. Therefore, in a *pn* junction the diffusion gradient moves electrons from *n* to *p* and holes from *p* to *n*, but as this happens a dipole of space charge is created between the uncompensated ionized dopants, and an induced electric field opposes the further diffusion of charge. When does equilibrium in the *pn* junction result? When the diffusion and the drift processes are perfectly balanced and the net current density is zero.

The *pn* junction in equilibrium consists of a neutral *n* region and a neutral *p* region, separated by a space charge region of width *W*. This structure forms a capacitor (conductor/insulator/conductor), and *pn* junctions have built-in capacitance which will partially dictate their switching speed. The electric field in the space charge region (for a step junction) is characteristically triangular shaped, with some peak value of electric field present. There is a built-in voltage drop across the junction, and, thus, from the energy band diagram we see that there is a potential barrier for any further movement of electrons and holes from side-to-side. This barrier to carrier transport maintains a net current density of zero, and the junction is by definition in equilibrium.

If one wanted to get current flowing again across the junction, how would this be done? Well, we must unbalance the drift and diffusion mechanisms by lowering the potential barrier to the electron and hole transport, and we can do this trivially by applying an external voltage to the *n* and *p* regions such that the *p* region (anode) is more positively biased than the *n* region (cathode). As shown in Figure 1.5, this effectively lowers the side-to-side barrier, drift no longer balances diffusion, and the carriers will once again start diffusing from side-to-side, generating useful current flow. This is called "forward bias". What happens if we apply a voltage to the junction of opposite sign? (i.e., *p* region more negatively biased than the *n* region). Well, the barrier the carriers experience grows, effectively preventing any current flow, a condition called "reverse bias" (Figure 1.5).

The *pn* junction thus forms a solid-state switch (a.k.a. the "diode"). Consider: Apply a voltage of one polarity and current flows. Apply a voltage of the opposite polarity and no current flows; an on/off switch. Shockley shared the Nobel Prize with Bardeen and Brattain largely for explaining this phenomenon, and of course by wrapping predictive theory around it which led to the demonstration of the BJT. The result of that particularly elegant derivation is the celebrated "Shockley equation" which governs the current flow in a *pn* junction

$$I = qA \left\{ \frac{D_n n_i^2}{L_n N_A} + \frac{D_p n_i^2}{L_p N_D} \right\} \left(e^{qV/kT} - 1 \right) = I_S \left(e^{qV/kT} - 1 \right) \tag{1.1}$$

where *A* is the junction area, *V* is the applied voltage, $D_{n,p}$ is the electron/hole diffusivity ($D_{n,p} = \mu_{n,p} kT$), $L_{n,p}$ is the electron/hole diffusion length, and I_S is the junction "saturation current" which collapses all of these factors into a single (measurable) parameter.

Observe, that all of the parameters in the Shockley equation refer to the minority carriers. If we build our junction with the *n* and *p* doping the same, then the relative contributions of the electron and hole minority carrier currents to the total current flowing will be comparable (to first order). Let us look closer at the operation of the junction. Under forward bias, electrons diffuse from the *n*-side to the *p*-side, where they become minority carriers. Those minority electrons are now free to recombine and will do so, on a length scale determined by L_n, and thus as we move from the center of the junction out into the neutral *p*-region, the minority electron population decreases due to recombination, inducing a concentration gradient as we move to the *p*-side, which drives a minority electron diffusion current. The same thing is happening with holes on the opposite side of the junction, and these two minority carrier diffusion currents add to produce the total forward bias current flow. What is the actual driving force behind the forward bias current in a *pn* junction? Recombination in the neutral regions, since recombination induces the minority diffusion currents. Alas, simple theory and reality are never coincident, and there a finite limits to the voltages that

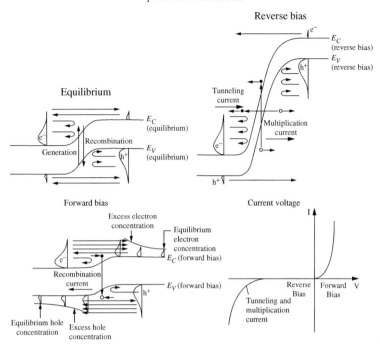

Figure 1.5   The pn junction under both forward and reverse bias, showing the resultant current–voltage characteristics. Reproduced with permission from Cressler, J. D.; *Silicon Earth: Introduction to the Microelectronics and Nanotechnology Revolution*; 2009, Cambridge University Press

can be applied to the diode, and how much current can be passed through it and how much voltage can be applied across it [3].

So, what makes the junction so useful? Well, as stated, it makes a nice on/off switch with low loss when forward biased, and it can provide very good electrical isolation when reverse biased. In power electronics the diode would be said to provide a "blocking" voltage, not allowing current flow in reverse bias up to some finite, and often huge, applied reverse voltage (hundreds to even thousands of volts). This is very useful. The diode can also function as a wonderful solid-state "rectifier". Rectifiers are ubiquitous in power generation, conversion, and transmission, (e.g., to turn AC voltage into DC voltage). Finally, the diode can also emit and detect light, which is also extremely useful as a transducer for converting optical to electrical energy, and vice versa (see Chapters 16 and 20).

All of this said, however, the diode does NOT possess gain, and, thus, is insufficient for realizing complex electronic systems. From a transistor perspective, however, the *pn* junction can be used to make a tunable minority carrier injector, which, if cleverly employed, can indeed produce gain when carefully implemented within a transistor. Importantly, one can trivially skew the relative magnitudes of the minority carrier injection from side-to-side in a *pn* junction by making the doping levels on one side of the junction much more heavily doped than on the other side. Let us imagine that the *n*-doping is far larger than the *p*-doping. Fittingly, this is referred to as a "one-sided" junction. In this scenario, it can be easily shown that electrons make up most of the total current flow in forward bias in such a junction. If we wanted to use a *pn* junction under forward bias to enhance the "forward-injection" of electrons into the *p*-region, and suppress the "back-injection" of holes into the n-region, we could simply use an $n^{++} - p^-$ junction as an "electron injector"! This will lead us directly to the BJT, a transistor with gain.

1.3 The Bipolar Junction Transistor and its Electronic Applications

The *pn* junction, as a two-terminal object, can be made to serve as an efficient minority carrier injector, but it does NOT possess inherent gain. This is the fundamental reason why we do not build microprocessors from diode-resistor logic. Diodes make excellent binary switches, but without a gain mechanism to overcome Nature's preference for attenuation, complex functions are not going to be achievable in practice. Let us imagine, however, that we add an additional third terminal to the device which somehow controls the current flow between the original two terminals. Let terminal 1 = the input "control" terminal, and terminals 2 and 3 have high current flow between them when biased appropriately by the control terminal. Then, under the right bias conditions, with large current flow between 2 and 3, if we could somehow manage to suppress the current flow to/from 1, we'd be in business. That is, small input current (1) generates large output current (from 2 to 3), and hence we have gain!

How do we do this in practice? Let us use two *pn* junctions, placed back-to-back, such that the control terminal (our #1; which we will call the "Base" terminal – B) is in the central *p* region, and the two high current flow path output terminals (our #2 and #3, which we will call the "Emitter" and "Collector" terminals – E, and C), are the two outside *n* regions (see Figure 1.6). Since the two central *p* regions are shared by both diodes, those can be coincident. That is, an *n* region separated from another *n* region by an intermediate *p* region actually contains two *pn* junctions.

BJT versus FET

At a deep level, the BJT and the FET are closely related devices. Both have two *pn* junctions which are integral to their functionality. In an FET, a "gate" electrode is capacitively coupled (through the gate oxide) to the charge conduction path, altering the current flow from source to drain. In the BJT, the "base" electrode is directly tied to the charge conduction path, altering the current flow from emitter to collector. Thus, the differences between BJTs and FETs lie with the how the control terminal is electrically tied to the charge conduction path.

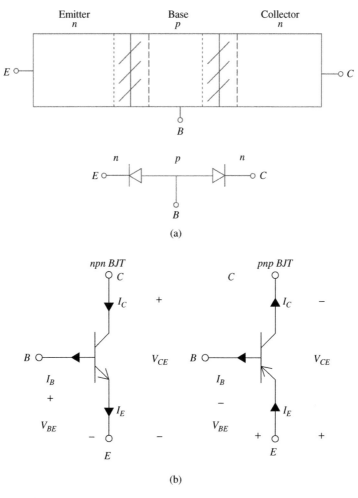

Figure 1.6   (a) Schematic of the two back-to-back pn junctions that form a bipolar junction transistor; (b) the circuit symbol of both doping polarity types are also shown. Reproduced with permission from Cressler, J. D.; *Silicon Earth: Introduction to the Microelectronics and Nanotechnology Revolution*; 2009, Cambridge University Press

Let us imagine forward biasing the emitter–base junction, and reverse biasing the collector–base junction, and then adding two more puzzle pieces: (1) We must dope the emitter very heavily with respect to the base, such that when we forward bias the emitter–base junction we have large electron flow from E to B and simultaneously suppress the hole flow from B to E (this is our tunable minority carrier injector!). (2) We must make the central base region VERY thin. Why? Well, if we don't, then the electrons injected from E to B will simply recombine in the base before they can reach the collector (to be collected and to generate the required large output current flow from E to C). Recall that the rough distance a minority carrier can travel before it recombines is given by the diffusion length ($L_{n,p}$). Clearly, we need the width of the p-type base region to be much, much less than this number; in practice, a few hundred nm is required for a modern BJT. The final result? We have created the *npn* BJT! (One could of course swap the doping polarities n to p and p to n and achieve the same result – a *pnp* BJT. We thus have two flavors of BJT, and this is often VERY handy in electronic circuit design.

Consider now how the BJT actually works: (1) The reverse-biased CB junction has negligible current flow. (2) The forward-biased EB junction injects (emits) lots of electrons from E to B, that diffuse across the base without recombining (because it is thin) and are collected at C, generating large electron flow from E to C (current). BUT, due to the doping asymmetry in the EB junction, while a large number of electrons get injected from E to B, very few holes flow from B to E. Forward electron current is large, but reverse hole current is small. That is: small input base current; large output collector. Gain! This is otherwise known in electronics as "current gain" (or β).

How do we make the BJT? Well, as might be imagined it is more complex than a *pn* junction, but even so, the effort is worth it. Figure 1.7 shows the simplest possible variant. Figure 1.7 also superposes both the equilibrium and forward-active bias energy band diagrams, with the carrier minority and majority carrier distributions, to help connect the *pn* junction physics to the BJT operation. Within the band diagram context, here is intuitively how the BJT works. In equilibrium, there is a large barrier for injecting electrons from the emitter into the base. Forward bias the EB junction and reverse bias the CB junction, and now the EB barrier is lowered, and large numbers of electrons are injected from E to B. Since B is very thin, and the CB junction is reverse biased, these injected electrons will diffuse across the base, slide down the potential hill of the CB junction, and be collected at C, where they generate a large electron current flow from E to C. Meanwhile, due to the doping asymmetry of the EB junction, only a small density of holes is injected from B to E to support the forward bias EB junction current flow. Hence, I_C is large, and I_B is small. Gain! A different visualization of the magnitudes of the various current contributions in a well-made, high gain, BJT, are illustrated in Figure 1.8.

Shockley's theory to obtain an expression for β is fairly straightforward from basic *pn* junction physics (although you have two different ones to contend with obviously), provided you make some reasonable assumptions on the thickness of the base (base width $W_b \ll L_{nb}$). For the output and input currents under forward-active (amplifier) bias, we obtain:

$$I_C \cong qA \left\{ \frac{D_{nb}n_i^2}{W_b N_{Ab}} \right\} e^{qV_{BE}/kT} = I_{CS} e^{qV_{BE}/kT} \tag{1.2}$$

$$I_B \cong qA \left\{ \frac{D_{pe}n_i^2}{L_{pe} N_{De}} \right\} e^{qV_{BE}/kT} = I_{BS} e^{qV_{BE}/kT} \tag{1.3}$$

where the "b" and "e", or "B" and "E", subscripts stand for base and emitter, respectively. Interestingly, the current gain does not to first-order depend on bias voltage, the size of the junction, or even the bandgap! We finally obtain,

$$\beta \cong \frac{I_C}{I_B} = \frac{I_{CS}}{I_{BS}} \cong \left\{ \frac{D_{nb}L_{pe}N_{De}}{D_{pe}W_b N_{Ab}} \right\} \tag{1.4}$$

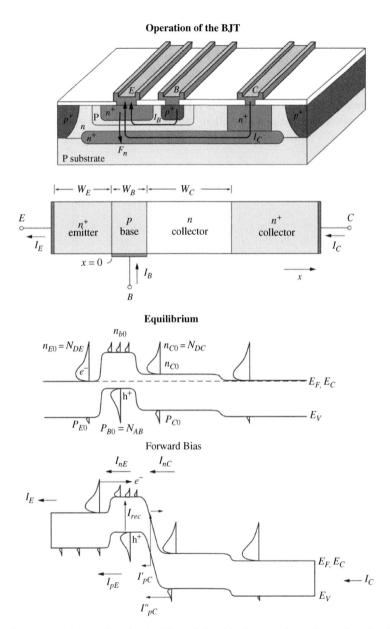

Figure 1.7 Basic structure and operational principles of the bipolar transistor. Reproduced with permission from Cressler, J. D.; *Silicon Earth: Introduction to the Microelectronics and Nanotechnology Revolution*; 2009, Cambridge University Press

Clearly, the current gain is a tunable parameter, giving us great flexibility in design. A common way to plot the BJT current–voltage characteristics is shown in Figure 1.8, where linear I_C is plotted versus linear V_{CE}, as a further function of I_B. Since I_C is larger than I_B, the gain is implicit here. This plot is known as the output "family" or "output characteristics". We use the output family to define the three regions of operation of the BJT: (1) "forward-active" (EB junction forward-biased; CB junction reverse-biased); (2) "saturation" (both EB and CB junctions forward-biased), and (3) "cut-off" (both EB and CB junctions reverse biased). As indicated, forward-active bias is typically for amplifiers, and as we will see, switching between cutoff and saturation will make an excellent regenerative digital switch!

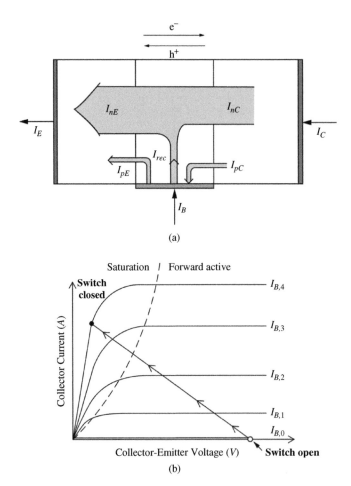

Figure 1.8   Sketch of (a) the relative current contributions of the bipolar transistor and (b) the resultant current–voltage characteristics. Reproduced with permission from Cressler, J. D.; *Silicon Earth: Introduction to the Microelectronics and Nanotechnology Revolution*; 2009, Cambridge University Press

How fast can transistors switch states (on to off)? The current speed record for a bipolar transistor digital switch is less than 10 picoseconds (0.000000000010 seconds – 10 trillionths of a second!). What limits that speed? Intuitively, the speed is limited by the time it takes the electrons to be injected from the emitter, transit (diffuse across) the base, and then be collected by the collector. In other words, a transistor can't be faster than it takes the charge to move through it. In most transistors, step two is the limiting one, and the so-called "base transit time" (τ_b) sets the fundamental speed limit on how fast the BJT can switch. A first-order base transit time expression can be easily derived,

$$\tau_b \cong \frac{W_b{}^2}{2D_{nb}} \tag{1.5}$$

Hence, the smaller τ_b is, the faster the BJT can switch. Clearly, making W_b as small as possible gives us a double benefit. It helps increase the current gain, yes, but even more importantly, it makes the transistor faster – quadratically!

So what does the BJT do for us? Let's restate some points for clarity. This beautiful three-terminal semiconductor device, if constructed correctly, will exhibit a (tunable) gain. Gain is the key to success in building any electronic system; hence the deserved fame of the BJT. This intrinsic gain will allow us to create a wide variety of amplifiers for use in a myriad of electronics applications. Amplifiers that take: (1) A small input current and turn it into a large output current (a.k.a., a "current amplifier"); (2) a small input voltage and turn it into a large output voltage (a.k.a., a "voltage amplifier"); (3) a small input current and turn it into a large output voltage (a.k.a., a "transconductance amplifier"); and (4) a small input voltage and turn it into a large output current (a.k.a., a "transimpedance amplifier"). Transconductance (g_m) in the electronics world just means the incremental change in current divided by the incremental change in voltage. As a real-world example of amplifiers-in-action, at the input of your cell phone you have a hand-crafted voltage amplifier that takes the tiny little RF signals and boosts them to a level sufficient to manipulate and decode them (see Chapter 14). In a receiver for a fiber optic link, you have a hand-crafted transimpedance amplifier that interfaces with the input photodetector, to change the in-coming photonic signals into electronic signals for processing (see Chapter 20).

In addition to building amplifiers, gain also allows us to construct nice regenerative binary switches. As can be seen in Figure 1.8, if the input base current I_B (or input voltage V_{BE}) is zero, the output current I_C is zero, the on/off switch is now open and the output voltage V_{CE} is thus high. Let us call that state a logical "1". Conversely, if the input current I_B (or input voltage V_{BE}) is large enough to turn on the transistor, the output current I_C is large, output voltage V_{CE} drops to a low value, and the on/off switch is now closed. Let's call that state a logical "0". A regenerative binary switch!

1.4 Optimization of Bipolar Transistors

There are two typical performance metrics (or figures-of-merit: FoM), which indicate how fast or how high a frequency a bipolar transistor can operate. The first is the so-called "cutoff frequency" (f_T), the frequency at which the AC (alternating current) current gain becomes unity. The f_T is simply given by the inverse of total transit time (τ_{ec}) from the emitter to the collector ($f_T = 1/2\pi\tau_{ec}$) and, thus, gives an estimate of the speed-limit of the BJT switch and is a good FoM for digital circuits. As described above, to improve f_T (make it larger) major attention must be paid to make the base width as narrow as possible.

1866   Alfred Nobel invents dynamite   ELECTRON DEVICES SOCIETY   1869   Opening of Suez Canal

Here, heavily doped polysilicon is introduced to form the emitter region. This idea is widely used even in the modern BJTs and is called a "poly-emitter". The poly-emitter is utilized to form a shallow out-diffusion for the emitter impurities, and thereby allows both a thin base and emitter design. The poly-emitter has an additional advantage; namely, that the very thin native interface oxide which naturally occurs between the polysilicon and the single-crystal silicon acts as an effective barrier to prevent the minority carrier (hole) back-injection from B to E. It is necessary to increase the base doping concentration for a narrower base (to avoid the disappearance of the neutral base, so-called device "punchthrough") but the emitter doping concentration already reaches its maximum value (limited by solid solubility), so the current gain in a scaled BJT naturally decreases due to the low emitter injection efficiency (the ratio of the injecting electrons from E to B to the injecting holes from B to E). Therefore, the poly-emitter interfacial oxide helps to increase the current gain and is a very useful secondary by-product. However, as the emitter scaling progresses, the interfacial oxide causes the problem of the high emitter resistance and, thus, must be carefully optimized.

The second important BJT FoM is the so-called "maximum oscillation frequency" (f_{max}), the frequency at which the unilateral power gain becomes unity. The unilateral power gain is the forward power gain in a feedback amplifier, so it is a suitable index for many analog and RF circuits. The f_{max} is approximately given by, $f_{max} = \sqrt{f_T/8\pi C_c r_b}$, where C_c is collector capacitance and r_b is base resistance. The critical difference between f_T and f_{max} is as follows. The f_T is a FoM determined from the one-dimensional (vertical) structure, but the f_{max} is a FoM which includes the two-dimensional (planar) structure of the device, because the parasitic C_c and r_b appear in the equation. This means, to improve f_{max}, it is essential to minimize the parasitic capacitances and resistances of the planar structure. As can be seen in Figure 1.7, the intrinsic region for BJT is a one-dimensional structure just under the emitter. The other areas of the transistor structure are provided mainly to lead the base and collector current to their electrodes, so they are non-essentially the operation of the device. To improve lateral parasitics, several important transistor structures and process sequences (e.g., the so-called "self-aligned transistor structure" or "self-aligned fabrication process") have been developed. Figure 1.9 shows a typical self-aligned BJT structure formed by using a self-aligned fabrication process. To reduce C_c, it is very important to reduce the junction area between the base and collector. Therefore, in this self-aligned transistor, the base electrode constructed by a polysilicon film is placed on a thick oxide layer, minimizing C_c in the extrinsic base region. To reduce r_b, the narrow (typically 100 nm wide or less) space between the polysilicon emitter and the polysilicon base is defined by the thickness of the insulator which is formed on the side of emitter or base polysilicon. This is the origin of the usage of the term "self-aligned", that is, the edge of the emitter and base is automatically defined by

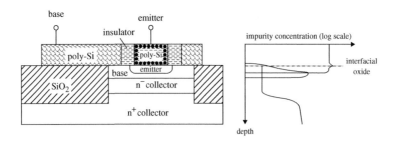

Figure 1.9 Self-aligned bipolar transistor structure and impurity profile under the emitter

the structure, independent of the lithography used. In the case of a non-self-aligned planar configuration defined by lithography, as shown in Figure 1.7, the base current must flow along a long (about 1 μm long or more) path, so it is difficult to decrease r_b. On the other hand, in a BJT formed using a self-aligned process, the distance separating from the emitter and base is very small, so this can be used to effectively reduce r_b.

Finally, the breakdown voltages (which set the maximum useful operating voltage of the BJT) are key transistor parameters for improving the high-speed and high-frequency characteristics of BJT. There is a fundamental trade-off between the speed (f_T) and the breakdown voltage (BV_{CEO}, the breakdown voltage between the collector and emitter when the base is open-circuited), often termed the "Johnson limit" [6]. The Johnson limit is derived only from considering fundamental issues associated with carrier transport, and predicts an achievable $f_T \cdot BV_{CEO}$ product of 200 GHzV, though in practice this value is significantly higher. The concept of a constant $f_T \cdot BV_{CEO}$ product in a BJT is useful for designing the collector region of the BJT, since it captures the tradeoff between achievable speed and operating voltage.

1.5 Silicon-Germanium Heterojunction Bipolar Transistors

The basic concept of the "heterojunction" bipolar transistor (HBT) was proposed by Shockley in the original BJT patent (refer to the history in [3]), and the basic theory of the HBT was published by Kroemer in 1957 [7]. Figure 1.10 shows the equilibrium energy band diagram, with the minority and majority carrier distributions, of the wide bandgap emitter HBT. The wide bandgap emitter creates a large barrier for injecting holes from the base into the emitter, thus increasing the current gain. Many III-V compound semiconductors (e.g., GaAs or InP) have been successfully applied in HBTs by virtue of their compositionally-adjustable growth technology which can tailor the bandgap for a specific need (called "bandgap engineering"). III-V HBTs benefit from this approach and can provide a large advance in performance over BJTs (see Chapter 14). However, bandgap engineering did not extend into the world of Si-based technologies for many decades, even though the basic idea was envisioned early-on for HBTs based on silicon-compatible silicon-germanium (SiGe) alloys. The lattice constants between Si and Ge differ by roughly 4.2%, so the SiGe films grown on Si are compressively strained. The criterion giving the stability of such pseudomorphically grown strained SiGe films on Si indicates a maximum "critical thickness"

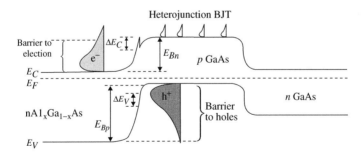

Figure 1.10   Basic idea behind the wide bandgap emitter heterojunction bipolar transistor. Reproduced with permission from Cressler, J. D.; *Silicon Earth: Introduction to the Microelectronics and Nanotechnology Revolution*; 2009, Cambridge University Press

of SiGe film for a given Ge content [8]. SiGe films which are device quality, meaning the SiGe films remain stable after thermal processing, were first epitaxially grown in the mid-1980s, and shortly thereafter the first SiGe HBTs were demonstrated [5].

The bandgap of Ge (0.66 eV) is smaller than that of Si (1.12 eV), so the SiGe HBT has a narrow bandgap base, differing from the wide bandgap emitter HBT. The compressive strain associated with sandwiched SiGe base layer between Si emitter and collector layers produces an additional bandgap shrinkage. As a result, a bandgap reduction of about 70–80 meV for each 10% of Ge content can be utilized in device engineering. Figure 1.11 shows the basic structure and forward-active bias energy band diagram of a SiGe

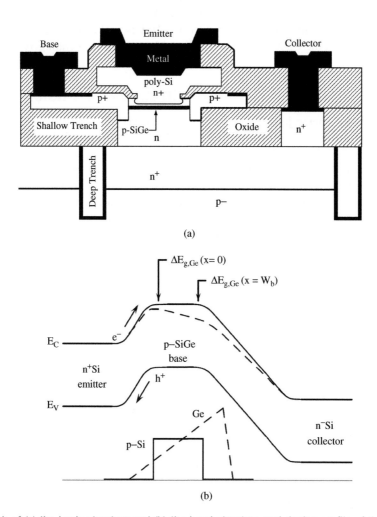

Figure 1.11   Sketch of (a) the basic structure and (b) the band structure and doping profile of the silicon-germanium heterojunction bipolar transistor (SiGe HBT). Reproduced with permission from Cressler, J. D.; *Silicon Earth: Introduction to the Microelectronics and Nanotechnology Revolution*; 2009, Cambridge University Press

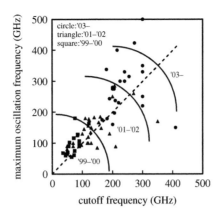

Figure 1.12 Evolutionary improvement in cutoff frequency and maximum oscillation frequency from 1999–2011 for SiGe HBTs

HBT. Similar to the wide bandgap emitter HBT, the emitter injection efficiency effectively increases due to the Ge-induced band offset occurred in the valence band.

After the first demonstration of functional SiGe HBT in 1987 [9], the development of SiGe HBTs evolved rapidly and their performance has dramatically improved from the mid-1990s to present. For Si BJTs, the peak f_T is limited to approximately 50 GHz. However, using a SiGe HBT, both f_T and $f_{\max}$ go rise above 300 GHz, as shown in Figure 1.12. In the early stages of evolution, the SiGe HBT had a non-self-aligned structure, so only f_T was improved by the shrinkage of the base width and bandgap engineering. However, SiGe HBTs soon incorporated self-aligned transistor structures, with rapid improvement in transistor $f_{\max}$. The schemes to fabricate self-aligned SiGe HBTs are roughly categorized into two types, depending on the SiGe epitaxial growth technologies used: selective or blanket epitaxial growth [10]. Recently, attention has been placed on achieving ultra-high $f_{\max}$ due to the emerging applications such as terahertz wireless systems.

One of the most important aspects of SiGe HBTs is that it can be easily combined with Si CMOS on the same wafer to enable highly-integrated systems. So-called SiGe BiCMOS (SiGe HBT + Si CMOS) technologies can be constructed using well-established Si-based processes and are 100% silicon manufacturing compatible. This represents a fundamental difference between SiGe HBTs technology and III-V HBTs (see also Chapter 14). The wide-spread application of SiGe HBTs in high-speed digital and RF/analog integrated circuits offer ample evidence to this crucial advantage enjoyed by SiGe BiCMOS technology (see examples in [10]).

References

[1] J. Bardeen and W. H. Brattain, "The transistor, a semiconductor triode", *Physical Review*, vol. 74, pp. 230–231, 1948.
[2] W. Shockley, M. Sparks, and G. K. Teal, "p-n junction transistors", *Physical Review*, vol. 83, p. 151, 1951.
[3] J. D. Cressler, *Silicon Earth: Introduction to the Microelectronics and Nanotechnology Revolution*, New York, NY, Cambridge University Press, 2009.
[4] K. K. Ng, *Complete Guide to Semiconductor Devices*, 2nd Edn, New York, NY, John Wiley & Sons, Inc., 2002.

[5] J. D. Cressler (ed.), *Silicon Heterostructure Handbook: Materials, Fabrication, Devices, Circuits, and Applications of SiGe and Si Strained-Layer Epitaxy*, Boca Raton, FL, CRC Press, 2006.

[6] E.O. Johnson, "Physical limitations on frequency and power parameters of transistors", *RCA Rev.*, vol. 26, pp. 163–177, 1965.

[7] H. Kroemer, "Theory of a wide-gap emitter for transistors", *Proc. IRE*, vol. 45, pp. 1535–1537, 1957.

[8] J. W. Matthews and A.E. Blakeslee, "Defects in epitaxial multilayers– I:misfit dislocations in layers", *J. Cryst. Growth*, vol. 27, pp. 118–125, 1974.

[9] S. S. Iyer, G. L. Patton, J. M. C. Stork, *et al.*, "Silicon-germanium base heterojunction bipolar transistors by molecular beam epitaxy", *Tech. Dig. IEEE Int. Elect. Dev. Meeting*, pp. 874–876, 1987.

[10] K. Washio, "Silicon-germanium (SiGe) heterojunction bipolar transistor (HBT) and bipolar complementary metal oxide semiconductor (BiCMOS) technologies", Chapter 18 in, *Silicon-Germanium Nanostructures* (eds Y. Shiraki and N. Usami), Cambridge, Woodhead Publishing, 2011.

1879 Edison invents the electric light bulb ELECTRON DEVICES SOCIETY 1880 Jaques and Pierre Curie discover the piezo electric effect in crystals

Chapter 2

MOSFETs

Hiroshi Iwai, Simon Min Sze, Yuan Taur and Hei Wong

2.1 Introduction

MOSFET (Metal-Oxide-Semiconductor Field-Effect Transistor) and the MOS-based NVSM (Non-Volatile Semiconductor Memory) are the two most manufactured electronic devices in modern information society and are definitely the most successful devices in the history of electronics. Although there is no official figure for estimating the numbers of MOSFET and NVSM per capita, considering the facts that the most advanced microprocessor has more than 2 billion MOSFETs in a chip, and that the biggest capacity NAND flash memory chip has more than tens of billions of NVSMs, we are quite sure that the number should exceed some billions for every person in the world. Because NVSMs will be considered in Chapter 3, this chapter aims at presenting a general overview on the development and evolution of the MOSFET, the problems encountered in downsizing the device to its ultimate dimension, and the technological options being available for the next couple of decades.

2.2 MOSFET Basics

A generic transistor is a three-terminal device where the current flow between two of the terminals is modulated by the bias of the third terminal (see Figure 2.1a). Among the different kinds of transistors being proposed in the last six decades, MOSFET has been the most widely investigated kind in terms

Guide to State-of-the-Art Electron Devices, First Edition. Edited by Joachim N. Burghartz.
© 2013 John Wiley & Sons, Ltd. Published 2013 by John Wiley & Sons, Ltd.

of applications and manufacturing. The unique structure of the MOSFETs is the third terminal, the gate, which is in fact a MOS capacitor consisting of a metal electrode, a thin oxide dielectric, and the substrate (see Figure 2.1b). The gate control terminal is normally used as an input with the MOS capacitor blocking the DC (direct) current flow, thus making this structure highly efficient in terms of extremely low power consumption.

The operation of the MOSFET is graphically illustrated in Figure 2.2. Considering a MOSFET with a positive threshold voltage (V_{th}), when the gate is biased at 0 V, that is, $V_g = 0$ V, the potential of p-type channel is 0 V provided that also the substrate is biased at 0 V. The heavily doped n^+ source and drain regions form pn junctions adjacent to the p-type channel. The built-in potential of the source junction, a negative potential barrier between the source and channel, suppresses the electrons from flowing out of the source (see Figure 2.2c). There is no current flow in the channel. Thus, the drain current $I_d = 0$, and the transistor is in the "off" state.

When the gate is biased with a sufficiently high voltage ($V_g > V_{th}$), the surface potential of the channel will be high enough to lower the potential barrier at the source-channel pn junction. For a positive drain bias, electrons can now travel from the source through the channel and enter the drain (see Figure 2.2d). This is the "on" state of the MOSFET. This describes the case of n-channel MOSFETs. We can make a different type of conduction channel by using an n-type substrate with p-type source and drain regions. In that case the channel current is composed of holes, and the drain current is now flowing from drain to source. This type of MOSFET is called the p-channel MOSFET. Details of a MOS device operation can be found in [1, 2].

There are many applications of MOSFETs. They can be used as a switch for logic and volatile memory (e.g., DRAM and SRAM; see Chapters 3 and 13) circuits; they can be used in an amplifier or a modulator to process the AC (alternating current) signal in an analog or a RF (radio frequency) circuit (see Chapters 12 and 14); and they can be used to convert a voltage in a power circuit (see Chapter 15). In this chapter, we focus on the issues regarding the digital applications of MOSFETs which is considered as the mainstream technology of modern microelectronics (see also Chapter 11).

For digital applications of MOSFETs, speed and power consumption are the major concerns. To achieve higher speed and lower power consumption, the dimensions of the MOSFETs have been shrunk continuously because the capacitance of logic nodes, C, decreases as the dimensions of MOSFETs decrease. A smaller capacitance results in a shorter switching time, t, and a smaller switching power, P,

Difference between 40 Years and 4 Billion Years of Evolution

Today, the performance of the most advanced microprocessors is tremendous and they are capable of controlling various systems. However, when comparing their performance to that of the brains of animals, we have to admit that their performance is still much inferior. For example, today's microprocessors cannot compete with the brain of a dragon y for its complicated three-dimensional ight control in terms of real-timeness and extremely low energy consumption. We are not even able to y an arti cial mosquito using a microprocessor, because the weight of a microprocessor and battery make it too heavy to lift the body. This is due to the difference between 40 years of integrated circuit history against 4 billion years of life evolution. However, we cannot use an insect's brain because it is not easy to sustain its life. At this moment, unfortunately we do not know how to implement such an amazing life algorithm into the integrated circuit architecture.

1884   First underground   ELECTRON   First article in Trans.   1885
             train runs in London   DEVICES   AIEE – notes on
                                        SOCIETY*   phenomena in
                                                        incandescent lamps

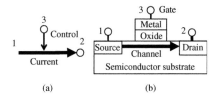

Figure 2.1 Schematic diagram illustrating (a) a generic transistor concept; and (b) the structure of a MOSFET

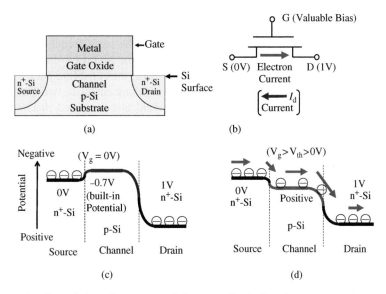

Figure 2.2 Structure, circuit symbol, and energy band diagrams illustrating the two operation modes of an n-channel Si MOSFET. (a) cross-section; (b) circuite configuration; (c) potential at Si surface at 'OFF'; and (d) Potential at Si surface at 'ON'

of a logic circuit. The relationship between capacitance, switching time, and switching power are given next:

$$t = CV_{supply}/I_d \qquad (2.1)$$

$$p = CV_{supply}^2 f/2 \qquad (2.2)$$

where V_{supply} is the supply voltage and f is the number of switching cycles per second or clock frequency.

In fact, the feature size of the MOSFETs, or minimum line width used for a MOSFET, has kept shrinking at an average rate of about 0.7 times about every two to three years over the past four decades [3–5] (see Figure 2.3), which has resulted in a line width reduction of a factor of 1/500 and a MOSFET area reduction of 1/250 000 since the 1970s.

Because of the downsizing of MOSFETs, the number of MOSFETs in an integrated circuit was doubled every two or three years. This is the well-known Moore's law which was proposed in 1966 by

Feature Size/Technology Node

(1970)10 µm→8 µm→6 µm→4 µm→3 µm→2 µm→1.2 µm →
0.8 µm→0.5 µm→0.35 µm→0.25 µm→180 nm→130 nm→90 nm→
65 nm→45 nm →32 nm→22 nm (2012)

Figure 2.3   The downsizing trend and technological nodes of CMOS technology

Gordon Moore [6]. Large scale integration (LSI) of MOSFETs started in early 1970s represented by the 1 kbit memory [7] and the integrated microprocessor [8] operated at 750 kHz clock frequency. Forty years later, the memory capacity has increased 128 million times, and the clock frequency of the microprocessor has increased 5000 times. It should be noted that the shrinking of the MOSFET dimensions also enables significant increase in the number of MOSFETs in a microprocessor, thus enabling parallel processing. Hence, the downsizing of MOSFETs contributes to the performance enhancement in two ways: shorter switching time of MOSFETs and parallel processing capability. It should also be noted that, with downsizing, the cost and power dissipation per MOSFET also become smaller [9]. Clearly, the downsizing of the MOSFET is critically responsible for the improvement of density, power, and performance of integrated circuits.

When the size of the MOSFET is reduced to the deep submicron range, the distance between the source and drain becomes so small that punchthrough will occur if the body doping remains the same [10]. Therefore, the biggest problem is how to switch off the MOSFETs. Figure 2.4 shows the cross-section of the device in the "off" state. There is an electron barrier near the source when $V_g = 0$ V. However, when a positive bias is applied to the drain, the depletion or space charge region near the drain pn junction becomes larger. This region may be insignificant in the long channel device. However, as the channel length reduces,

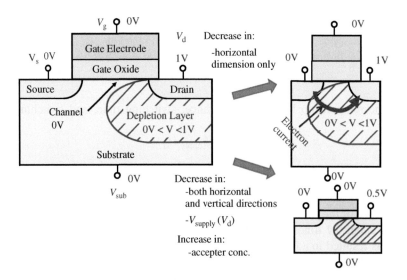

Figure 2.4   Problems for MOSFET downsizing and scaling method

the depletion region may extend over most of the channel region and lower the electron barrier near the source region. Under this situation, electrons can transport from the source to the drain even under zero gate bias; that is, the "off" current is no longer negligible or the device is not exactly in the "off" mode.

To solve this problem, the gate oxide thickness should be decreased accordingly in order to enhance the gate controllability of the channel potential [8]. In addition, the depth of the source and drain junctions should also be reduced so as to suppress the drain depletion layer depth and its lateral extension underneath the gate. These principles were first highlighted by Dennard and co-workers who developed them into a constant-field scaling method in 1974 [11]. In this method, all the horizontal and vertical dimensions, as well as the supply voltage, are multiplied by the same factor, k (< 1), whereas the dopant (acceptor or donor) concentrations of the channel are multiplied by $1/k$ (> 1). The typical value of k is about 0.7 in the scaling trend of last several decades [12]. It is noted that the device size did scale according to this trend for over 40 years, but the supply voltage and doping concentrations did not scale accordingly. As the downsizing approaches the decananometer scale, and V_{supply} and V_{th} have become smaller, the "off" current or subthreshold conduction cannot be suppressed effectively with the aforementioned scaling method. Regardless of the gate length, even when the gate is biased below the threshold voltage, there is still a small fraction of source electrons with enough energy to surmount the source potential barrier, transport across the channel region, and arrive at the drain. This current is known as subthreshold leakage current. Unlike the "on" state drain current which is a linear or quadratic function of $V_g - V_{th}$, the subthreshold drain leakage current is an exponential function of gate bias because of the presence of potential barrier [13]. Figure 2.5 depicts the log $I_d - V_g$ plot of current–voltage characteristics of a MOSFET from the "off" state to "on" state. The straight line portion represents the exponential V_g dependence in the subthreshold region. When the supply voltage, V_{supply}, is scaled down along with the device size, V_{th} needs to be reduced as well. Otherwise, the on-state current I_d, which changes with $V_{supply} - V_{th}$, will be overproportionally reduced, or the device may not even be turned on at all. In Figure 2.5, the decrease of V_{th} gives rise to a parallel shift

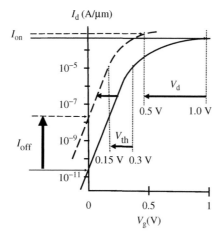

Figure 2.5 Illustration of the effect of threshold voltage on the "off" current of a MOSFET

of the log $I_d - V_g$ curve to the left. The slope of the log $I_d - V_g$ plot is governed by the electron thermal voltage kT/q from Boltzmann statistics of electron energy distribution, which remains unchanged regardless of the transistor's channel length. Figure 2.5 further indicates that the I_{off} leakage increases exponentially with V_{th}. Although I_{off} in the order of 10^{-8} A is insignificant for a single MOSFET, the overall "off" current of the entire chip may exceed 100 A in future integration of trillions of transistors. This is a serious issue! The problem is alleviated to some extent by adopting various smart power management schemes to cut off certain leakage paths. Nevertheless, subthreshold leakage remains a critical issue for further device downsizing.

The actual device structure in use today has evolved in many aspects to overcome the problems encountered in downsizing. Figure 2.6 shows a typical MOSFET structure fabricated for its use in modern logic circuits. The structure becomes more complex with many new materials introduced. The basic planar structure, however, remains identical until the 32 nm node.

In conventional bulk MOSFET scaling, the depletion depth has been scaled by increasing the doping concentration of the p-type body, that is, the region of substrate right underneath the channel. By the 28 nm node, the body doping level has reached 10^{19} cm^{-3}. This leads to several problems: band-to-band tunneling at the reverse biased drain-body junction, high threshold voltage even with low work function gates, and dopant number fluctuation in minimum size devices. It is possible to circumvent these problems by making a structural change to the MOSFET, namely, a double-gate (DG) MOSFET or fin-FET in which the two-dimensional short-channel effect is controlled by the thickness of the silicon film instead of the depletion region depth. A thin substrate, which makes the back gate (substrate) control of the channel possible, will be a quite promising approach. However, the substrate cannot be too thin as the substrate electrode may be in short-circuit with the drain. This issue can be solved by making an isolated bottom electrode resembling to the top MOS gate. Controlling the channel potential from both sides of the silicon is the idea of the double-gate MOSFET or fin-FET [14, 15] (which was first named as Delta FET in 1989 [16]). The DG-FET is categorized as one of so-called multiple-gate MOSFETs. The ultimate multiple-gate MOSFET structure can be a nanowire in which the gate surrounds all surfaces of the wire channel. Such an ideal structure, compared to a bulk MOSFET in terms of suppressing the drain depletion depth, is shown in Figure 2.7. There are many variations of the multiple-gate nanowire FET [17–23], as depicted in Figure 2.8. Even a fin-FET could be categorized to be a nanowire FET in a broad sense as the fin width approaches decananometer dimensions.

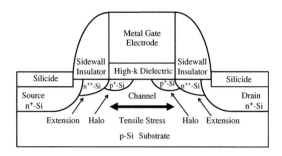

Figure 2.6   Cross-sectional view of the MOSFET structure used in recent technology nodes

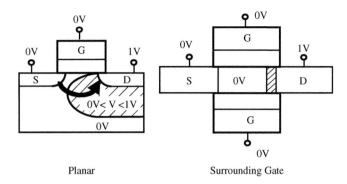

Figure 2.7 Comparison of a planar MOSFET (*left*) with a nanowire FET (*right*) from the drain depletion layer control point of view

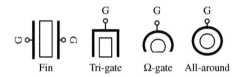

Figure 2.8 Conceptual layouts of a nanowire MOSFET

2.3 The Evolution of MOSFETs

The evolution of MOSFETs from the production point of view can be roughly divided into six stages as described next.

Stage I (1925–1960): From ideas to experiments

The original idea of a field effect transistor can be dated back to 1925 [24, 25]. Back in the 1920s and 1930s, MOS-like insulated gate field effect transistors were invented [25–27]. There might have been some primitive experiments during this period, but no report of success was published. Later, in the 1940s, Shockley's group at Bell Laboratories strove to realize MOSFET operation, but they did not succeed, because of the very poor interface between the semiconductor and the gate insulator [25].

Stage II(1960–1970): The first MOS transistor and modeling

The first functional MOSFET was made in 1959 by using a Si substrate, a silicon dioxide (SiO_2) gate dielectric, and an Al gate electrode [28]. However, as its driving current was smaller than that of a bipolar junction transistor (BJT), and the threshold voltage was unstable due to mobile ion like sodium contamination in the gate SiO_2, MOSFETs were not considered to be a competitor to the mainstream bipolar technology at that time. In this period, the MOSFET drain current, I_d, was successfully modeled analytically in terms of the drain voltage V_d and the gate voltage V_g [29, 30].

Stage III (1970–1985): Integration and downsizing

The gate instability problem [31, 32] was solved in the late 1960s by introducing impurity gettering methods and control of the cleanliness of the process environment [33, 34]. MOSFETs were then used in the fabrication of large scale integrated (LSI) circuits, such as memories and microprocessors. The typical feature size of the devices fabricated in early 1970s was about 10 μm. P-channel MOSFETs were used in these early LSIs because of the more forgiving Si/SiO$_2$ interfacial requirement for V_{th} control. Then, just one or two generations later, N-channel MOSFETs, with their higher carrier mobility and therefore higher drain current, replaced the PMOS technology. NMOS LSIs were then used for the fabrication of microprocessor and many other logic devices and memories, such as ROM (Read-Only Memory), and RAM (Random-Access Memory) (see also Chapter 3). However, BJTs were still the mainstream technology for high performance mainframe computers.

In the late 1960s, the polysilicon (poly-Si) gate electrode was developed for MOSFETs [35]. The introduction of poly-Si gate made the self-alignment of source/drain to the gate electrode possible. In this technology, the poly-Si gate electrode was used as a mask for the high-temperature dopant diffusion to form the source-drain region. In the Al gate technology, the source/drain diffusion had to be done with a separate photolithographic mask before the gate electrode formation as the Al melting point is far below the diffusion temperature for impurities in silicon. To guard against worst-case misalignment, the Al gates had a large overlap with the source-drain regions in the nominal case, which led to both significant area increase and excessive parasitic capacitance and resistance, resulting in severe loss of performance. The poly-Si gate is a very important milestone in the downsizing of MOSFETs.

In early 1970s, short-channel effects that cause V_{th} lowering and increase of "off" leakage current due to the drain depletion layer were investigated [36]. As described in the previous chapter, the constant-field scaling method was proposed in order to suppress the short channel effect in 1974 [11]. However, it is not necessary for all the device parameters to be scaled with the same factor, k. Consequently, more general guidelines for the suppression of the short channel effects were proposed using simple empirical relation of device vertical parameters: gate oxide thickness, source/drain junction depths, and source/drain depletion widths [37]. Later on, the guideline was further refined by analytically solving a simplified two-dimensional potential problem, leading to the concept of "scale length" [38–40].

Stage IV (1985–2005): CMOS, and structure evolution

In the late 1970s, the hot carrier effect was one of the major concerns. Channel hot carriers generated at the drain edge can be injected into the gate oxide and cause threshold voltage and mobility degradation [41–49]. This is because channel electrons (or holes) are accelerated by the strong lateral electric field near the drain in a short-channel MOSFET to sufficiently high energy to overcome the Si/SiO$_2$ interface barrier. The hot carriers injected into SiO$_2$ caused damage in terms of interface state and fixed charge generation, which then led to V_{th} and I_d variations over time. To solve this problem, the lightly-doped drain (LDD) structure (see Figure 2.6) was proposed in the late 1970s [50, 51]. LDD region near the drain mitigates the electric field hence the hot carrier effects. Note, that the source uses the LDD structure also because of the symmetrical fabrication process. The supply voltage had remained unchanged at 5 V for almost 20 years since the 6 μm generation. Finally, the supply voltage was decreased to 3.3 V at the 0.5 μm technology generation. This relieves the lateral electric field near the drain. The source/drain structure was further modified in the late 1990s. The LDD-like region became ultra-heavily doped and ultra-shallow, called the source-drain "extension". The purpose is to decouple the shallow junction needed for suppression of short-channel effect from the deep junction needed for source-drain contact.

The downscaling of the source/drain junction depth led to a significant increase in the source/drain resistance. The dopant concentration at the source/drain edge was made as high and abrupt as possible to minimize the resistance. However, this made the short-channel effects and hot-carrier reliability worse. Several additional measures were introduced. Hot-carrier effects were relieved by further reducing the supply voltage. Short-channel effects were controlled by adding a higher doped region of the same type as the body adjacent to the drain, called halo [52–54]. The effect of the halo is to suppress the drain depletion layer. By symmetry, the halo is formed on both drain and source (see Figure 2.5).

Since the mid-1980s, CMOS LSIs started to replace NMOS LSIs because of the extremely low standby power of CMOS circuits, even though the fabrication process of CMOS is more complex with the additional PMOS transistor. CMOS LSIs had been used since the 1970s for very low-power and lower-speed applications such as digital watches and pocket calculators as well as for specific use for aerospace. When CMOS became a mainstream technology, the n^+ poly-Si gate electrode was used for both n- and p-channel MOSFETs. However, although the process is simpler, n^+ poly-Si gate is not suitable for p-channel MOSFETs, as its low work function resulted in a V_{th} far too negative for p-channel MOSFETs. The n^+ poly-Si gate in p-channel MOSFETs was eventually changed to p^+ poly-Si, while n^+ poly-Si was still used for n-channel MOSFETs. This was called dual gate CMOS process [55–57]. However, as the gate oxide thickness was reduced for control of short-channel effects, the boron dopant in p^+ poly-Si can diffuse across the gate oxide to the channel during the high-temperature annealing process. This is called boron penetration [58–60]. To suppress boron penetration, some amount of nitrogen atoms were introduced into the SiO_2, by high-temperature annealing of SiO_2 in nitrogen-containing ambient such as NH_3 and N_2O [59, 61–66]. With this process, the gate SiO_2 becomes oxynitride (SiO_xN_y). Thus, the gate insulator material was changed from SiO_2 to the oxynitride 30 years after the fabrication of first MOSFET. There were so many problems regarding the Si/SiO_xN_y interfacial property initially that it took many years to resolve them.

To decrease the resistance of the gate electrode, a refractory metal silicide layer was deposited on top of the poly-Si gate electrode. The poly-Si/silicide stacked structure was called polycide [67, 68]. $MoSi_2$ and WSi_2 were used in the first generation. Source/drain resistance also needed to be reduced, thus a technique called salicide [69], that formed silicide layers over the source/drain regions as well as the gate electrode in a self-aligned manner, was developed. In the first generation, $TiSi_2$ [70] and $CoSi_2$ [71] were used. As the junction depth became shallower, NiSi was introduced [72–76]. Many expected and unexpected problems arose when implementing these new materials into production and, thus, it took 10 years to completely resolve them (see also Chapters 6 and 10).

Regarding the gate oxide thickness, it had been thought that 3 nm would be the limit below which the oxide ceased to function as an insulator because of the onset of direct-tunneling [77]. It was later shown that even in the presence of direct-tunneling, individual MOSFETs still operate quite well in the circuit as long as the gate leakage current is negligible compared to the drain current. It turns out that a short-channel device can tolerate a higher gate current density than a long channel device because the drain current increases whereas the gate current decreases with shorter channel length. [78–80]. The limit of

1897 First reported splitting of atomic spectra in a magnetic field by Pieter Zeeman (Anomalous Zeeman Effect)

ELECTRON DEVICES SOCIETY® Ferdinand Braun invents the cathode ray tube (CRT)

gate current density is set by the chip standby power due to the accumulated leakage current of all the turned-on MOSFETs. A SiO_xN_y gate oxide of 1.2 nm thick was used for microprocessor production [81].

To reduce active power consumption, the silicon-on-insulator (SOI) structure was proposed [82–84]. SOI MOSFETs have low source and drain junction capacitance, therefore are attractive for low-power applications.

Stage V (2005–2012): New materials and new structures

As the feature size of the transistor is scaled towards the atomic scale, it becomes increasingly difficult to achieve performance gains through downsizing. To further enhance the drive current, several mobility enhancement techniques were introduced [85–89]. It was found that the electron and hole mobilities can be enhanced under process-induced tensile and compressive stresses, respectively.

Once the gate oxide thickness reaches 1 nm, it cannot be scaled any further because of the huge leakage current by direct-tunneling [90]. To get around this issue, high dielectric constant (high-k) material [81, 89, 91–99] was introduced to replace the SiO_xN_y gate insulator. Meanwhile, poly-Si gate electrode was replaced by metal electrode such as TaN, TiN, W, in order to eliminate the problem of poly-Si gate depletion. The depletion layer capacitance connected in series to the gate oxide capacitance decreases the MOSFET on-current. High-k with metal gate stack with EOT (equivalent oxide thickness) of 1 nm was successfully introduced by Intel in 2007 using HfO_2-based gate insulator [89].

When the gate length reaches 20 nm, suppression of the short-channel effect by higher body doping is no longer practical. Tri-gate MOSFETs are now being introduced to replace bulk MOSFETs for the 22 nm node and beyond [100]. This is a monumental change of the MOSFET structure in its 40-year history. For the same gate oxide thickness, tri-gate or double-gate MOSFETs can be scaled to shorter channel lengths than bulk MOSFETs [101, 102].

Outlook (2013 and beyond): Nanowire MOSFETs

What would be the ultimate limit for downsizing? The operation of a single MOSFET will be limited by direct tunneling between the source and the drain. This happens at a channel or gate length less than 3 nm where a single MOSFET will operate with a high level of "off" leakage. So the limitation is around 3 nm [102]. For VLSI systems, gate length shortening will be limited by the total amount of subthreshold leakage current of the chip [103]. There is no universal number as it depends on the applications and the circuit configurations [104].

It is noted that the subthreshold current depends on V_{th}. In terms of the subthreshold leakage current suppression, V_{th} should be as high as possible. However, a large V_{th} will decrease the drive current in the "on" state. To obtain both low subthreshold leakage and high drive current, we need to keep V_{supply} high to allow higher V_{th}. But high V_{supply} will increase the dynamic power consumption and enhance the short-channel effect [80]. At this moment, the MOSFET for large scale integrated circuits is expected to be further scaled to 8 nm or maybe even below. To suppress the short-channel effect with high drivability, nanowire FETs appear to be a good candidate [105–107]. Thinning the EOT of high-k gate insulator is still required to suppress the short-channel effect for further downsizing of MOSFETs. Note, that the benefit of on-current increase by EOT reduction gradually diminishes as the intrinsic gate oxide capacitance becomes comparable to the inversion layer capacitance at EOT = 0.4 nm. An EOT of sub-0.4 nm could be possible by choosing proper gate stack materials [108, 109].

Another practical issue with ultra-short channel MOSFETs is the dopant diffusion from the source/drain into the channel. The dopant diffusion from the source/drain to the channel becomes a serious issue as the channel region may disappear as a result of the dopant diffusion. This can be circumvented by replacing the heavily doped source/drain with metal or silicide [110]. Such a Schottky source-drain is inherently abrupt and is governed by the crystal boundary between the metal/silicide and semiconductor. The resistance of metal/silicide source/drain is significantly lower than that of diffusive source/drain. It should be noted that the silicide can encroach the portion of nanowire under the sidewall to form the source/drain at that portion [111]. This will significantly decrease the resistance. The channel-to-source/drain contact resistance could be also smaller provided that low electron- and hole-barrier Schottky junctions can be made on n-channel and p-channel nanowire FETs, respectively.

Further performance improvement and power reduction may potentially be achieved with high-mobility, new channel materials such as Ge and/or III-V semiconductors in the future [112–116]. At present, there are still many technological challenges to overcome before they can be considered for billions- and trillions-transistor ULSI products (see also Chapters 5 and 21).

2.4 Closing Remarks

The MOSFET is the most fundamental and the smallest electronic device ever manufactured. It is truly amazing that it has gone through so many generations of evolution with unprecedented impact to everyday life. Nevertheless, the downsizing of MOSFET will eventually end in the next one or two decades because many of the device parameters are now approaching fundamental physical and manufacturing limits. We believe CMOS technology will still remain to be the mainstream IC technology for a long period after that, as no other device technology can conceivably be developed into as large an integration scale as the present level in the foreseeable future.

References

[1] S. M. Sze and K. K. Ng, *Physics of Semiconductor Devices*, 3rd edn, John Wiley & Sons, Inc., Hoboken, 2007.
[2] Y. Taur and T. H. Ning, *Fundamentals of Modern VLSI Devices*, 2nd edn, Cambridge University Press, Cambridge, 2009.
[3] R. H. Dennard, F. H. Gaensslen, V. L. Rideout, E. Bassous, A. R. LeBlanc, "Design of ion-implanted MOSFETs with very small physical dimensions", *IEEE J. Solid-State Circuits*, vol. 9, pp. 256–268, 1974.
[4] G. A. Sai-Halasz, M. R. Wordemen, D. P. Kern, *et al.*, "Design and experimental technology for 0.1-μm gate length low temperature operation FETs", *IEEE Electron Devices Lett.*, vol. 8, pp. 463–466, 1987.
[5] M. Ono, M. Saito, T. Yoshitomi, *et al.*, "Sub-50 nm gate length N-MOSFETs with 10 nm phosphorus source and drain junction", *Tech.Dig. IEEE Int. El. Dev. Meeting (IEDM)*, pp. 119–122, December, 1993.
[6] G. E. Moore, "Cramming more components onto integrated circuits", *Electronics Magazine*, vol. 38(8), 1965.
[7] W. Regitz, and J. Karp, "A three transistor-cell, 1024-bit, 500 ns MOS RAM", *Dig. Techn. P. IEEE International Solid-State Circuits Conference (ISSCC)*, pp. 42–43, 1970.

1900   First test flight of Zeppelin   ELECTRON DEVICES SOCIETY   Porcelain capacitors come up

[8] F. Faggin, and M. E. Hoff, "Standard parts and custom design merge in four-chip processor kit", *Electronics*, pp. 112–116, 1972

[9] H. Wong and H. Iwai, "The road to miniaturization", *Phys. World*, vol. 18, pp. 40–44, 2005.

[10] S. Ogura, P. J. Tsang, W. W. Walker, *et al.*, "Design and characteristics of the lightly doped drain-source (LDD) insulated gate field-effect transistor", *IEEE Trans. Electron Devices*, vol. 27, pp. 1359– 1367, 1980.

[11] H. Wong and H. Iwai, "On the scaling issues and high-k replacement of ultrathin gate dielectrics for nanoscale MOS transistors", *Microelectron. Engineer.* vol. 83, pp. 1867–1904, 2006.

[12] H. Iwai, "CMOS downsizing toward sub-100 nm", *Solid–State Electron.* vol. 48, pp. 497–503, 2003

[13] R. R. Troutman, and S. N. Chakravarti, "Subthreshold characteristics of insulated-gate field effect transistors", *IEEE Trans. Circuit Theory*, vol. CT-20, pp. 656–665, 1973.

[14] D. Hisamoto, W.–C. Lee, J. Kedzierski, *et al.*, "FinFET-a self-aligned double-gate MOSFET scalable to 20 nm", *IEEE Trans. Electron Devices*, vol. 47, pp. 2320–2325, 2000.

[15] A. Keshavarzi, D. Somasekhar, M. Rashed, *et al.*, "Architecting Advanced Technologies for 14 nm and beyond with 3D FinFET transistors for the future SoC applications", *Tech. Dig. IEEE Int. El. Dev. Meeting (IEDM)*, pp. 67–70, 2011.

[16] D. Hisamoto, T. Kaga, and E. Takeda, "Impact of vertical SOI "DELTA" structure on planar device technology", *IEEE Trans. Electron Devices*, vol. 38, pp. 1419–1424, 1991.

[17] J. Appenzeller, J. Knoch, M. T. Bjork, H. Riel, H. Schmid, and W. Riess, "Toward nanowire electronics", *IEEE Trans. Electron Devices*, vol. 55, pp. 2827–2845, 2008.

[18] L. Wei, X. Ping, and C. M. Lieber, "Nanowire transistor performance limits and applications", *IEEE Trans. Electron Devices*, vol. 55, pp. 2859–2876, 2008.

[19] M. Saitoh, Y. Nakabayashi, K. Uchida, and T. Numata, "Short-channel performance improvement by raised source/drain extensions with thin spacers in trigate silicon nanowire MOSFETs", *IEEE Electron Device Lett.* vol. 32, pp. 273–275, 2011.

[20] J. Nah, E.-S. Liu, K. M. Varahramyan, D. Shahrjerdi, S. K. Banerjee, and E. Tutuc, "Scaling properties of – core–shell nanowire field-effect transistors", *IEEE Trans. Electron Devices*, vol. 57, pp. 491–495, 2010.

[21] Y. Lee, K. Kakushima, K. Shiraishi, K. Natori, and H. Iwai, "Size dependent properties of ballistic silicon nanowire field effect transistors", *J. Appl. Phys.*, vol. 107, pp. 113705, 2010.

[22] K. Tachi, M. Casse, D. Jang, *et al.*, "Relationship between mobility and high-k interface properties in advanced Si and SiGe nanowires", *Tech. Dig. IEEE Int. El. Dev. Meeting (IEDM)*, pp. 313–316, 2009.

[23] S. Sato, Y. Lee, K. Kakushima, P. Ahmet, K. Ohmori, K. Natori, K. Yamada, and H. Iwai, "Gate semi-around Si nanowire FET fabricated by conventional CMOS process with very high drivability", *Proc. European Solid-State Device Research Conference* (ESSDERC), pp. 361–364, 2010.

[24] J. E. Lilienfeld, "Method and apparatus for controlling electric current", US Paten 1,745,175, Canadian application Oct. 22 1925, Filed October 8, 1926.

[25] C. T. Sah, "Evolution of the MOS transistor - From conception to VLSI", *Proc. IEEE*, vol. 76, No. 10, pp. 1280–1285, 1988.

[26] J. E. Lilienfeld, "Device for controlling electric current", U.S. Patent No. 1,900,018, Filed March 28, 1928.

[27] O. Heil, "Improvements in or relating to electrical amplifiers and other control arrangements and devices", British Patent No. 439, 457, Filed March 5, 1935.

[28] D. Kahng and M. M. Atalla, "Silicon-silicon dioxide field induced surface device", *IRE-AIEE Solid-State Device Research Conference*, MIT, 1960.

[29] H. C. Pao and C. T. Sah, "Effects of diffusion current on characteristics of metal-oxide (insulator) semiconductor transistors", *Solid-State Electron.* vol. 9, pp. 927–937, 1966.

[30] Y. A. El-Mansy and A. R. Boothroyd, "A simple two-dimensional model for IGFET operation in the saturation region", *IEEE Trans. Electron Devices*, vol. 24, pp. 254–262, 1977.

[31] E. H. Snow, A. S. Grove, B. E. Deal, and C. T. Sah, "Ion transport phenomena in insulating films", *J. Appl. Phys.*, vol. 36, p. 1664, 1965.

1901   Maybach builds first Mercedes car   ELECTRON DEVICES SOCIETY   Marconi transmits telegraphic messages

[32] E. Yon, W. H. Ko, and A. B. Kuper, "Sodium distribution in thermal oxide on silicon by radiochemical and MOS analysis", *IEEE Trans. Electron Devices*, vol. 13, p. 276, 1966.

[33] D. R. Kerr, J. S. Logan, P. J. Burkhardt, and W. A. Pliskin, "Stabilization of SiO_2 passivation layers with P_2O_5", *IBM J.*, vol. 8, p. 376, 1964.

[34] B. E. Deal, "A scientist's perspective on the early days of MOS technology", *ECS Interface*, pp. 42–45, 2007.

[35] F. Faggin, T. Klein, and L. Vadasz, "Insulated gate field effect transistor integrated circuits with silicon gates", *IEEE Trans. Electron Devices*, vol. 16, p. 236, 1969.

[36] L. D. Yau, "A simple theory to predict the threshold voltage of short-channel IGFET", *Solid-State Electron.*, vol.17, pp. 1059–1063, 1974.

[37] J. R. Brews, W. Fichtner, E. H. Nicollian, and S. M. Sze, "Generalized guide for MOSFET miniaturization", *IEEE Electron Device Lett.*, vol. 1, pp. 2–4, 1980.

[38] T. Toyabe and S. Asai, "Analytical models of threshold voltage and breakdown voltage of short-channel MOSFETs derived from two-dimensionalvanalysis", *IEEE Trans. Electron Devices*, vol. 26, pp. 453–461, 1979.

[39] W. R. Bandy, and R. S. Winton, "A new approach for modeling the MOSFET using a simple, continuous analytical expression for drain conductance which includes velocity-saturation in a fundamental way", *IEEE Trans. Comput. Aided Design Integrated Circuits and Systems*, vol. 15, pp. 475–483, 1996.

[40] Q. Xie, J. Xu, and Y. Taur, "Review and critique of analytic models of MOSFET short-channel effects in Subthreshold", *IEEE Trans. Electron Devices*, vol. 59, pp. 1569–1579, 2012.

[41] S. A. Abbas, and R. C. Dockerty, "Hot carrier instability in IGFETs", *Appl. Phys. Lett.*, vol. 27, pp. 147–148, 1975.

[42] T. H. Ning, C. M. Osburn, and H. N. Yu, "Emission probability of hot electron from silicon to silicon dioxide", *J. Appl., Phys.*, vol. 48, pp. 286–293, 1979.

[43] E. Takeda and N. Suzuki, "An empirical model for device degradation due to hot-carrier injection", *IEEE Electron Device Lett.*, vol. 4, pp. 111–113, 1983.

[44] G. Sodini. P.-K. Ko, and J. L. Moll, "The effect of high fields on MOS device and circuit performance", *IEEE Trans. Electron Devices*, vol. ED-31, pp. 1386–1393, 1984.

[45] C. Hu, S. C. Tam, F.-C. Hsu, *et al.*, "Hot-electron-induced MOSFET degradation – model, monitor, and improvement", *IEEE J. Solid-State Circuits*, vol. 20, pp. 295–305, 1985.

[46] F. Matsuoka, H. Iwai, H. Hayashida, *et al.*, "Analysis of hot carrier induced degradation mode on PMOSFETs", *IEEE Trans. Electron Devices*, vol. 37, pp. 1487–1495, 1990.

[47] Y. Toyoshima, H. Iwai, F. Matsuoka, *et al.*, "Analysis of gate oxide thickness dependence of hot carrier induced degradation in thin gate oxide nMOSFETs", *IEEE Trans. Electron Devices*, vol. 37, pp. 1496–1503, 1990.

[48] H. Wong, "A physically based drain avalanche breakdown model for MOSFETs", *IEEE Trans. Electron Devices*, vol. ED-42, pp. 2197–2202, 1995.

[49] H. Wong and Y.C. Cheng, "Generation of interface states at the silicon/oxide interface due to hot electron injection", *J. Appl. Phys.*, vol. 74, pp. 7364–7368, 1993.

[50] S. Ogura, P. J. Tsang, W. W. Walker, *et al.*, "Design and characteristics of the lightly doped drain-source (LDD) insulated gate field-effect transistor", *IEEE Trans. Electron Devices*, vol. 27, pp. 1359–1367, 1980.

[51] H. C.-H. Wang, C.-C. Wang, C. H. Diaz, *et al.*, "Arsenic/phosphorus LDD optimization by taking advantage of phosphorus transient enhanced diffusion for high voltage input/output CMOS devices", *IEEE Trans. Electron Devices*, vol. 49, pp. 67–71, 2002.

[52] C. F. Codella, and S. Ogura, "Halo doping effects in submicron DI-LDD device design", *Tech. Dig. IEEE Int. El. Dev. Meeting (IEDM)*, pp. 230–233, 1985.

[53] Y. Taur, S. Wind, Y. Mii, *et al.*, "High performance 0.1 um CMOS devices with 1.5 V power supply", *Tech. Dig. IEEE Int. El. Dev. Meeting (IEDM)*, pp. 127–130, December 1993.

[54] B. Yu, H. Wang, O. Milic, Q. Wang, W. Wang, J.-X. An, and M.-R. Lin, "50 nm gate-length CMOS transistor with super-Halo: Design, process and reliability", *Tech. Dig. IEEE Int. El. Dev. Meeting (IEDM)*, pp. 653–657, 1999.

[55] J. R. Pfiester, and L. C. Parrillo, "A novel dual p⁻/p⁺ poly gate CMOS VLSI technology", *IEEE Trans. Electron Devices*, vol. 35, pp. 1305–1310, 1988.

[56] C. Y. Wong, Y. C. Sun, Y. Taur, C. S. Oh, R. Angelucci, and B. Davari, "Doping of N+ and P+ polysilicon in a dual-gate CMOS process", *Tech. Dig. IEEE Int. El. Dev. Meeting (IEDM)*, pp. 238–241, 1988.

[57] H. Hayashida, Y. Toyoshima, Y. Suizu, K. Mitsuhashi, H. Iwai, K. Maeguchi, "Dopant redistribution in dual gate W-polycide CMOS and its improvement by RTA", *Dig. Symp., VLSI Technology*, pp. 29–30, May, 1989, Kyoto, Japan, 1989.

[58] Y. C. Sun, C. Y. Wong, Y. Taur, and C. Hsu, "Study of boron penetration through thin oxide with P+ polysilicon gate", *Tech. Dig., Symp. VLSI Technology*, pp. 17–18, May 1989.

[59] T. Morimoto, H. S. Momose, Y. Ozawa, K. Yamabe, and H. Iwai, "Effects of boron penetration and resultant limitations in ultra-thin pure-oxide and nitrided-oxide gate-films", *Tech. Dig. IEEE Int. El. Dev. Meeting (IEDM)*, pp. 429–432, 1990.

[60] J. R. Pfiester, F. K. Baker, T. C. Mele, *et al.*, "The effects of boron penetration on p⁺ polysilicon gated PMOS devices", *IEEE Trans. Electron Devices*, vol. 37, pp. 1842–1851, 1990.

[61] T. Hori and H. Iwasaki, "Improved hot-carrier immunity in submicrometer MOSFET's with reoxidized nitrided oxides prepared by rapid thermal processing", *IEEE Electron Device Lett.*, vol. 10, pp. 64–66, 1989.

[62] H. S. Momose, S. Kitagawa, K. Yamabe, and H. Iwai, "Hot carrier related phenomena for n- and p-channel MOSFETs with nitrided gate oxide by RTA", *Tech. Dig. IEEE Int. El. Dev. Meeting (IEDM)*, pp. 267–270, 1989.

[63] B. S. Doyle and A. Philipossian, "Role of nitridation/reoxidation of NH₃-nitrided gate dielectrics on the hot-carrier resistance of CMOS transistors", *IEEE Electron Device Lett.*, vol. 18, pp. 267–269, 1997.

[64] H. Wong, B. L. Yang, and Y. C. Cheng, "Chemistry of silicon oxide annealed in ammonia and its influence on interface traps", *Applied Surface Science*, vol. 72, pp. 49–54, 1993.

[65] H. Wong, and V. A. Gritsenko, "Defects in oxynitride gate dielectric", *Microelectron. Reliab.*, vol. 42, pp. 597–605, 2002.

[66] V. Gritsenko and H. Wong, "Atomic and electronic structures of traps in silicon oxide and silicon oxynitride", *Crit. Reviews in Solid State and Materials Sciences*, vol. 36, pp. 129–147, 2011.

[67] B. L. Crowder, and S. Zirinsky, "1 μm MOSFET VLSI technology: Part VII-metal silicide interconnection technology-A future perspective", *IEEE J. Solid-State Circuits*, vol. SC-14, pp. 291–293, 1979.

[68] S. P. Murarka, "Refractory silicides for low resistivity gates and interconnects", *Tech. Dig. IEEE Int. El. Dev. Meeting (IEDM)*, pp. 454–457, 1979.

[69] T. Shibata, K. Hieda, M. Sato, M. Konaka, R. L. M. Dang and H. Iizuka, "An optimally designed process for submicron MOSFETs", *Tech. Dig. IEEE Int. El. Dev. Meeting (IEDM)*, pp. 647–650, 1981.

[70] M. E. Alperin, T. C. Holaway, R. A. Haken, C. D. Gosmeryer, R. V. Karnaugh, and W. D. Parmantie, "Development of self-aligned titanium salicide process for VLSI applications", *IEEE Trans. Electron Devices*, vol. 32, pp. 141–149, 1985.

[71] J. B. Laskey, J. S. Nakos, O. J. Chan, and P. J. Geiss, "Comparison of transformation of low resistivity phase and agglomeration of TiSi₂, and CoSi₂", *IEEE Trans. Electron Devices*, vol. 38, pp. 262–269, 1991.

[72] T. Morimoto, H. S. Momose, T. Iinuma, *et al.*, "A NiSi salicide technology for advanced logic devices", *Tech. Dig. IEEE Int. El. Dev. Meeting (IEDM)*, pp. 653–656, 1991.

[73] T. Ohguro, S. Nakamura, M. Koike, *et al.*, "Analysis of resistance behavior in Ti- and Ni- salicided polysilicon films", *IEEE Trans. Electron Devices*, vol. 41, pp. 2305–2317, December 1994.

[74] H. Iwai, T. Ohguro, and H. Ohmi, "NiSi salicide technology for scaled CMOS", *Microelectron. Engineer.*, vol. 60, pp. 157–169, 2002.

[75] M. C. Poon, M. Wong, F. Deng, S. S. Lau, and H. Wong, "Thermal stability of cobalt and nickel silicides", *Microelectron. Reliab.*, vol. 38 pp. 1495–1498, 1998.

[76] M. Sun, M. Kim, J.-H. Ku, *et al.*, "Thermally robust Ta-doped Ni salicide process promising for sub-50 nm CMOSFETs", *Tech. Dig. Symp. VLSI Technology*, pp. 81–82, 2003.

1903   First motor flight by Orville and Wilbur Wright   ELECTRON DEVICES SOCIETY

[77] B. Honestain and S. C. Mead, "Fundamental limitations in microelectronics – I", *Solid-State Electronics*, vol. 15, pp. 819–892, 1972.

[78] H. S. Momose, M. Ono, T. Yoshitomi, T. Ohguro, S. Nakamura, M. Saito, and H. Iwai, "Tunneling gate oxide approach to ultra-high current drive in small-geometry MOSFETs", *Tech. Dig. IEEE Int. El. Dev. Meeting (IEDM)*, pp. 593–596, 1994.

[79] H. S. Momose, S. Nakamura, T. Ohguro, T. Yoshitomi, E. Morifuji, T. Morimoto, Y. Katsumata, and H. Iwai, "Uniformity and reliability of 1.5 nm direct-tunneling gate oxide Si MOSFETs", *Symp. on VLSI Technology*, pp.15–16, 1997.

[80] Y. Taur, D. A. Buchanan, W. Chen, *et al.*, "CMOS scaling into the nanometer regime", *Proc. IEEE*, vol. 85, No. 4, pp. 486–504, April 1997.

[81] R. Chau, S. Datta, M. Doczy, J. Kavalieros, and M. Metz., "Gate dielectric scaling for high-performance CMOS: from SiO$_2$ to high-k", *Ext. Abst. of International Workshop on Gate Insulator (IWGI)*, pp. 124–126, 2003.

[82] R. Tsuchiya, M. Horiuchi, S. Kimura, *et al.*, "Silicon on thin BOX: A new paradigm of the CMOSFET for low-power and high-performance application featuring wide-range back-bias control", *Tech. Dig. IEEE Int. El. Dev. Meeting (IEDM)*, pp. 631–634, 2004.

[83] F.-L. Yang, C.-C. Huang, Chi.-C. Huang, *et al.*, "45 nm node planar-SOI technology with 0.296 μm^2 6T-SRAM cell", *Symp. on VLSI Technology*, pp. 8–9, 2004.

[84] D. Esseni, and E. Sangiorgi, "Low field electron mobility in ultra-thin SOI MOSFETs: experimental characterization and theoretical investigation", *Solid-State Electron*, vol. 48, pp. 927–936, 2004.

[85] I. Aberg, C. N. Chleirigh, O. O. Olubuyide, X. Duan, and J. L. Hoyt, "High electron and hole mobility enhancements in thin-body strained Si/strained SiGe/strained Si heterostructure on insulator", *Tech. Dig. IEEE Int. El. Dev. Meeting (IEDM)*, pp. 173–176, 2004.

[86] K. Ismail, S. F. Nelson, J. O. Chu, and B. S. Meyerson, "Electron transport properties of Si/SiGe heterostructures: Measurements and device implications", *Appl. Phys. Lett.*, vol. 63, pp. 660–662, 1993.

[87] F. Ren, M. Hong, W. S. Hobson, *et al.*, "Demonstration of enhancement-mode p- and n-channel GaAs MOSFETS with Ga$_2$O$_3$(Gd$_2$O$_3$) as gate oxide", *Solid-State Electron.*, vol. 41, pp. 1751–1753, 1997.

[88] T. Mizuno, N. Sugiyama, T. Tezuka, Y. Moriyama, S. Nakaharai, and S. Takagi, "(110)-surface strained-SOI CMOS devices with higher carrier mobility", *Tech. Dig. VLSI Technology Symp.*, pp. 97–98, 2003.

[89] K. Mistry, C. Allen, C. Auth, *et al.*, "A 45 nm logic technology with high-k+metal gate transistors, strained silicon, 9 Cu interconnect layers, 193 nm dry patterning, and 100% Pb-free packaging", *Tech. Dig. IEEE Int. El. Dev. Meeting (IEDM)*, pp. 247–250, 2007.

[90] S.-H. Lo, D. A. Buchanan, Y. Taur and W. Wang, "Quantum mechanical modeling of electron tunnelling current from the inversion layer of ultra-thin-oxide nMOSFETs", *IEEE Electron Device Lett.*, vol. 18, pp. 209–211, 1997.

[91] W. J. Qi, R. Nieh, B. Hun Lee, L. Kang, Y. Jeon, K. Onishi, T. Ngai, S. Banerjee, and J. C. Lee, "MOSCAP and MOSFET characteristics using ZrO$_2$ gate dielectric deposited directly on Si", *Tech. Dig. IEEE Int. El. Dev. Meeting (IEDM)*, pp. 145–148, 1999.

[92] Y. Ma, Y. Ono, L. Stecker, D. R. Evans, and S. T. Hsu, "Zirconium oxide based gate dielectrics with equivalent oxide thickness of less than 1.0 nm and performance of submicron MOSFET using nitride gate replacement process", *Tech. Dig. IEEE Int. El. Dev. Meeting (IEDM)*, pp. 149–151, 1999.

[93] B. H. Lee, L. Kang, W. J. Qi, R. Nieh, Y. Jeon, K. Onishi, and J. C. Lee, "Ultrathin hafnium oxide with low leakage and excellent reliability for alternative gate dielectric application", *Tech. Dig. IEEE Int. El. Dev. Meeting (IEDM)*, pp. 133–136, 1999.

[94] E. P. Gusev, E. Cartier, D. A. Buchanan, M. Gribelyuk, M. Copel, H. Okorn-Schmidt, and C. D'Emic, "Ultrathin high-k metal oxides on silicon: processing, characterization and integration issues", *Microelectron. Eng.*, vol. 59, pp. 341–349, 2001.

[95] G. D. Wilk, R. M. Wallace, and J. M. Anthony, "High-k gate dielectrics: Currnet status and materials properties considerations", *J. Appl. Phys.*, vol. 89, pp. 5243–5276, 2001.

1904 — Otto Lehmann publishes on fundamentals of liquid crystals — ELECTRON DEVICES SOCIETY

[96]   G. Bersuker, P. Zeitzoff, G. Brown and H. R. Huff, "Novel dielectric materials for future transistor generations", *Mater. Today*, pp. 26–33, 2004.

[97]   H. Wong, K. L. Ng, N. Zhan, M. C. Poon, and C. W. Kok, "Interface bonding structure of hafnium oxide prepared by direct sputtering of hafnium in oxygen", *J. Vac. Sci. Technol. B*, vol. 22, pp. 1094–1100, 2004.

[98]   A. A. Rastorguev, V. I. Belyi, T. P. Smirnova, L. V. Yakovkina, M. V. Zamorynskaya, V. A. Gritsenko, and H. Wong, "Luminescence of intrinsic and extrinsic defects in hafnium oxide films", *Phys. Rev. B*, vol. 76, 235–315, 2007.

[99]   D. Misra, H. Iwai, and H. Wong, "High-k gate dielectrics", *ECS Interface*, vol. 14, pp. 30–32, 2005.

[100] B. Doyle, B. Boyanov, S. Datta, M. Doczy, S. Hareland, B. Jin, J. Kavalieros, T. Linton, R. Rios, and R. Chau, "Tri-gate fully-depleted CMOS transistors: fabrication, design and layout", *Symp. VLSI Technology*, pp. 133–134, 2003.

[101] D. J. Frank, S. E. Laux, and M. V. Fischetti, "Monte Carlo simulation of a 30 nm dual-gate MOSFET: How short can we go?", *Tech. Dig. IEEE Int. El. Dev. Meeting (IEDM)*, pp. 553–556, 1992.

[102] D. J. Frank, Y. Taur, and H.-S. P. Wong, "Generalized scale length for two-dimensional effects in MOSFETs", *IEEE Electron Dev. Lett.*, vol. 19, No. 10, pp. 385–387, October 1998.

[103] H. Iwai, "Roadmap for 22 nm and beyond", *Microelectron. Engineer.*, vol. 86, pp. 1520–1528, 2009.

[104] T. Skotnicki, C. F.-Beranger, C. Gallon, *et al.*, "Innovative materials, devices, and CMOS technologies for low-power mobile multimedia", *IEEE Trans. Electron Devices*, vol. 55, pp. 96–130, 2008.

[105] B. Yu, L. Wang, Y. Yuan, P. M. Asbeck, and Y. Taur, "Scaling of nanowire transistors", *IEEE Trans. Electron Devices*, vol. 55, pp. 2846–2858, 2008.

[106] H. Iwai, "CMOS technology after reaching the scale limit", *Proc. IEEE Workshop on Junction Technology*, pp. 1–2, 2008.

[107] H. Iwai, K. Natori, K. Shiraishi, J. Iwata, A. Oshiyama, K. Yamada, K. Ohmori, K. Kakushima, P. Ahmet, "Si nanowire FET and its modeling", *Science China*, vol. 54, pp. 1004–1011, 2011.

[108] K. Kakushima, K. Okamoto, T. Koyanagi, *et al.*, "Selection of rare earth silicates for highly scaled gate dielectrics", *Microelectron. Engineer.* vol. 87, pp. 1868–1871, 2010.

[109] T. Kawanago, Y. Lee, K. Kakushima, *et al.*, "EOT of 0.62 nm and high electron mobility in La-silicate/Si structure based nMOSFETs achieved by utilizing metal-inserted Poly-Si stacks and annealing at high temperature", *IEEE Trans. Electron Devices*, vol. 59, p. 269, 2012.

[110] M. P. Lepselter and S. M. Sze, "SB-IGFET, An insulated-gate field-effect transistor using Schottky barrier contacts as source and drain", *Proc. IEEE*, vol. 56. pp. 1400–1402, 1968.

[111] N. Shigemori, S. Sato, K. Kakushima, *et al.*, "Suppression of lateral encroachment of Ni silicide into Si nanowires using nitrogen incorporation", *ECS 218th Meeting*, 2010.

[112] T. Hoshii, M. Deura, M. Sugiyama, *et al.*, "Epitaxial lateral overgrowth of InGaAs on SiO_2 from (111) Si micro channel areas", *Phys. Stat. Solidi*, vol. 5, pp. 2733–2735, 2008.

[113] S. H. Kim, M. Yokoyama, N. Taoka, *et al.*, "High performance extremely-thin body III-V-on-insulator MOSFETs on a Si substrate with Ni-InGaAs metal S/D and MOS interface buffer engineering", *Symp. VLSI Technology*, pp. 58–59, 2011.

[114] R. Zhang, T. Iwasaki, N. Taoka, M. Takenaka, and S. Takagi, "High mobility Ge pMOSFETs with ∼ 1nm thin EOT using Al_2O_3/GeO_x/Ge gate stacks fabricated by plasma post oxidation", *Symp. VLSI Technology*, pp. 56–57, 2011.

[115] A. Toriumi, C. H. Lee, S. K. Wang, *et al.*, "Material potential and scalability challenges of germanium CMOS", *Tech. Dig. IEEE Int. El. Dev. Meeting (IEDM)*, pp. 646–649, 2011.

[116] M. V. Fischetti, L. Wang, B. Yu, C. Sachs, P. M. Asbeck, Y. Taur, and M. Rodwell, "Simulation of electron transport in high-mobility-MOSFETs: density of states bottleneck and 'source starvation'", *Tech. Dig. IEEE Int. El. Dev. Meeting (IEDM)*, pp. 109–112, 2007.

1905   Albert Einstein publishes his Theory of Relativity   ELECTRON DEVICES SOCIETY $E = mc^2$

Chapter 3

Memory Devices

Kinam Kim and Dong Jin Jung

3.1 Introduction

It is very important in the evolution of memory devices to reminisce about an automatic machine that can perform a series of well-defined steps for solving a problem at a very fast speed with no error. This device can be called a calculator. In that sense, Atanasoff and Berry invented the first (electronic) calculator between 1937–1942. This machine stood out as a very important milestone because of conceiving three critical concepts that still hold in modern computers. First, it used punched cards for data input and output, which means that data processing was carried out in a binary format. Secondly, the data processing was electronic rather than mechanic due to use of both vacuum tubes and rotating drums for memory via regenerative capacitors, which were similar to today's DRAM (Dynamic Random Access Memory) capacitors. Thirdly, the memory part was separated from the computational part. Today's elements of computation consist of logic gates and flip-flops, both of which can execute any Boolean operation and read and write data in connection with memory devices. Thanks to computers and memory devices, global sales on semiconductors have been more than quadrupled over the past two decades, more precisely between 1988–2008, leading to a 9% compound annual growth rate [1]. These advances were initially foreseen in 1965 when G. Moore published his famous prediction about the constant growth rate of chip complexity; in fact, it has repeatedly been proven that the number of transistors integrated into silicon chips has indeed doubled every

Guide to State-of-the-Art Electron Devices, First Edition. Edited by Joachim N. Burghartz.
© 2013 John Wiley & Sons, Ltd. Published 2013 by John Wiley & Sons, Ltd.

John Ambrose Fleming invents oscillation valve as a detector of wireless transmission

This invention is often viewed as the birth of electronics

ELECTRON DEVICES SOCIETY

18 months [2]. Such remarkable achievements owe much to successful inventions pioneered in the past by many researchers. Back in the 1940s, scientists desperately sought fundamental alternatives to vacuum tubes, devices capable of rectifying, amplifying and switching electrical signals mostly for telephonic applications. This is because the vacuum tubes raised several fundamental issues: huge power consumption, the heating up of the cathode filament due to poor controllability, and a long set-up time for a proper operation – all associated with the thermionic emission of electrons in the vacuum tube. Therefore, in 1947, J. Bardeen, W. Brattain and W. Schockley at Bell Laboratories succeeded in demonstrating the first solid-state electronic device capable of replacing vacuum tubes [3]. The device could control large changes in current flow by comparably small input signals, leading to a huge gain of the electronic signals. Therefore, its name 'transistor' came from the words 'transfer' and 'resistor' according to its operational principle that transfers an input current signal from a low resistance circuit to a high resistance one at the output (see also Chapter 1). Without MOSFET (the Metal-Oxide-Semiconductor Field Effect Transistor), neither Moore's law would have been realized nor conventional *shrink technology* been feasible. This is because of two important parameters for device scaling: oxide thickness and junction depth. In 1959, such MOS devices were demonstrated by Atalla and Khang, who were able to overcome surface states that block the electric field from penetrating into the semiconductor through the thermally grown silicon-dioxide [4, 5]. Moreover, in 1963, Wanlass and Sah introduced CMOS (Complementary MOS) logic, virtually providing zero leakage-current, which means that cross-coupled CMOS inverters can hold a bi-stable bit with any access transistors [6]. In other words, the SRAM (Static Random Access Memory) became feasible in a reasonable packing density even with low standby power. (Note: This is not necessarily true because the stand-by current of SRAM begins to exceed the DRAM's at the deep sub-micron scale due to the high electric field at the junction.)

The advent of the computer in the 1940s, together with scalable MOS transistors appearing in the 1960s, led to the emergence of a new era in which massive and speedy memory devices evolved for high performance computation. From the point of view of functionality, none of the MOS memory devices satisfy both aspects 'massive and speedy' at the same time. For example, the read and write speed of a SRAM (cache memory) is extremely fast but only provides a moderate memory density (typically several hundreds of megabytes). Communicating data between the CPU (Central Processing Unit) and memory itself, DRAM requires a high speed (not much faster than the cache memory) and its memory size is relatively large, amounting

Cache Memory

File system cache is kind of physical memory that stores recently used data as long as possible to allow quicker access to that data rather than retrieving the data from the hard disk. Data throughput rate of the latest HDD is still orders of magnitude slower than the processor/system-memory clock speed. For optimum performance of a computer it is important to bridge any performance gap in between the components. To improve the gap between CPU and main memory, a CPU cache is used. Computers use two types of cache memory: Level-1 cache integrated with the processor, and Level 2 cache consisting of very fast SRAMs and placed outside of the processor.

1906 Lee De Forest receives the patent on the tri-gate vacuum tube (triode)

ELECTRON DEVICES SOCIETY

today to several gigabytes. Both DRAM and SRAM are volatile memories, that is, the bit-storing capability vanishes when power is turned off. Auxiliary memories such as HDD (Hard Disk Drive) and SSD (Solid-State Disk) are generally much more capacious than most others. Code memory such as the EEPROM (Electrically Erasable and Programmable Read Only Memory) has also to be taken into account when considering auxiliary memories. They are non-volatile because the information is preserved even without any power supply. In this chapter, we first illustrate how both volatile (Section 3.2) and non-volatile MOS (Section 3.3) memories have penetrated technological barriers to meet the expectations set by Moore's law. In these two categories we focus our discussion on the DRAM evolution, the volatile memory with the highest on$-$off ratio of array transistors, and on the NAND (Not AND) flash, the non-volatile memory that is very similar to a ROM (Read-Only Memory). Though most experts believe that silicon technology will maintain its leadership down to the critical 20 nm dimension, beyond this node a number of fundamental and application-specific obstacles may prevent further shrinkage of established memory devices. Therefore, another aim of this chapter is to present various possible paths to overcome these obstacles and eventually to maintain the technology-scaling trend beyond 20-nm node in Section 3.4.

3.2 Volatile Memories

For volatile memory devices, off-leakage current (I_{OFF}) is regarded as a considerably important factor and thus volatile memory technologies tend to be developed with a greater emphasis on reducing I_{OFF}. Since the first 1024-bit DRAM was demonstrated by Intel^{TM} in the early 1970s [7], DRAM technology has now reached the 20-nm node and 4 Gb (gigabit) in packing density. In DRAM with only one transistor and one capacitor per bit, charging and discharging properties of cell capacitors depend strongly on performance of cell array transistors (CATs). On-current (I_{ON}) of the CAT plays a critical role in its charging behaviours, while CAT's I_{OFF} is a decisive factor to determine the data retention time under standby condition. On the one hand, I_{ON} needs to be greater than 10^{-6} Amps to achieve a reasonable read and write speed. On the other, I_{OFF} has to be kept below a level of 10^{-16} Amps to minimize charge loss after having charged up the cell capacitors to ensure adequate sensing-signal margin, thus resulting in adequate

A Negative Word-Line (NWL) Scheme in DRAM

In a ground word-line (GWL) scheme, back-bias voltage (V_{BB}) has in general been utilized to reduce junction capacitance; to improve cutoff characteristics; and to enhance latch-up endurance. Amid a trend of lowering V_{TH} down for more I_{ON}, another external bias has been needed to govern sub-threshold leakage current, by providing CATs' word-lines with negative voltage. Since a level of *dc* (direct current) bias at unselected word-lines is negative, the sub-threshold leakage current becomes extremely low because its channel has never the chance to be at the on-set of inversion, in order to maintain a reasonable level of off-leakage current despite the low V_{TH}.

data retention time. In spite of relentless technology migration, the $I_{ON}-I_{OFF}$ ratio has continued to be 10^{10}, approximately. DRAM CATs have evolved to meet such requirements. However, the higher density memory requires the shorter in transistor's channel length and results in a higher level of the channel doping to block the short-channel-effect (SCE). The increased electric field in CATs is therefore causing a higher I_{OFF} which increases even more due to contributions of leakage current: sub-threshold leakage current; gate-induced drain leakage (GIDL) current; and also junction leakage current from the storage node. Therefore, the degradation of the data retention time becomes significant below the 100 nm node. Since the mid-2000s this issue has been overcome by introducing 3D (three-dimensional) cell transistors, where the junction electric-field can be greatly reduced due to lightly doped channel. One example is the RCAT (Recess Channel Array Transistor) structure whose channel surrounds a part of silicon substrate so that the elongated channel can be embodied in the array transistor [8]. Further innovations for the $I_{ON}-I_{OFF}$ ratio have been pursued in a way of a negative word-line (NWL) scheme.

The NWL scheme allows to reduce CAT's threshold voltage (V_{TH}), and thus to achieve more I_{ON}. However, since CAT's gate potential turns more negative during holding data stored at the storage junction, the GIDL current increases as a function of gate-to-storage-node voltage. Many device engineers have made a lot of effort to tackle this problem and finally ended up with exploiting different gate-oxide thickness [9].

Furthermore, a body-tied finFET (fin field-effect-transistor) appeared as a CAT because of wrapping the gate for maximizing gate controllability (low V_{TH} and sub-threshold swing), providing more I_{ON} (see Figure 3.1a). Note, that the basic structure of the body-tied finFET was firstly proposed in 1990 by D. Hisamoto et al. to reduce the SCE [10].

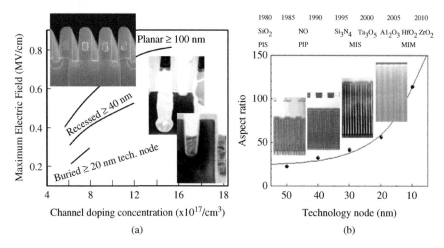

Figure 3.1 (a) Changes in maximum electric field as a function of channel doping concentration, together with evolution trends of DRAM's cell array transistors (CATs). *Insets*: Cross-sectional views of planar-type (PCAT) upper, recess-type (RCAT) middle and buried-type (BCAT) lower. (b) A trend of changes in aspect ratios of cell capacitors as a function of technology node. *Insets:* Cell capacitor's micrographs (S.J. Hong, *IEDM*, 2010). (a) and (b) are reproduced with permission from S. K. Park, "The future memory technologies," SEMATECH Symposium Korea, 27–28 October 2011, © Hynix Semiconductor Inc.

Meanwhile, DRAM's capacitor has changed to a stacked 2D (two-dimensional) structure, integrated under the bit-line in the process architecture until the mid-1990s. The stacked capacitor was first proposed, in 1978, by M. Koyanagi *et al.* [11] and was commercially realized in 1985 [12]. Since then, DRAM has changed in structure to have an integration scheme of cell-capacitors placed over bit-line (COB) though there was an attempt to use trench-type capacitors, which are buried deeply into the silicon substrate. To maintain almost non-scalable requirement of cell capacitance, more or less 20 fF per cell, DRAM's cell capacitors have evolved in two ways: One is to expand the capacitor's area as much as possible, while the other is to use high-κ dielectrics. In the 1990s, the dielectric materials of the cell capacitors have adopted silicon-based dielectrics, SiO_2/Si_3N_4, whose dielectric constant lies in between 3.9–7.0. With these relatively low-κ dielectrics, a cell capacitor has headed for expanding its area. Thus, its structure has been transformed in substantially complex ways, from a simple stack to a hemi-spherical-silicon-grain (HSG) stack, and finally to a HSG cylinder until the late 1990s [13]. High-κ dielectrics since the beginning of twenty-first century have opened a new way of building cell capacitors. In high-κ dielectrics, increase in cell capacitance can be achieved simply by increasing the height of the cylinder. Such an increase in height gives rise to skyrocketing of aspect ratios of cell capacitors when technology scales due to dramatic decrease in footprint. Typically, an aspect ratio ranges from 6–9 until the 100-nm technology node. A higher aspect ratio has brought up another issue, that is, in building cell capacitors mechanically robust. As a result of mechanical instability of OCS (one-cylinder-stack) structures, the silicon industry has introduced a novel type of capacitor structure, that is, the supporter-added OCS, such as mesh type cell capacitors, which can increase the cell capacitor height along with the desired mechanical stability [14]. Taking into account the recent advances of DRAM cell capacitor technology, the aspect ratio reaches 40 to 50, which is far beyond those of the world tallest skyscrapers, ranging from 8–10, as seen in Figure 3.1(b).

3.3 Non-Volatile Memories

When turning on a computer, a set of basic instructions is needed for starting up the system. This instruction set checks first whether all input/output states in the system are in order, next looks for a storage device containing a copy of operating system (OS) and then loads the OS program into the RAMs. Thus, these instruction codes need to be stored in some specific memory that is never interfered by any power failure. That specific memory should be ROM-type and definitely non-volatile because information loss during power interruption must have no effect. The first non-volatile MOS memory was suggested in 1967 by D. Khang and S. Sze [15]. Due to critical drawbacks of tunnel-oxide imperfections at that stage of process technology, this first concept was not realized until Frohman-Bentchkowsky introduced a FAMOS (Floating gate Avalanche injection MOS) device in 1971 [16]. The FAMOS device has evolved into a class of memory products called EPROM (Electrically Programmable ROM). However, the original EPROM had no mechanism of electrical erasure due to lack of an external gate. Therefore, the erasure function was performed by UV (ultra-violet) or X-ray irradiation. Next, by adding an external gate, electrical erasure became possible via field emission through the top dielectric due to poly-oxide conduction. The first nonvolatile memory device relying on Fowler-Nordheim (FN) tunneling for both write and erase steps was proposed in 1978 [17]. Incorporating the thin oxide region overlapped between floating gate and drain, this FN device was introduced in 1980 as an EEPROM transistor. Struggling against a pertinent issue of how effectively to erase and then program a ROM, researchers finally found a solution of global erase in one step (i.e. a flash) with this EEPROM.

Today's NAND flash memory has the smallest cell size among silicon-memory devices commercially available due to its simple one-transistor configuration per bit and a serial connection of multiple cells in a string. Because of this, NAND flash has carved out a huge market for itself, as was expected since it first appeared in 1984 by F. Masuoka *et al.* [18]. As a consequence of the rise of the mobile communication era, NAND flash has evolved toward an ever-smaller cell size in two ways: by increasing string size and by developing two-bit-per-cell technology which was suggested in 1995 by M. Bauer *et al.* [19]. Since the later 1990s flash memory has continued to migrate the technology node to 70 nm until the beginning of 2000s, using a floating gate (FG) [20, 21]. Then, MLC (Multi-Level Cell) technology has driven flash memory to much higher packing density during the mid-2000s [22, 23].

In the meantime, from a scaling point of view, flash memories have faced a serious problem at and beyond the 50 nm technology node: Cell-to-cell separation becomes so small that influence of adjacent cells cannot be ruled out. This is often posed not only by physical aspects of cell structures but also by certain aspects of its performance. To circumvent the cell-to-cell interference, the width of a floating gate tends to be more aggressively squeezed than the space between floating gates. This results in a high aspect ratio of a gate stack. Such a high aspect ratio can provoke fabrication difficulty of memory cells due to its mechanical instability. And stored charge (e.g., electrons) in a floating gate can redistribute easily in operating conditions, leading to vulnerability of poor data retention. Since the interference originates from another type of coupling between floating gates (FGs), it is desirable to find innovative structures, where charge storage media do not have a continuum of charge like the floating gate style but have a discrete sort such as charge traps (CTs) in a nitride layer. The typical examples are non-volatile memories with non-floating gate, for example, SONOS (silicon-oxide-nitride-oxide-silicon) [24], SANOS (silicon-alumina-nitride-oxide-silicon) [25], TANOS (TaN-alumina-nitride-oxide-silicon) [26] or nanocrystal dots [27, 28]. Recently, a 32 Gb flash memory has been reported at the 40-nm technology node [29]. This pioneering achievement is based on a novel structure with a high-κ dielectric of Al_2O_3 as the top oxide and TaN as a top electrode. With this approach, it is possible to achieve several essential properties for the NAND flash memory: reasonable programming/erasing characteristics, an adequate window for multi-bit operation and robust reliability. As technology migrates, a body-tied finFET could be a very promising candidate because it can increase storage electrons effectively by expanding channel width of cell transistors, similar to 3D CATs in the DRAM. In this pursuit, a research group has successfully developed flash memory with a TANOS structure based on a 3D, body-tied finFET [30], where they can obtain excellent performance of NAND-flash cells with robust reliability. If there are much higher-κ dielectrics compared to Al_2O_3, then further downscaling of the finFET CT-NAND flash memory should be feasible. Beyond the 20 nm node, we believe that the most likely way to increase memory density is to stack cells vertically. In line with this, there has been a report of a vertical NAND flash memory [31], where a string of memory cells is built vertically by stacking up the memory cells rather than horizontally by squeezing the lateral dimensions. As such, the vertical memory cells via 3D technology looms on the horizon, in particular, for the NAND flash memory (see Figure 3.2).

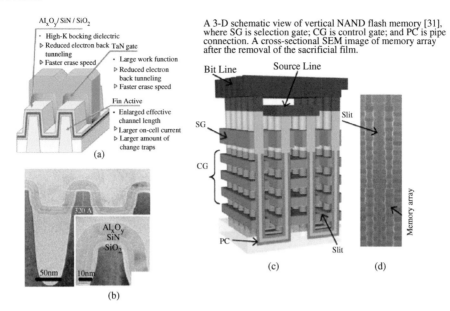

Figure 3.2 (a) A schematic diagram of 3D, body-tied FinFET NAND cells [30]. (b) A micrograph of the 3D TANOS NAND flash cell [30]. (c) A 3D schematic view of vertical NAND flash memory [31], where SG is selecting gate; CG is control gate; and PC is pipe connection. (d) A cross-sectional SEM image of memory array after the removal of the sacrificial film. Parts (a) and (b) are reproduced with permission from © 2006 IEEE. Parts (c) and (d) are reproduced with permission from © 2009 Toshiba Co. Ltd.

3.4 Future Perspectives of MOS Memories

As the era of 2D, planar-based *shrink technology* is coming to an end, semiconductor institutions face enormous hurdles to keep up with the pace of miniaturization according to the Moore's Law and to meet the requirements of the highly demanding applications in mobile products. They have attempted to tackle those barriers by smart and versatile approaches of 3D technology in integration hierarchy. One strand of the responses is to modify structures of elementary constituents such as DRAM's CATs and storage transistors of flash memory, as shown in Figure 3.2(a). A second thread revisits these modifications to a higher level of integration: memory stacking [32, 33]. And yet another move is to upgrade this into a system in a way of fusing of each device in functionality by utilizing smart CMOS technology, for example, *through-silicon-via* (TSV) (see Figure 3.3c). As the use of 3D silicon technology has great potential to migrate today's information technology (IT) devices into a wide diversification of multi-functional

Memory Stacking

Along with 3D nFETs, one of the memory manufacturers has implemented 3D integration technology with SRAM to reduce large cell-size [32]. Since transistors stacked onto a given area do not need to isolate p-well to n-well, SRAM-cell size of $85-100$ F^2, where the minimum feature size F is de ned as half pitch of bit-line or word-line, was reduced to 25 F^2. Encouraged by this successful approach, stacked ash memory has also been demonstrated successfully [33].

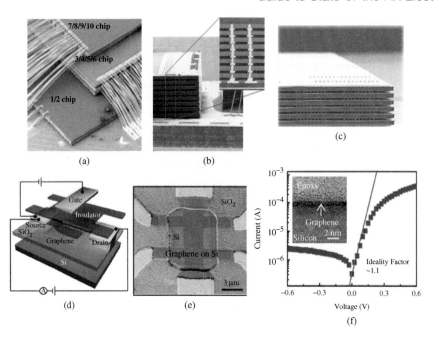

Figure 3.3   (a) A bird's eye view of multi-chip-package (MCP) by wire bonding. (b) A wafer-level stack package with *through-via-hole*. (c) A micrograph of 3D integrated circuits using *through-silicon-via* (TSV). (d) A schematic drawing of a graphene-silicon barristor [36]. (e) A False-coloured SEM image of the barristor before the top gate fabrication. (f) A current-voltage characteristic of the barristor. Parts (d), (e) and (f) are reproduced from H. J. Yang *et al.*, p. 2. *Sciencexpress,* May 17, 2012, © 2012 American Association for the Advancement of Science (AAAS)

electronics, it can also stimulate a trend that merges one technology with another, ranging from *new materials* through *new devices* to *new concepts* [34−36]. In that sense, there has been a recent report of a very innovative approach to improve an on-off ratio by adopting new materials, such as the graphene-silicon barristor with a gate-controlled Schottky barrier [36]. We believe that this invention has great potential in the memory evolution because of the followings: Firstly, high-mobility materials, for example, electron/hole mobility in graphene is several orders of magnitude higher than that in silicon, which means it behaves metallic, were successfully implemented in conventional CMOS technology, even on a 6-inch wafer. This shows that bringing high-functional materials to mass production is not in a dream any more. Secondly, the graphene-silicon barristor has been controlled accurately by a gate, giving rise to an $I_{ON}-I_{OFF}$ ratio of $\sim 10^5$ despite the early stage of the development for this device (see Figure 3.3d−f).

This suggests that it could eventually perform at a level required for memory array switches. Third is that fabrication has been carried out in a vertical way. This means that elemental 3D devices in integration technology can allow us to have massive capability both in memory size and in performance. Therefore, introduction of very powerful personal gadgetry is waiting for us in a not too distant future.

1916   Czochralski crystal growth process invented (Jan Czochralski)   ELECTRON DEVICES SOCIETY   1917   1918

3.5 Closing Remarks

Not only are there many challenges and obstacles to be met and overcome by the silicon industry as technology enters the deep nanometer-dimension era but there are also promising opportunities. Supported by new technologies such as 3D devices and by a wealth of new materials, we will overcome many hurdles ahead of us and respond to technological challenges in the years to come. All plausible solutions described earlier tell us that planar-based technology will reach an impassable limit. Three-dimensional technology begins to provide clear signs of serving as a foundation for refueling the silicon industry. The advantages of 3D integration are numerous. They include: elimination of variability in the electrical characteristics of deep nanometre-scale transistors; extendable use of silicon infrastructures (e.g., optical lithography tools); and formation of a baseline for multi-functional electronics, facilitating to implement a hierarchical architecture, where each layer is dedicated to a specific functional purpose. Over the next decade, we will see endeavors in numerous areas that will greatly stimulate the semiconductor business. Successful evolutions of device structures will continue and even accelerate at a greater pace in the not-too-distant future. In addition, device designs will converge onto a single mobile platform, covering many different capacities and services from telecommunication through broadcasting and a much higher degree of data processing. In line with this, silicon technology will still play a critical role in realizing merged solutions for functionality. All of these will allow us to have invaluable clues not just on how to prepare future silicon technology, but also on how to positively influence the entire silicon industry. This will enable us to attain an even more sophisticated fusing of technologies. As seen in the past, silicon technology will continue to provide our society with versatile solutions and as-yet unforeseen benefits in much more cost-effective ways. Today's prosperity of silicon industries came mostly from MOS memories based on MOSFETs that can control surface states via a thermally grown silicon-dioxide. Very recently, there has been a very interesting report that a graphene-silicon barristor can have the dramatic $I_{ON}-I_{OFF}$ ratio by achieving work-function controls through graphene-silicon contacts. This might suggest another marked milestone in semiconductor technology.

Performance Bottleneck

Chip performance increases annually by 70% coming mostly from device scaling (50%) and transistor's speed (15%), while pin bandwidth is improved at an annual rate by almost 30% as pin counts on PCB (Printed Circuit Board) is slowly increased (10%) due to lack of interconnect resources [34]. This discrepancy between the two used to be overcome by increase in data rates. Raising the data rate is now reaching the limit and causing rapid increase in input/output power dissipation. It is hence desired to nd an effective way of chip stacking. One example of this is through-silicon-via (TSV) and another is inductive coupling, that is, through-chip interface (TCI).

End of World War I | 1919 | ELECTRON DEVICES SOCIETY | 1920 | Term "Electronics" appears

References

[1] R. Chitkara and W. Ballhous, *"A Change of Pace for Semiconductor Industry?"*, Available at http://www.pwc.com/
     en_GX/gx/technology/pdf/change-of-pace-in-the-semiconductor-industry.pdf (accessed October 15, 2012), PriceWa-
     terhouseCoopers, p. 19, 2009.

[2] G. Moore, "Cramming more components onto integrated circuits", *Electronics Magazine*, 19, pp. 24–27, April 1965.

[3] J. Bardeen and W. H. Brattain, "The transistor, a semiconductor triode", *Phys. Rev.*, 74, p. 230, 1948.

[4] M. M. Atalla, E. Tannenbaum and E. J. Scheibner, "Stabilization of silicon surfaces by thermally grown oxides",
     Bell System Technical Journal, 38, pp. 749–783, May 1959.

[5] D. Khang, "Silicon-silicon dioxide field induced surface devices", Technical memorandum issued by Bell Labs (16
     January, 1961) reprinted in Sze, S. M. (ed.) *Semiconductor Devices: Pioneering Papers*, pp. 583–596, Singapore:
     World Scientific Publishing Co., 1991.

[6] F. M. Wanlass, and C. T. Sah, "Nanowatt Logic Using Field-Effect Metal-Oxide Semiconductor Triodes", *Dig.
     Techn. P. IEEE International Solid State Circuits Conference (ISSCC)*, pp. 32–33, February 20, 1963.

[7] R. H. Dennard, "Evolution of the MOSFET dynamic RAM - A personal view", *IEEE Trans. on Electron Devices*, 31,
     Issue 11, pp. 1549–1555, November 1984; R. H. Dennard, "Field-effect transistor memory", *U. S. Patent 3,387,286*,
     Filed 14 July, 1967. Issued 4 June, 1968.

[8] J. Y. Kim, C. S. Lee, S. E. Kim, *et al.*, "The breakthrough in data retention time of DRAM Using Recess-Channel-
     Array Transistor (RCAT) for 88 nm feature size and beyond", Dig. Tech. Papers *VLSI Technology Symposium*,
     pp. 11–12, June 2003.

[9] D. J. Jung, E. S. Lee, H. W. Seo, *et al.*, "Data-retention time in 40 nm of technology node", *Proc. 32nd Conference
     of Semiconductor in Samsung*, December 2009; J. Lee, J. W. Lee, J. Y. Lee *et al.*, "GiDL engineering in 50 nm of
     technology node", *31st. Conference of Semiconductor in Samsung*, December 2008.

[10] D. Hisamoto, T. Kaga, Y. Kawamoto and E. Taketa, "A Fully Depleted Lean-Channel Transistor (DELTA)-A Novel
     Vertical Ultrathin SOI MOSFET", *IEEE Electron Device Lett.*, 11, pp. 36–38, January 1990.

[11] M. Koyanagi, H. Sunami, N. Hashimoto and M. Ashikawa, "Novel high density, stacked capacitor MOS RAM",
     Techn. dig. IEEE International Electron Devices Meeting (IEDM), pp. 348–351, December 1978.

[12] Y. Takemae, T. Ema, M. Nakano, *et al.*, "A 1-Mb DRAM with 3-dimensional stacked capacitor cell", *Dig. Techn.
     P. IEEE International Solid State Circuits Conference (ISSCC)*, pp. 250–251, February 1985.

[13] M. Sakao, N. Kasai, T. Ishijima, *et al.*, "A capacito-over-bit-line (COB) cell with a hemishperical-grain storage node
     for 64 Mb DRAMs", *Techn. Dig. IEEE International Electron Devices Meeting (IEDM)*, pp. 655–658, December
     1990.

[14] D. Kim, J. Y. Kim, M. Huh, *et al.*, "A mechanically enhanced storage node for virtually unlimited height (MESH)
     capacitor aiming at sub 70 nm DRAMs", *Techn. Dig. IEEE International Electron Devices Meeting (IEDM)*, pp.
     69–72, December 2004.

[15] D. Khang and S. M. Sze, "A floating gate and its application to memory devices", *Bell System Technical Journal*,
     46, p. 1288, May 1967.

[16] D. Frohman-Bentchkowsky, "Memory behaviour in a floating gate avalanche injection MOS (FAMOS) structure",
     Appl. Phys. Lett., 18, p. 332, 1971.

[17] E. Harari, L. Schmitz, B. Troutman and S. Wang, *et al.*, "A 256bit non-volatile static RAM", *Dig. Techn. P. IEEE
     International Solid State Circuits Conference (ISSCC)*, pp. 108–109, 1978.

[18] F. Masuoka, M. Asano, H. Iwahashi, *et al.*, "A new Flash EEPROM cell using triple polysilicon technology", *Techn.
     Dig. IEEE International Electron Devices Meeting (IEDM)*, pp. 464–467, December 1984.

[19] M. Bauer, R. Alexis, G. Atwood, *et al.*, "A Multilevel-Cell32Mb Flash Memory", *Dig. Techn. P. IEEE International
     Solid State Circuits Conference (ISSCC)*, pp. 132–133, 1995.

[20] S. Keeny, "A 130nm Generation High Density Etox Flash Memory Technology", *Techn. Dig. IEEE International Electron Devices Meeting (IEDM)*, pp. 2.5.1–4, December 2001.

[21] Y. Yim, K. S. Shin, S. H. Hur, *et al.*, "70nm NAND flash technology with 0.025 μm^2 cell size for 4Gb flash memory", *Techn. Dig. IEEE International Electron Devices Meeting (IEDM)*, pp. 34.1.1–4, December 1995.

[22] J. Park, H. Hur, J. Lee, *et al.*, "8Gb MLC (Multi-Level Cell) NAND flash Memory using 63 nm Process Technology", *Techn. Dig. IEEE International Electron Devices Meeting (IEDM)*, pp. 873–876, December 2004.

[23] D. Byeon, S. Lee, Y. Lim, *et al.*, "An 8 Gb multi-level NAND flash memory with 63 nm STI CMOS process technology", *Dig. Techn. P. IEEE International Solid-State Circuits Conference (ISSCC)*, pp. 46–47, February 2005.

[24] S. Mori, E. Sakagami, H. Araki, *et al.*, "ONO inter-poly dielectric scaling for nonvolatile memory applications", *IEEE Transactions on Electron Devices*, 38, No. 2, pp. 386–391, August 1991.

[25] C. Lee, S. Hur, Y. Shin, *et al.*, "Charge-trapping device structure of SiO_2/SiN/high-*k* dielectric Al_2O_3 for high-density flash memory", *Appl. Phys. Lett.*, 86, pp. 152908-1–152908-3, April 2005.

[26] Y. Shin, J. Choi, C. Kang, *et al.*, "A Novel NAND-type MONOS Memory using 63 nm Process Technology for Multi-Gigabit Flash EEPROMs", *Techn. Dig. IEEE International Electron Devices Meeting (IEDM)*, pp. 327–330, December 2006.

[27] S. Tiwari, F. Rana, K. Chan, *et al.*, "Volatile and nonvolatile memories in silicon with nano crystal storage, *Electron Devices Meeting*", *Techn. Dig. IEEE International Electron Devices Meeting (IEDM)*, pp. 521–524, December 1995.

[28] A. Nakajima, T. Futatsugi, H. Nakao, *et al.*, "Microstructures and electrical properties of Sn nanocrystals in thin thermally grown SiO_2 formed via low energy implantation", *Journal of Applied Physics*, 84, no 3, pp.1316–1320, August 1998.

[29] Y. Park, J. Choi, C. Kang, *et al.*, "Highly manufacturable 32 Gb multi-level NAND flash memory with 0.0098 nm^2 cell size using TANOS (Si-oxide-nitride-Al_2O_3-TaN)", *Techn. Dig. IEEE International Electron Devices Meeting (IEDM)*, pp. 1–4, December 2006.

[30] S. Lee, J. Lee, J. Choe *et al.*, "Improved post-cycling characteristic of FinFET NAND flash", *Techn. dig. IEEE International Electron Devices Meeting (IEDM)*, pp. 1–4, December 2006.

[31] R. Katsumata, M. Kito, Y. Fukuzumi, *et al.*, "Pipe-shaped BiCS Flash Memory with 16 Stacked Layers and Multi-Level-Cell Operation for Ultra High Density Storage Devices", *Dig. Tech. Papers VLSI Technology Symposium*, pp. 136–137, June 2009.

[32] S. Jung, J. Jang, W. Cho, *et al.*, "The Revolutionary and Truly 3-Dimensional $25F^2$ SRAM Cell Technology with the smallest S^3 (Stacked Single-crystal Si) Cell, 0.16 um^2, and SSTFT (Stacked Single-crystal Thin Film Transistor) for Ultra High Density SRAM", *Tech. Papers VLSI Technology Symposium*, pp. 228–229, June 2004.

[33] S. Jung, J. Jang, W. Cho, *et al.*, "Three Dimensionally Stacked NAND flash Memory Technology Using Stacking Single Crystal Si Layers on ILD and TANOS Structure for Beyond 30 nm Node", *Techn. Dig. IEEE International Electron Devices Meeting (IEDM)*, pp. 1–4, December 2006.

[34] T. Kuroda, "Perspective of Low-Power and High-Speed Wireless Inter-Chip Communications for SiP Integration", *Proc. European Solid-State Device Research Conference (ESSDERC)*, pp. 3–6, September 2006.

[35] K. Kim, J. Choi, T. Kim, *et al.*, "A Role for Graphene in Silicon-Based Semiconductor Devices", *Nature*, 479, pp. 338–344, November 2011.

[36] H. J. Yang, J. Heo, S. Park, *et al.*, "Graphene Barristor, a Triode Device with a Gate-Controlled Schottky Barrier", pp. 1–5. Sciencexpress, May 2012.

1921   1922   ELECTRON DEVICES SOCIETY   1923   US crossed non-stop in a Fokker T2

Chapter 4

Passive Components

Joachim N. Burghartz and Colin C. McAndrew

4.1 Discrete and Integrated Passive Components

Passive components were used by electrical engineers as early as 1745 (the Leyden jar, the first capacitor), which was 160 years before the first active device, the vacuum tube, marked the birth of radio electronics. Resistors and coils (inductors) were then developed through the early 1800s as experimental apparatus to unravel the basic science of electromagnetic phenomena, culminating in Maxwell's equations, and were later used in electronic circuits. In early radios the passives, such as the mechanically tunable capacitor of the *LC* resonator in the receiver, were as important as the rectifying vacuum diode.

Discrete passive components (Table 4.1) could be fabricated more accurately than the early active devices, and often tunable passive components were adjusted to compensate for the variability of active

Passive Component or Device?

Active electronic devices require a source of energy for their operation, exhibit power gain, are non-linear, and have three or more terminals. In contrast, passive components do not require energy for operation, do not have power gain, are often linear, and have two (intended) terminals. Inductors, capacitors, resistors and their variable variants fall into that category. Obviously, there are no "passive devices."

Guide to State-of-the-Art Electron Devices, First Edition. Edited by Joachim N. Burghartz.
© 2013 John Wiley & Sons, Ltd. Published 2013 by John Wiley & Sons, Ltd.

1924 | Airship ZR-3 crosses the Atlantic Ocean | ELECTRON DEVICES SOCIETY | 1925 | Kronig, Uhlenbeck and Goudsmit postulate the spin model for electrons

Table 4.1   Examples of discrete, thin and thick film, and integrated passive components

		Resistor	Capacitor	Inductor
	Symbol	─\/\/\─	─┤├─	─⌒⌒⌒─
	Unit	Ohm Ω	Farad F	Henry H
Discrete Passive Components	Example			
	Value Range	mΩ – MΩ	pF – mF	nH – mH
Thin and Thick Film Passive Components	Example			
	Value Range	Ω – MΩ	pF – μF	pH – μH
Integrated Passive Components	Back-end Example			
	Front-end Example			n.a.
	Value Range	mΩ – kΩ	fF – pF	pH – nH

Image sources: top left, © Leo Blanchette/Fotolia.com; top center, © Gavin/Fotolia.com; top right, © Ionescu Bogdan/Fotolia.com; bottom left, © International Manufacturing Services (IMS); bottom center, © IMS CHIPS; bottom right, © IEEE, reprinted with permission from IEEE Journal of Solid-State Circuits, September, 2003

devices (through trimming resistances, capacitances, and inductances, see Table 4.2) as well as for frequency tuning. In the vacuum electronics era the large tubes, rather than the discrete passive components, were the limiting factor in miniaturization.

With the advent of transistor electronics in the late 1940s the focus shifted to development of more compact passives, leading to thick and thin film process technologies and ultimately to integrated microelectronics. Integrated resistors were a crucial part of the PMOS and NMOS process technologies of the

| 1926 | Air ship "Norge" flies over the North Pole | ELECTRON DEVICES SOCIETY | Patent of the field effect transistor issued to Julius Edgar Lilienfeld | |

Table 4.2 Examples of variable/trimmable passive components

	Resistor	Capacitor	Inductor
Symbol	⎓⋀⋀⋀⟋	⫲⟋	⟋⟋⟋⟋⟋
Discrete			
Thin/Thick Film			
Integrated			

Image sources: top row left, © Gdead; top row center, © Dmitriy Syechin/Fotolia.com; top row right, © Electronics DIY; middle row left, LS Laser Systems GmbH; middle row center, LS Laser Systems GmbH; middle row right, IMS CHIPS; bottom row left, Keyence Corporation; bottom row center, IMS CHIPS; top row right, IMS CHIPS

late 1960s and early 1970s but die size (i.e., cost) minimization spurred the development of all-transistor circuits. Resistors continued to be part of high-speed digital bipolar technologies until the late 1980s.

After the virtual disappearance of digital bipolar technologies and the ascendance of CMOS in the early 1990s passives became unnecessary for volume digital applications, except for the storage capacitor in a DRAM (Chapter 3), although high quality (linear, well-matched) resistors and capacitors remained important for precision analog circuits, such as data converters and switched capacitor circuits. When high-performance BiCMOS emerged in the mid-1990s as the technology of choice for integrated RF systems, and portable RF communications became the driving force behind the semiconductor industry, passives returned to the limelight. Not only were resistors and capacitors important, but for frequency tuning it became necessary to place inductors on silicon (see Section 4.3), even though it was a significant challenge to develop sufficiently high-quality inductors in standard CMOS and

Passive Components – the Intended and the Unintended

Passive components within an integrated circuit are prone to parasitic capacitive or inductive coupling to neighboring components and to the chip substrate, through which signal cross-talk may occur. Both the intended discrete components and the unintended parasitics have to be considered in circuit design and layout.

1927

ELECTRON DEVICES SOCIETY®

Charles Lindbergh is first to cross the Atlantic Ocean by plane

Table 4.3   Basic distributed passive components

	Wave Guide	Microstrip	Coplanar Wave Guide (CPW)
Schematic			
Example			
Preferred Modes	TEM	Quasi TEM	Quasi TEM
Bandwidth	Low	High	Medium
Losses	Low	High	High
Power Capacity	High	Low	Low
Physical Size	Very Large	Small	Medium
Ease of Fabrication	Medium	Easy	Very Easy
Integrability	Very Hard	Easy	Very Easy

Image sources: left, IMS CHIPS; center, Henrique Miranda; right Michael Forman

BiCMOS processes. Voltage tunable capacitors (varactors), both *PN*-junction and then MOS, were developed to allow for implementation of on-chip variable frequency oscillators, and the on-chip RF component suite was extended to include transformers, *PiN*-diodes, and MOS or MEMS switches [1]. At present, even distributed passive components are used to enable integrated millimeter wave systems [2], see Table 4.3. However, for the design of integrated RF, microwave and millimeter wave systems both the intended passive components and the unintended parasitics need to be taken into consideration (see Section 4.4 and sidebar).

4.2 Application in Analog ICs and DRAM

Prior to the predominance of CMOS technology resistors were widely used as loads in bipolar technologies, and that is still the case for communications circuits built in modern SiGe processes (Chapters 1 and 14). Any resistive layer in an integration process can be described by its sheet resistance (see sidebar) and be used

to build a resistor. In double-poly bipolar technologies the extrinsic-base (200–400 Ω/sq.) and the emitter polysilicon (100–200 Ω/sq.) layers, the pinched intrinsic base layer (1–10 kΩ/sq.), and the buried layer structure (10–50 Ω/sq.) are exploited for that purpose. Higher sheet resistances (~1000 Ω/sq.) are achievable by building a resistor in the collector epitaxial layer.

Analog functions require passive circuit components which have small temperature and voltage coefficients, good matching, and can be seamlessly integrated with the active devices [3]. Polysilicon resistors (Figure 4.1a) are preferred since their temperature coefficient (500–1500 ppm/°C) is lower than that of diffused resistors (2500–3000 ppm/°C) and they are usually much more linear. *P*-type polysilicon can exhibit a negative temperature coefficient of resistance, which can be useful for compensating resistors with positive temperature coefficients, thereby enabling the design of circuits with minimum variation over temperature.

The nominal sheet resistance of heavily doped polysilicon is on the order of 200 Ω/sq., which is sufficient for reasonable area load resistors for Emitter-Coupled Logic (ECL) bipolar circuits. Far higher sheet resistances are required for Static Random Access Memory (SRAM) cells if load resistors are used instead of PMOS load transistors as in the common six-transistor SRAM cell (Chapters 3 and 13).

In that case polysilicon resistor structures are very lowly doped, having highly doped regions only for the metal contacts (Figure 4.1a). Since

Sheet Resistance

The resistance of any layer in a silicon integration process, having a de ned geometry, can be de ned by its resistivity (ρ), its thickness (t) and by its lateral dimensions, length (l) and width (w),

$$R = \frac{\rho}{t}\frac{l}{w} = R_S \frac{l}{w}$$

The ratio ρ/t is de ned as sheet resistance R_S, given in ohms/square (Ω/sq.), which applies to the resistance of a square shape of the layer ($l = w$).

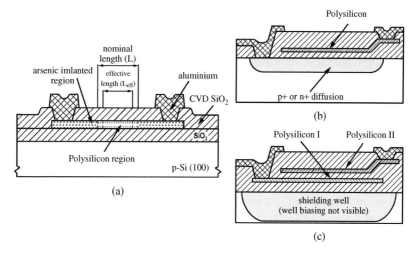

Figure 4.1   Passive components for analog circuit integration: (a) polysilicon thin-film resistor [3], (b) capacitor having one polysilicon and one diffused electrode and (c) capacitor with two polysilicon electrodes

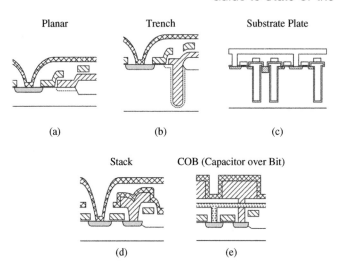

Figure 4.2 High-density capacitor structures used for the different DRAM nodes [4]

the active transistors in SRAM are typically NMOS, arsenic implantation with doses of 10^{13} cm^{-2} to 10^{15} cm^{-2} in combination with an adjusted polysilicon layer thickness are used to adjust the sheet resistance to values ranging from 10^{12} Ω/sq. to 10^{4} Ω/sq. [3]. High-performance capacitors are also important elements in analog CMOS and BiCMOS data converter and switched capacitor filter circuits. The most commonly used structures have either a highly doped contact in silicon and a heavily doped poly contact separated by a well-defined insulator layer (Figure 4.1b), or the structure is built above the silicon substrate by using two heavily doped polysilicon layers (Figure 4.1c). Thermal oxidation of silicon can be exploited in double-poly capacitors to achieve a well-controlled, reproducible capacitance per unit area, with small parasitics. Even lower parasitic, more linear capacitors can be built as metal-insulator-metal (MIM) capacitors in the back-end processing, although these increase process complexity and cost. In contrast, Dynamic Random Access Memory (DRAM) structures require a very high capacitance/area value to be realized [4]; details of these DRAM capacitors are explained in Chapters 3 and 13.

4.3 The Planar Spiral Inductor – A Case Study

Forming a useful inductor above a conductive silicon substrate with only a thin metallization layer available, such as in a planar silicon process technology, is a challenging task. The most popular planar spiral coil structure is modeled by an inductance L_S, a metal series resistance R_S, a parasitic capacitance C_{Ox} between metal and substrate, and a bulk resistance R_B (inset of Figure 4.3). The quality factor (Q, see sidebar) is limited at low frequency by R_S and at high frequency by C_{Ox} (see Figure 4.3). The self-resonance frequency $f_{SR} = 1/\left(2\pi\sqrt{L_S \cdot C_P}\right)$, which for typical layouts depends mainly on the inter-wire capacitance C_P, should be considerably higher than the operating frequency $f\left(Q_{max}\right)$ at the maximum Q(Q_{max} in Figure 4.3). Obviously, both Q_{max} and $f\left(Q_{max}\right)$ can be adjusted by tailoring the terms $\omega L_S/R_S$ and $1/\omega C_B \cdot R_B$.

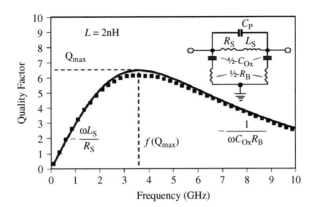

Figure 4.3 Quality factor of an inductor versus frequency, indicating a maximum (Q_{max}) at $f(Q_{max})$, which can be tailored by adjusting the parasitic capacitances C_{Ox} and C_P and resistances R_S and R_B [9]

The magnetic field extending into the substrate can cause eddy currents in the conductive silicon, thus reducing the apparent inductance L_S and lowering Q. Inductors on silicon date to 1990 [5] but it took until 1995 to develop inductors with useful values of Q [6]; there multi-level inter-connect technology was exploited by connecting the uppermost metal layers together in parallel, thereby minimizing R_S, and omitting the lowermost metal layers, thereby increasing the separation of the inductor from the substrate and minimizing C_{Ox}. Later, in 1997, Q values beyond 20 were demonstrated through the use of copper metallization and a high-resistivity silicon substrate [7].

Increasing the resistivity of the substrate can either increase or decrease Q, depending on the operating regime (Figure 4.4). For substrate resistivities above about 10 $\Omega \cdot$ cm both eddy current and capacitive losses decrease as the resistivity increases, thereby increasing Q_{max} (Inductor Mode). For resistivities below 10 $\Omega \cdot$ cm f_{SR} drops and Q_{max} increases with decreasing resistivity because the inductor acts as an *LC* resonator with $Q = Q_L \cdot Q_C/(Q_L + Q_C)$ where $Q_L = \omega L_S/R_S$ and $Q_C = 1/\omega C_{Ox} \cdot R_B$ (Resonator Mode). Below about 1 $\Omega \cdot$ cm eddy currents in the substrate rise so high that L_S becomes noticeably reduced (Eddy Current Mode). Insertion of a uniform metal shield between the spiral coil and the substrate prevents eddy currents in the substrate, although strong eddy currents then appear within the shield and the structure operates as a resonator rather than an inductor. Patterning the metal shield significantly reduces eddy currents in the metal [8] though there is still some substrate loss [9]. Although the substrate back-side contact is often assumed to be an ideal ground (insert of Figure 4.3) the

Quality Factor (Q)

This gure-of-merit for passive components describes the ability to store inductive or capacitive energy relative to ohmic losses. Q is the ratio of the imaginary part im(Z) to the real part re(Z) of the component's impedance Z: Q = |im(Z)|/re(Z). The sign of im(Z) is positive if a component is inductive and negative if it is capacitive. At resonance im(Z) = 0 and the sign of im(Z) changes. Q is also used to denote the ratio of center frequency f_0 to bandwidth Δf of resonators and lters, Q = $f_0/\Delta f$.

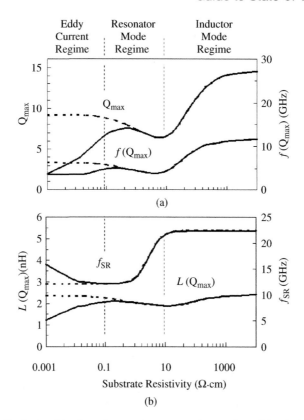

Figure 4.4 Maximum quality factor (Q_{max}), optimum frequency of operation $f(Q_{max})$, inductance $L(Q_{max})$ and self-resonance frequency (f_{SR}) versus substrate resistivity for uniformly conducting (solid lines) and patterned (dashed lines) substrates, which minimize eddy current losses [9]

actual placement of substrate contacts can have a pronounced effect on the inductor characteristics and needs to be taken into consideration with circuit layout [10]. This is of particular concern if the lossy silicon underneath the coil is removed in order to increase Q [11]. Moreover, such structures lack silicon as an effective heat conductor, leading to possible thermal effects [12]. Another concern is that coils for high inductances are relatively large, leading to higher cost, and have lower values of Q. Approaches to implement integrate inductors with minimum area have exploited solenoid-like stacked coil structures [9] or ferromagnetic cores [13].

Besides inductor optimization for maximum Q and minimum area, development of accurate inductor models has also been a challenge. While inductor modeling started more than 60 years ago [14], significant attention to accurate models began only after the first practical integrated inductors were demonstrated [6], leading to a wide variety of models today [15].

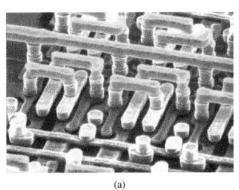 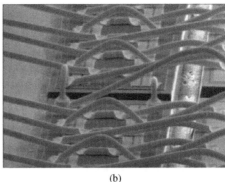

(a)   (b)

Figure 4.5   On-chip interconnect (a) and package-to-chip wire-bonds (b). (a) Reproduced by permission of IBM and (b) Courtesy of Palomar Technologies, Inc. © 2012 www.palomartechnologies.com

4.4 Parasitics in Integrated Circuits

Figure 4.5 shows on-chip interconnect and package-to-chip wire-bonds. Every ideal "wire" in reality has parasitic inductance and resistance associated with it, and every wire is capacitatively coupled to every other piece of wire. As the linewidths and spacings in integrated circuits have shrunk, the parasitic wire resistances and capacitances have unavoidably increased. These parasitics – unintended passives – decrease the overall performance level of integrated circuits (ICs), and since about 2004 have been one of the main factors that have stalled the historic trend in the increase of clock speeds of large digital circuits.

Handling parasitics in IC design is not conceptually difficult, apart from the skin effect for inductance (which is often less important than capacitance or resistance) there are no complex dependencies on bias as there are for MOS varactors. However, in modern microprocessors there are up to several billion individual transistors, and the sheer magnitude of the task of determining all of the parasitics is daunting. The final interconnect patterning is not known until after layout, however, design of such large ICs is done hierarchically by partitioning the circuit into blocks, so circuit performance in the presence of parasitics can be done at the block level. Accurate accounting for parasitics can be critical for high precision analog circuits and high frequency RF circuits, which are often done at the schematic level. Final verification simulations must include parasitics that are extracted from the physical layout.

References

[1]   J. N. Burghartz, "Passive Components", in *The Heterostructure Handbook*: *Materials, Fabrication, Devices, Circuits, and Applications of SiGe and Si Strained-Layer Epitaxy*, J. D. Cressler (ed.), Boca Raton, FL, CRC Press, pp. 249–263, 2005.

[2]   D. Beck, D. M. Herrmann, and E. Kasper, "CMOS on FZ-high resistivity substrate for monolithic integration of SiGe-RF-circuitry and readout electronics", *IEEE Trans. Electron Devices*, vol. 44, no. 7, pp. 1091–1101, Jul. 1997.

[3]   S. Wolf, "*Silicon Processing for the VLSI Era, Volume 2 – Process Integration*", Lattice Press, Sunset Beach, CA, 1990.

[4] K. Nakamura, "DRAM Cell and Technology Trends", Short Course at the *1996* International Electron Devices Meeting *(IEDM)*.

[5] N. M. Nguyen and R.G. Meyer, "Si IC-compatible inductors and LC passive filters", *IEEE J. Solid-State Circuits*, vol. 25, no. 4, pp. 1028–1031, Aug. 1990.

[6] J. N. Burghartz, M. Soyuer, K. A. Jenkins, and M. D. Hulvey, "High-Q inductors in standard silicon interconnect technology and its application to an integrated RF power amplifier", *Techn. Dig. IEEE International Electron Devices Meeting (IEDM)*, in 1995, pp. 1015–1018.

[7] J. N. Burghartz, D. C. Edelstein, K. A. Jenkins, and Y. H. Kwark "Spiral inductors and transmission lines in silicon technology using copper-damascene interconnects and low-loss substrates", *IEEE Trans. Microwave Theory Tech.*, vol. 45, no. 10, pp. 1961–1968, Oct. 1997.

[8] C. P. Yue and S. S. Wong, "On-chip spiral inductors with patterned ground shields for Si-based RF ICs", *IEEE J. Solid-State Circuits*, vol. 33, no. 5, pp. 743–752, May 1998.

[9] J. N. Burghartz and B. Rejaei, "On the design of RF spiral Inductors on silicon", *IEEE Trans. Electron Devices*, vol. 50, no. 3, pp. 718–729, Mar. 2003.

[10] J. N. Burghartz, A. E. Ruehli, K. A. Jenkins, M. Soyuer, and D. Nguyen-Ngoc, "Novel substrate contact structure for high-Q silicon-integrated spiral inductors", *Techn. Dig. IEEE International Electron Devices Meeting (IEDM)*, pp. 55–58, 1997.

[11] J. Y.-C. Chang, A. A. Abidi, and M. Gaitan, "Large suspended inductors on silicon and their use in a 2-μm CMOS RF amplifier", *IEEE Electron Device Lett.*, vol. 14, no. 5, pp. 246–248, May 1993.

[12] H. Sagkol, B. Rejaei, and J. N. Burghartz, "Thermal issues in micromachined spiral inductors for high-power applications", *IEEE Trans. Electron Devices*, vol. 55, no. 11, pp. 3288–3294, 2008.

[13] Y. Zhuang, M. Vroubel, B. Rejaei, and J. N. Burghartz, "Ferromagnetic RF inductors and transformers for standard CMOS/BiCMOS", *Techn. Dig. IEEE International Electron Devices Meeting (IEDM)*, pp. 475–478, 2002.

[14] F. W. Grover, *Inductance Calculations, Princeton, New Jersey: Van Nostrand, 1946*. Reprinted by Dover Publications, New York, NY, 1962.

[15] J. R. Long and M. A. Copeland, "The modeling, characterization, and design of monolithic inductors for silicon RF ICs", *IEEE J. Solid-State Circuits*, vol. 32, no. 3, pp. 357–369, Mar. 1997.

1937 — Hindenburg accident at Lakehurst, New Jersey — ELECTRON DEVICES SOCIETY — Sigurd and Russell Varian invent the Klystron

Chapter 5

Emerging Devices

Supriyo Bandyopadhyay, Marc Cahay and Avik W. Ghosh

5.1 Non-Charge-Based Switching

The primitive unit of all digital circuitry is the bistable "switch" whose two stable states encode the binary bits 0 and 1. When the circuit performs a useful function (e.g., signal processing or computing), the switches transition between their two states many times, and each time they dissipate a certain amount of energy. While the amount dissipated by a single switch in a single event is tiny, it may quickly add up to a very large amount when billions of switches toggle a few billion times per second.

The quintessential switch is the metal-insulator-semiconductor field-effect-transistor (MISFET) whose two conductance states ("on" and "off") are associated with different amounts of charge stored in the transistor's channel. When the MISFET switches, the channel charge is changed by an amount ΔQ in an amount of time Δt, resulting in current $I = \Delta Q / \Delta t$ and accompanying energy dissipation

Why Charge-Based Devices are Dissipative

Charge is a scalar quantity that has *magnitude*, but no *direction*. If digital bit states 0 and 1 are encoded in "charge", they must be encoded in two different *amounts* or magnitudes of charge. Switching bits will invariably require changing the amount of charge in the device, which will cause current ow and associated energy dissipation. Excessive dissipation is the primary impediment to downscaling devices in accordance with the celebrated Moore's law.

Guide to State-of-the-Art Electron Devices, First Edition. Edited by Joachim N. Burghartz.
© 2013 John Wiley & Sons, Ltd. Published 2013 by John Wiley & Sons, Ltd.

1938   Planet Pluto discovered

$E_d = \Delta Q V = I^2 R \Delta t = (\Delta Q)^2 R / \Delta t$, where R is the resistance in the path of the current. This dissipation can be considerable when the switch operates in a noisy environment, or in an interconnect-driven circuit, where ΔQ has to be large, or when it is switched fast (Δt is small). Today's best transistors dissipate about 4 fJ of energy to switch in a circuit in ~1 ns. Assuming, that a future chip (around 2020) will have a transistor density of 10^{11}/cm^2 and still dissipate a few fJ per bit flip per clock cycle at 1 GHz, the energy dissipation will be ~170 kW/cm^2 if the clock rate goes up to about 4 GHz! That is comparable to the energy dissipated in a rocket nozzle and is sufficient to overwhelm any heat sinking technology. This highlights the need for low energy switches.

If digital bits can be encoded in two mutually anti-parallel spin orientations of a single electron ("up-spin" = 1 and "down-spin" = 0), then they can be switched by merely flipping the spin without moving the electron in space and causing current flow ($I = 0$). Some energy will still be dissipated since the up- and down-spin states are usually not energetically degenerate, but that energy is typically much less than the energy dissipated in moving charges around. This realization motivates the research in employing alternate state variables – different from charge – to store, process and communicate information in a digital format. Spin has been a favorite alternate state variable, which has given birth to the field of *Spintronics*. A genre of devices, loosely referred to as *Spin Field Effect Transistors* (SPINFET), were the earliest semiconductor spintronic switches proposed in 1990 by Supriyo Datta and Biswajit Das of Purdue University [1]. The basic device structure is identical to that of a MISFET, except that the source and drain contacts are ferromagnetic and usually magnetized parallel to each other. The source preferentially injects electrons with a particular spin polarization into the channel. The gate voltage does *not* modulate the charge concentration in the channel ($\Delta Q = 0$), but instead rotates the spins' polarization by modulating the spin-orbit interaction in the channel. If the rotation angle is an odd multiple of 180°, then the drain blocks the electrons from transmitting through the channel and the channel current ideally falls to zero. On the other hand, if the angle of rotation is an even multiple of 180°, then the drain transmits the channel electrons and the current is highest. Thus, by modulating the spin-orbit interaction, and therefore the angle of spin precession, with a gate voltage, one can modulate the drain current and realize switching action (see Figure 5.1).

Clearly, this device is very different from a traditional field effect transistor and its transfer characteristic (drain current versus gate voltage) is oscillatory. That has no bearing on its switching ability, but what does have bearing is that no ferromagnet can selectively inject spin of only one polarization. This makes the conductance on/off

Single Spin Logic (SSL)

An electron's spin orientation becomes "bistable" in a magnetic field since the spin can orient only parallel or anti-parallel to the magnetic field. These two orientations encode logic bits 0 and 1. By placing semiconductor quantum dots, each containing a single free electron, in specific geometric patterns on a wafer and then applying a global magnetic field on the dots, one implements an array of interacting binary switches since each spin acts like a switch with two stable states. The interaction between them is due to exchange coupling between spins. Using such arrangements of binary switches, it is possible to implement Boolean logic gates. The spins in certain quantum dots (designated as input bits) are forcibly aligned (with very strong local magnetic fields generated in those dots with current lines) along one or the other orientation to conform to the input bits, while the interactions between all the spins will make the spin in another quantum dot (designated as the output bit) automatically align in a direction that represents a pre-determined Boolean function of the input bits. Thus a gate is realized. Realization of a NAND gate is shown in Figure 5.2. This idea was proposed by Supriyo Bandyopadhyay, Biswajit Das, and Albert E. Miller, then at the University of Notre Dame, in 1994.

Walter Schottky first decribes the metal-semiconductor junction - the Schottky diode | First IRE sponsored conference on electron tubes, New York | ELECTRON DEVICES SOCIETY | E.E. Simmons and A.C. Ruge invent the strain gauge | First nuclear reaction

ratio – perhaps the most important figure of merit for a "switch" – rather small. Furthermore, when switching action is realized in this fashion, the channel charge is not modulated, but the dissipation does not necessarily decrease. After all, the two states of this switch are still demarcated by different amounts of current flowing through the channel (current "on" and "off"), so that dissipation is still significant. The innate advantage of spin, that is, electrons need not move in space to switch, was not exploited.

The *Single Spin Logic* (SSL) idea (see the sidebar) is different in this respect [2]. Here, each electron acts as a switch and there is no charge movement when the device switches. This reduces the dissipation dramatically. However, single spins are very unstable at room temperature, which is why they are not all that practical. On the other hand, a collection of spins, such as a single domain nanoscale ferromagnet, has a stable magnetization at room temperature. By making the magnet's shape anisotropic (e.g., an elliptical cylinder), its magnetization can be made bistable so that it can act as a binary switch and encode logic bits 0 and 1. Because of strong exchange interaction between the spins, all the $\sim 10^4$ spins in a single domain ferromagnet tend to rotate together in unison (even at room temperature), behaving like a giant classical spin, when the magnetization flips. Because of this remarkable feature, a striking example of correlated switching, the energy dissipated in switching a magnet containing 10^4 spins is not 10^4 times more than that dissipated in flipping a single spin, but instead it is only a few times more!

Since single-domain nanomagnets can be almost as efficient as single spins and additionally they can work at room temperature, Boolean logic gates have been fashioned out of them. They work in essentially the same way as SL, with dipole interactions between magnets replacing exchange interactions between spins. These architectures have been termed *Magnetic Quantum Cellular Automata* by Russell Cowburn and Mark Welland of Cambridge University, who demonstrated its operation in 2000 [3]. They are really Boolean logic and not true cellular automata, which is why a more appropriate term for them is *Nanomagnetic Logic* (NML). One problem that plagues both SSL and NML is that these are *wireless* architectures where bit information is transmitted between nearest neighbors through nearest neighbor exchange coupling or dipole coupling instead of through wires. Since these interactions are bidirectional, information cannot be propagated one-way from the input stage to the output stage. Unidirectionality is, however, imposed in time, instead of space, using sequential clocking of a particular type known as *Bennett clocking*, first postulated by Charles Bennett of the IBM T. J. Watson Research Center in the early 1980s [4].

Hybrid Spintronics/Straintronics

A two-phase *multiferroic* nanomagnet is a bilayer structure comprising a $\sim$6 nm thick magnetostrictive layer like Terfenol-D elastically coupled with a $\sim$ 40 nm thick piezoelectric layer like lead-zirconate-titanate. The lateral dimensions of the nanomagnet are $\sim$100 nm. A tiny voltage of $\sim$10 mV applied across the piezoelectric layer generates a strain in it that is transferred to the magnetostrictive layer and rotates its magnetization. Less than 1 aJ of energy is dissipated to ﬂip the magnetization and thus switch a multiferroic magnet in this fashion in $\sim$1 ns. The static bit error probability at room temperature is $\sim 10^{-14}$. The NAND gate shown with fan-in and fan-out in Figure 5.2 (right) is based on this paradigm and dissipates $\sim$2 aJ of energy per gate operation while operating at a clock rate of 1 GHz with a sinusoidal four-phase clock.

1939   Beginning of World War II   ELECTRON DEVICES SOCIETY   Discovery of pn barrier by Russell Ohl

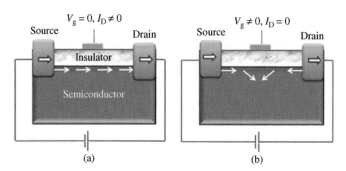

Figure 5.1   The structure of the Spin Field Effect Transistor. (a) When the gate voltage is zero, there is ideally no spin-orbit interaction in the channel and the electrons injected by the source arrive at the drain with their spin orientations intact. They are transmitted, resulting in non-zero current. When the gate voltage is turned on, spin-orbit interaction rotates the spins injected by the source through 180° and the drain blocks them from transmitting. The current then falls to zero. Spin polarization is shown with the arrowheads

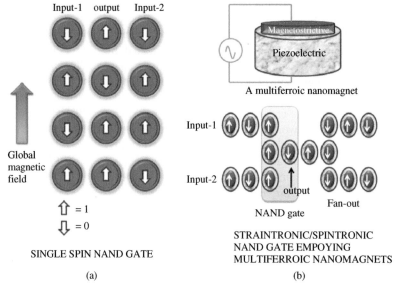

Figure 5.2   (a) A single spin NAND gate shown for all possible combinations of the two input bits. The peripheral spins are input bits and the central spin is the output bit. The output is always the NAND function of the inputs because exchange coupling enforces anti-ferromagnetic ordering along a row. (b) A two-phase multiferroic nanomagnet is shown on top and a NAND gate is shown below. Dipole coupling enforces anti-ferromagnetic ordering along a row and ferromagnetic along a column

How energy-efficient NML is, and whether it can eclipse conventional transistor-logic, is pivotally dependent on how much energy is expended in Bennett clocking. The latter requires rotating the magnetization through large angles. If this is done with a local magnetic field generated with a current line, then the energy dissipated is about *1000 times more* than what is dissipated to switch a transistor. If it is done by passing a spin-polarized current through the magnet, which delivers a spin-transfer torque on the magnetization, then the energy dissipated is still about *100 times more* than in a transistor. However, a new scheme, known as hybrid straintronics/spintronics, proposed by Jayasimha Atulasimha and Supriyo Bandyopadhyay of Virginia Commonwealth University last year, appears to be very energy-efficient and should be able to switch a magnet while dissipating roughly *5000 times less* energy than a transistor [5]. In this scheme, the magnetization of a multiferrroic nanomagnet, consisting of a piezoelectric layer elastically coupled with a magnetostrictive layer, is flipped by applying a tiny voltage of $\sim$10 mV across the structure, resulting in a strain in the magnetostrictive layer that ends up rotating the magnetization. This electrical control of magnetization is essential for practical device applications.

The notion of switching spins with an electric field has a number of precedents [6, 7] and is rapidly burgeoning into an intensely active research area. Even in traditional semiconductor spintronics, ways must be found for the creation, manipulation, and detection of spin-polarized currents by *purely electrical* means. Efforts towards that goal are sometimes referred to as "spintronics without magnetism" [8]. Since spin-orbit coupling (SOC) couples the electron orbital motion to its spin, SOC has been envisioned as a possible tool for all-electrical spin control and generation of spin-polarized current without ferromagnetic contacts and applied magnetic fields. It has been shown that SOC can be used to modulate spin polarized currents by taking advantage of symmetry-breaking factors such as interfaces, electric fields, strain, and crystalline directions. There is a recent theoretical and experimental study of a hitherto unexplored SOC to realize electrical spin injection and detection. This SOC is induced by the transverse in-plane electric field due to the gradients of the lateral confining potential of a quantum wire and is referred to as *lateral spin-orbit* coupling (LSOC). LSOC has been shown to lead to a net spin polarization in side-gated quantum point contacts (QPCs) – short quantum wires – when the confining potential of the QPC is made sufficiently asymmetric [9]. The net spin polarization is triggered by an asymmetric LSOC and becomes robust in the presence of strong electron–electron (e-e) interaction. In the last few years, another approach to semiconductor spintronics has emerged based on topological methodologies for generating spin polarization in semiconductors [10, 11].

5.2 Carbon as a Replacement for Silicon and the Rise of Graphene Electronics and Moletronics

In parallel with developments in unconventional switching, there has been an avalanche of activity exploring unconventional materials for switching, in particular, carbon (see also Chapters 2 and 21). Carbon is unique in that its bond-energies for 3D tetrahedral and 2D hexagonal lattices are comparable, making it ubiquitous in biology and also a fascinating candidate for electronics. Carbon-based electronics has taken several routes – from organic molecules to fullerenes, nanotubes and graphene. The most commonly studied organic molecules are the alkanes, usually attached to gold electrodes using thiol (sulfur) -based "alligator clips" and interrogated from the other side using another metal electrode such as an STM (scanning tunneling microscope) tip or a deposited top contact.

1941 · Konrad Zuse builds the Z3- the first programmable computer, an entirely mechanical machine · ELECTRON DEVICES SOCIETY® · 1942

The alkanes (e.g., dodecane, octadecane) consist of a chain of successive zigzagged, tetrahedral bonds that make electron flow very difficult along the axis and generate a large band-gap ($\sim$8–10 eV). The current through alkanes has a characteristic activated behavior describable using the classical Simmons model for tunneling [15], with a decay constant that has now been calibrated across multiple experiments. But a more interesting class of molecules is the aromatics, which consist of interlinked benzene rings (e.g., phenyl, xylyl, styrene, anthracene). The π-bonds of benzene sticking out of their plane create a delocalized, conjugated electron pool that can conduct current and hop between rings, increasing the overall mobility and decreasing the band-gap of benzene based compounds to $\sim$2–3 eV. Depending on the workfunction of the metal electrode, the electrons can now access either the molecular conduction band or valence band (in chemical parlance, lowest unoccupied molecular orbital or LUMO, and highest occupied molecular orbital or HOMO, respectively) at relatively modest voltages, creating a rise in the current and a corresponding peak in the conductance-voltage spectrum. By periodically interrupting these molecular rings with chemical moieties and multiple bonds, chemists hope to create molecular "bridges" and "bus-stops," and even rotate individual rings electrically with dipolar side-groups, disrupting the flow of electrons as needed.

Considerable challenges exist in fabricating stable metal-molecule-metal junctions given the sensitivity and fragility of the overall measurement set-up. Nonetheless, a lot of experiments have now been reproduced across laboratories, and calibrated with "first-principles" theories on molecular transport [16]. A simple Landauer theory can now explain many of the alkane and aromatic currents, typically with an injection barrier $\sim$1.3 eV. The resulting resistance is $\sim$$10^4$ MΩ for C12 chains with doubly chemisorbed contacts, and for aromatics about 1.3 MΩ for phenyl dithiol and $\sim$20 MΩ for xylyl dithiol. These numbers are consistent with theoretical estimates [17].

While theory and experiment seem to be converging, the currents through small molecules clearly seem inadequate to allow fast switching, limited by the low density of states ("quantum capacitance") of the tunneling electrons across a small cross section, as well as by the poor contacts between metals and molecules that leads to large quantized junction resistances and series resistances. A deeper problem arises with the gateability of ultra-short molecules, Simply put, the extreme molecular aspect ratio makes it hard to bring the gate close enough to turn on the electrons without substantially leaking them through the gate itself. Furthermore, electronic states from the electrodes tunnel into the

Single Molecular Electronics

Single molecule electronics (or moletronics), as a eld of study, traces back to the seminal 1974 work by Ari Aviram and Mark Ratner [12] who postulated that a molecular donor-bridge-acceptor array can act as a rectifying diode. As molecular structures and bandstructures can be varied" 'bottom-up" using synthetic chemistry, it allows us to imagine a range of candidates for passive and active electronics, perhaps even exploiting properties of recognition and dynamical stereochemistry. The added possibility of self-assembling ordered molecular arrays at dimensions where lithography is prohibitive makes moletronics an ultimate (if perhaps distant) ideal for nanoelectronics [13]. Most importantly, measuring currents through single molecules provide unparalleled insight into the fundamentals of quantum charge transport. Organic molecules also play a major role in spintronics since the spin polarization of an electron is extremely long-lived in organic molecules (primarily hydrocarbons) owing to the weak spin-orbit interaction in organics [14].

molecular gap, making it hard to turn the molecule off (in fact, hydrogen becomes metallic when chemisorbed on platinum electrodes!). As we shall see, this is a problem that persists with larger molecular nets such as graphene.

Perhaps the real promise of molecular electronics involves our ability to gate molecular degrees of freedom *other than charge*, in the spirit of the introduction. For instance, by engineering a dipole into a molecular ring and rotating that ring physically away from the conducting path with a gate field, one can precipitate a nonlinear phase transition that switches the electrons sub-thermally [18]. As discussed earlier, the energy dissipated in binary switching equals $\Delta QV = NqV$, where q is the electronic charge, N is the number of electrons and V is the switching voltage. N is $\sim 10^4$ due to thermal robustness and drivability constraints in a circuit (i.e., the need to charge the interconnect capacitance [19]), while V is limited by the thermal energy, as typical switching events involve thermally exciting electrons over a barrier. But the ability to gate the *conformation* of a molecule, instead of its charge, would allow us to turn its current off with much less voltage, as discussed by Ghosh *et al.* [18]. This has, interestingly, been seen in biology since the 1950s. In fact, the classic Nobel Prize-winning paper by Hodgkin-Huxley shows ion channels that turn on at voltages substantially lower than the Boltzmann limit [20]. (More recently, nanoscale mechanical relays or NEMFETs are purported to operate on this same principle.) The physics is analogous to magnetic logic, in that many charges (rather than spins) act as a single giant charge and rotate close to the price of one.

This mechanism actually exploits a property that organic molecules excel at, namely, their conformational flexibility. A different property that also distinguishes molecules, but less explored for switching applications, is that molecules have strong electron–electron and electron–phonon correlations, naturally tending to localize electrons and vibrations into tight compartments, with observable many-body excitations [21, 22]. Such excitations lead to sharp hysteretic switching between conducting and non-conducting (or "dark") electronic and conformational many-body states [17], creating a wealth of signatures such as Coulomb Blockade, Coulomb staircase, inelastic tunneling spectra, phonon sidebands, negative temperature anomaly and random telegraph signals [17]. In other words, an appealing role of organic molecules seems to be as generators of *scattering*, rather than *conducting* states in a transistor channel – applications such as resistive random access memory (RRAMs) and surface modulated field effect transistors (SurFETs) [17] (see Figure 5.3).

Proceeding beyond single molecules, an array of benzene rings leads to a hexagonal 2D structure called *graphene* that is a single carbon atom thick – the winner of the 2010 Nobel Prize for physics (in fact, the edges of patterned graphene nanoribbons actually resemble benzene) [23]. The 2D network creates a stiffer set of bonds that allows much superior (in fact world-record) electronic and thermal conduction. Graphene can now be fabricated by a variety of methods such as CVD growth and sublimation in SiC, mechanical exfoliation, catalytic growth on metals, as well as chemical techniques such as unzipping carbon nanotubes and growing them straight out of benzene. The conical bandstructure of graphene resembles photons, with an ultralow effective mass. In addition, the two bands derive from two distinct sets of "pseudospins" (bonding-antibonding combinations of carbon dimers) whose orthogonality suppresses 1D backscattering, leading to ultrahigh mobilities for suspended samples (orders of magnitude larger than silicon). This has led to the great interest in graphene for high-speed RF applications and interconnects. Ironically, the gaplessness of the graphene bands, responsible for the high mobility, also leads to high OFF currents (similar to

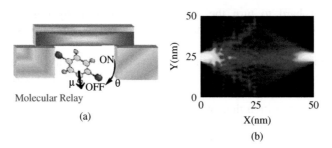

(a)   (b)

Figure 5.3 Non-charge-based switching in carbon electronics. (a) A molecular relay can be rotated with a gate tunable electric field coupled to a dipolar group to create lower OFF currents than the Boltzmann limit [17, 18]. Reproduced with permission from Ghosh, A. W.; *Comprehensive Semiconductor Science and Technology*, Vol. 5; © 2011 Elsevier Science. (b) Pseudospin dynamics in a graphene PN junction allows the gate to chaperone electrons along various trajectories analogous to Snell's law in optics, with a "refractive index" that can be negative, thus creating ultra-sharp tunneling-dominated switching events [25]. Figure courtesy Sajjad, R. *et al.,* unpublished

benzene). Aside from fabrication challenges such as transferring large area graphene from a metallic to an insulating substrate, a significant focus has been the opening of a bandgap in graphene by breaking its sublattice symmetry (e.g., selective doping of dimer atoms), or by quantization (e.g., patterning narrow graphene nanoribbons). However, such a bandgap invariably destroys the mobility and increases band-to-band tunneling and non-saturating currents [24] – partly due to extrinsic reasons associated with fabrication and morphology, but more significantly due to inherent bandstructure reasons such as an increase in effective mass with increasing parabolicity, and a concomitant increase in density of states that increases the probability of phonon scattering.

It is possible in theory to gate non-electronic degrees of freedom in graphene, specifically, its pseudospins [25]. The additional symmetry of the pseudospin states allows us to enhance or suppress transmission of electrons at a graphene PN junction, in seeming agreement with experiments [26]. By exploiting the phase evolution of the pseudospins, it is theoretically possible to turn on a graphene switch at much lower voltages than the Boltzmann limit, and realize a gate tunable pseudospin-FET somewhat analogous to the spin-FET described earlier.

5.3 Closing Remarks

There is some speculation about the true merits of emerging fields such as moletronics and spintronics. It is probably clear by now that as simple substitutes of the silicon transistor, the use of molecules or spins is likely to significantly degrade performance in many cases, particularly in conventional applications. Whether novel applications come out of these topics remains to be seen, although there is no doubt that we have learned an enormous amount about the fundamentals of electron transport at atomic scales, and the predictive power of "first-principles" theories describing these systems. The real promise of moletronics and spintronics could well lie in our creative ability to exploit their inherent strengths towards non-charge based switching, such as the strong exchange interaction among the vector spin variables in a magnet, conformational flexibility and strong electron correlations in short organic molecules, or the photon-like spectrum and pseudospin structure in graphene.

References

[1] S. Datta and B. Das, "Electronic analog of electro-optic modulator", *Appl. Phys. Lett.*, Vol. 56, 665 (1990).

[2] S. Bandyopadhyay, B. Das and A. E. Miller, "Supercomputing with spin-polarized single electrons in a quantum coupled architecture", *Nanotechnology*, Vol. 5, 113 (1994).

[3] R. P. Cowburn and M. E. Welland, "Room temperature magnetic quantum cellular automata", *Science*, Vol. 287, 1466 (2000).

[4] C. H. Bennett. "The thermodynamics of computation: A review", *Int. J. Theor. Phys.*, Vol. 21, 905 (1982).

[5] K. Roy, S. Bandyopadhyay, and J. Atulasimha, "Hybrid spintronics and straintronics: A magnetic technology for ultralow energy computing and signal processing", *Appl. Phys. Lett.*, Vol. 99, 063108 (2011).

[6] T. Brintlinger, S.-H. Lim, K. H. Baloch, P. Alexander, Y. Qi, J. Barry, J. Melngailis, L. Salamanca-Riba, I. Takeuchi and J. Cummings, "In Situ Observation of Reversible Nanomagnetic Switching Induced by Electric Fields", *Nano Lett.*, Vol. 10, 1219 (2010).

[7] T. Chung, S. Keller, and G. P. Carman, "Electric-field-induced reversible magnetic single-domain evolution in a magnetoelectric thin film", *Appl. Phys. Lett.*, Vol. 94, 132501 (2009).

[8] D. Awschalom and N. Samarth, "Spintronics without magnetism", *Physics*, Vol. 2, 50 (2009).

[9] P. Debray, S. M. S. Rahman, J. Wan, R. S. Newrock, M. Cahay, A. T. Ngo, S. E. Ulloa, S. T. Herbert and M. Muhammad, "All-electric quantum point contact spin-polarizer", *Nature Nanotech.*, Vol. 4, 759 (2009).

[10] J. Zaanen, " Fast electrons tie quantum knots", *Science*, Vol. 323, 888 (2009).

[11] M. König, S. Wiedmann, C. Brüne, A. Roth, H. Buhmann, L. W. Molenkamp, X.-L. Qi and S.-C. Zhang, "Quantum Spin Hall insulator state in HgTe quantum wells", *Science*, Vol. 318, 766 (2007).

[12] A. Aviram, M. A. Ratner, "Molecular rectifiers", *Chem. Phys. Lett.*, Vol. 29, 274 (1974).

[13] C. Joachim, J. Gimzewski and A. Aviram, "Electronics using hybrid-molecular and mono-molecular devices", *Nature*, Vol. 408, 541 (2000).

[14] S. Pramanik, C-G Stefanita, S. Patibandla, S. Bandyopadhyay, K. Garre, N. Harth and M. Cahay, "Observation of extremely long spin relaxation times in an organic nanowire spin valve", *Nature Nanotech.*, Vol. 2, 216 (2007).

[15] J. G. Simmons, "Generalized formula for the electric tunnel effect between similar electrodes separated by a thin insulating film", *J. Appl. Phys.*, Vol. 34, 1793 (1963).

[16] A. W. Ghosh, P. S. Damle, S. Datta and A. Nitzan, "Molecular electronics: Theory and device prospects", *MRS Bulletin*, Vol. 29, 391 (2004).

[17] A. W. Ghosh, "Electronics with Molecules", in Bhattacharya, P., Fornari, R., and Kamimura, H. (eds) *Comprehensive Semiconductor Science and Technology*, Vol. 5, 383 (2011), Amsterdam: Elsevier.

[18] A. W. Ghosh, T. Rakshit, S. Datta, "Gating of molecular transistors: electrostatic and conformational", *Nano Lett.*, Vol. 4, 565 (2004).

[19] V. V. Zhirnov, R. Calvin, J. Hutchby and G. Bourianoff, "Limits to binary logic switch scaling – a Gedanken model", *Proc. IEEE*, Vol. 91, 1934 (2003).

[20] A. L. Hodgkin and A. F. Huxley, "Currents carried by sodium and potassium ions through the membrane of the giant axon of loligo", *J. Physiol.*, Vol. 116, 449 (1952).

1946   ENIAC - first electronic computer   ELECTRON DEVICES SOCIETY   Patent of semiconductor junction solar cell (Russell Ohl)

[21] B. Muralidharan, A. Ghosh, S. Pati, and S. Datta, "Theory of high bias coulomb blockade in ultrashort molecules", *IEEE Trans. Nanotech.*, Vol. 6, 536 (2007).

[22] B. Muralidharan, A. W. Ghosh, and S. Datta, "Probing electronic excitations in molecular conduction", *Phys. Rev. B*, Vol. 73, 155410 (2006).

[23] F. Tseng, D. Unluer, M. R. Stan, and A. W. Ghosh, "Graphene Nanoribbons: From Chemistry to Circuits", in Raza, Hassan (ed.) *Graphene Nanoelectronics, Metrology, Synthesis, Properties and Applications Series: NanoScience and Technology*, p. 555 (2012).

[24] F. Tseng, D. Unluer, K. Holcomb, M. R. Stan, and A. W. Ghosh, "Diluted chirality dependence in edge rough graphene nanoribbon field-effect transistors", *Appl. Phys. Lett.*, Vol. 94, 223112 (2009).

[25] R. N. Sajjad and A. W. Ghosh, "High efficiency switching using graphene based electron optics", *Appl. Phys. Lett.*, Vol. 99, 123101 (2011).

[26] R. J. Sajjad, S. Sutter, J. U. Lee, and A. W. Ghosh, "Manifestation of chiral tunneling at a tilted graphene p-n junction", *Phys. Rev. B.*, Vol. 86, 155412 (2012).

1947   P.R. Wallace develops band theory for single layer graphene   ELECTRON DEVICES SOCIETY   First supersonic flight

Part II

ASPECTS OF DEVICE AND IC MANUFACTURING

John Bardeen and Walter Braittain demonstrate the point contact transistor …

… under direction of William Shockley - the first transistor

ELECTRON DEVICES SOCIETY®

ASPECTS OF DEVICE AND IC
MANUFACTURING

Chapter 6

Electronic Materials

James C. Sturm, Ken Rim, James S. Harris and Chung-Chih Wu

6.1 Introduction

The progress in electronic devices over the past 60+ years has been made possible in large part by our ability to make more complex structures involving more materials – semiconductors, insulators, and metals. This has been true across the device spectrum, from silicon-based integrated circuits and power devices to compound semiconductor electronic and optoelectronic devices to solar cells to flat panel displays. In this section we review the increasing range of materials used in devices over this period, summarized by the elements of the periodic table used in each class of devices. For brevity, we include only materials in the active device structures themselves – not in packaging or the chemicals/materials used in processing – and for the most part focus on materials in structures that have reached production.

6.2 Silicon Device Technology

6.2.1 1960–1970s: Early Integrated Circuits

Although transistors were first demonstrated on germanium substrates, most early discrete transistors and integrated circuits (ICs) were based on silicon bipolar transistors (Figure 6.1; see also Chapter 1). This was due to the larger bandgap of silicon and the ability of silicon dioxide to provide a stable insulator, to mask

Guide to State-of-the-Art Electron Devices, First Edition. Edited by Joachim N. Burghartz.
© 2013 John Wiley & Sons, Ltd. Published 2013 by John Wiley & Sons, Ltd.

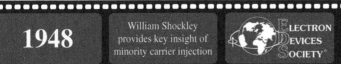

1948 — William Shockley provides key insight of minority carrier injection — ELECTRON DEVICES SOCIETY — — William Shockley presents the pn junction theory

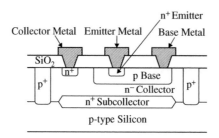

Figure 6.1   Cross section of double-diffused (base and emitter) silicon bipolar transistor technology, with n$^+$ subcollector formed before n-type epitaxy, and p$^+$ sinker isolation rings

dopant diffusion, and to passivate surface states. The advent of the planar process enabled by junction isolation and the use of silicon dioxide to mask dopants was an important milestone for early ICs. In the 1970s, ICs based on silicon MOSFETs also became popular, especially for digital logic and memory, and came to dominate the IC industry (see Chapters 2 and 10). Early ICs (both bipolar or FET) typically had one layer of metal interconnect and device dimensions of several microns to tens of microns. Common dopants for silicon devices were boron, phosphorus, and later arsenic because of its lower diffusion coefficient compared to phosphorus. Silicon dioxide was the dominant insulator in the field region and gate insulator in MOSFETs, and aluminum was the main interconnect metal as well as the gate metal of early MOSFETs. Often a silicon-nitride passivation layer was added as well. The industry was still learning to control unwanted impurities, and doped oxides such as phosphosilicate glass (PSG) and borophosphosilicate glass (BPSG) saw increasing usage for their gettering properties, as well as their superior reflow characteristics for smoothing steps.

More Materials

Progress in silicon device technology was for many years enabled by scaling, with a nearly fixed set of materials. However, as scaling has become more difficult, in the last two decades the materials set has rapidly expanded.

6.2.2   1980–1990s: CMOS Scaling Accelerates

During the 1980s and 1990s, the geometric scaling of silicon CMOS ICs to reduce dimensions accelerated and drove innovations in materials for electronic devices. In the front end of silicon CMOS, additional dopant species such as indium and antimony were introduced to form more abrupt junction profiles and help control short channel effects of FETs [1]. In the latter part of the 1990s, advanced technologies experimented with co-implantation of species such as carbon, fluorine, and nitrogen along with the traditional dopants. Nitrogen was added to gate oxides to form oxynitrides (SiON) for reliability purposes.

William Shockley publishes homo- and heterojunction bipolar transistor concepts

ELECTRON DEVICES SOCIETY

1949

IRE committee on electron tubes and solid state devices

Short channel FETs required shallow junctions and thus a metallization which prevented aluminum interaction with the silicon and which had low parasitic resistance. A Self-ALigned Silicide ("salicide") became popular and quickly became the industry standard. Tungsten silicide and titanium silicide were used early on, while cobalt silicide and eventually nickel silicide became the dominant materials for contacts due to their low contact resistance to silicon. The salicide process involved blanket metal deposition, with the silicide itself being self-aligned to the edge of the gate region, so silicide contacts achieved excellent parasitic resistance properties. The silicides, including PtSi for p-type, served as a barrier between the aluminum and the silicon.

During the 1980s and 1990s, the need for high performance multi-level metallization drove advances in interconnects and dielectrics, such as copper metallization for low resistance [2]. In contrast to aluminum that was deposited by physical deposition techniques, copper was electroplated into pre-etched trenches for the wires, and then polished back (chemical–mechanical polishing) in a "damascene" process. Since it was essential to keep copper from diffusing into the silicon region, much of the innovation in copper integration was focused on barrier metals to encapsulate the copper. Tungsten contact plugs were used between the copper wires and the silicide. For barriers to the silicon, metals and metal nitrides such as Ti, TiN, and TaN were dominant.

Outside of mainstream silicon CMOS technology, the advancement of Si/SiGe Heterojunction Bipolar Transistor (HBT) technology was another significant development in the IC industry during this period (see also Chapters 1 and 14). Towards the turn of the century, addition of carbon to the SiGe base ($Si_{1-x-y}Ge_xC_y$) became prevalent to further control the boron out-diffusion into emitter and collector regions. In these two decades, memories technology, particularly DRAM and Flash ROM continued to scale, but the process and materials for flash memories technology closely tracked the mainstream logic technology. DRAM storage capacitors utilized novel geometries (trenches and stacks) but materials similar to those of digital CMOS.

6.2.3 2000s: CMOS in the Nanometer Regime

As the technology moved below the 100-nm node, geometric scaling alone was insufficient to sustain the performance trend, and the IC industry turned to material innovation to boost performance. Strained Si technology was developed to take advantage of changes in the silicon electronic band structure and carrier transport properties to increase MOSFET performance. The most prominent example was so-called embedded SiGe, where epitaxial strained SiGe was deposited in the source and drain region of the MOSFETs to impart strain in the channel region (Figure 6.2a). The deposition of nitride films with inherent stress (both compressive and tensile) were also used to increase both NFET and PFET currents due to a strained Si channel.

Beyond the 45-nm node, gate tunneling current inhibited further scaling of oxide thickness. In response, the next prominent material innovation in CMOS technology was high K/metal gate technology to achieve an effective gate oxide thickness near and even under 1 nm. The use of gate insulators with a higher K (dielectric constant) than silicon dioxide enabled further scaling of gate capacitance, while maintaining the physical thickness of the gate dielectric to suppress tunneling current. Hf-based oxides, Al_2O_3, and other rare earth oxides such as lanthanum and zirconium oxides were intensely studied, but by the 32-nm node, the industry converged on HfO_2 and related materials for their superior material stability during processing and high resulting channel mobility.

1950   From
                              Tubes to Transistors   **1950**

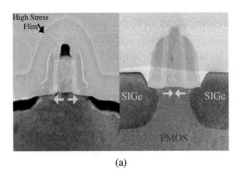 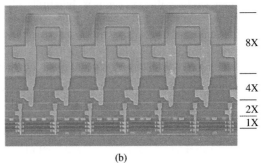

(a)   (b)

Figure 6.2 (a) Intel's strained Si MOSET transistors in 90-nm technology [1]. © 2003 IEEE. Reprinted with permission from Ghani, T. *et al.*, A 90nm high volume manufacturing logic technology featuring novel 45nm gate length strained silicon CMOS transistors; *Tech. Dig. IEEE International Electron Devices Meeting*, 2003. (b) Cross-section of advanced multi-level interconnects in IBM 65-nm technology, with up to ten levels of copper metallization with W plugs, SiCOH and F-doped TEOS as low-K interlevel dielectrics, and relaxed dimensions in the upper layers [2]. © 2005 The Japan Society of Applied Physics. Reprinted from the Symposium on VLSI Technology, *Digest of Technical Papers*, 2005

For low interconnect capacitance, the first material to replace conventional SiO_2 was fluorinated SiO_2, which reduced K from 3.9 to ~3.5. Today, carbon-doped porous silica-glass deposited by chemical vapor deposition is in wide use, with K reduced to 2.5–3 (Figure 6.2b).

Scaling also drove material innovation in memory technology as well. In DRAM technology, to achieve a high capacitance in a small area, many materials were evaluated in the research lab, including tantalum pentoxide (Ta_2O_5), barium strontium titanate ($SrBaTiO_3$), lead zirconate titanate ($PbZrTiO_3$), and $SrBi_2Ta_2O_9$, etc. In the end, Al_2O_3, $Hf_xAl_yO_z$, and related multi-layer structures were the most popular high K choices. Phase-change memories have emerged recently as a nonvolatile memory structure which can be scaled to very small geometries. A promising material today for this application is the chalcogenide alloy GeSbTe. A second emerging nonvolatile technology is magnetoresistive random access memory (MRAM), based on magnetic tunnel junctions to enable the switching of the magnetization in a multilayer thin film structure. It offers the possible combination of high speed and high reliability (many read/write cycles). Prototype magnetic materials in the thin film stack include the CoFeB-MgO or FePt systems and Co/Pt multilayers [3].

6.2.4 Power Devices and Solar Cells

Silicon has long been the leading material for power transistors and commercial solar cells. Power devices include low resistance versions of MOSFETs and insulated gate bipolar transistors (IGBTs). IGBT fabrication technology generally involves a combination of MOS and bipolar approaches, with the material set generally covered earlier (see also Chapter 15). In power devices gold may be used for low resistance and contact reliability. Silicon is also the leading semiconductor for the production of solar cells (see also Chapter 16). The silicon may be single crystalline, polycrystalline, or even amorphous in thin-film cells. The contacts of solar cells add some extra materials to those previously listed. Metallization may be made of silver because of its low resistivity and its ability to be screen printed at low cost. In thin-film cells, such as those of hydrogenated amorphous silicon, silicon-germanium and/or silicon-carbon alloys, unpatterned transparent conductors such

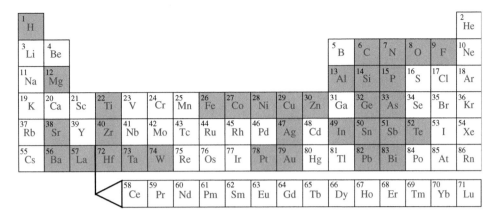

Figure 6.3 Summary of elements (shaded) used in silicon device technology

as indium tin oxide (ITO) or even zinc oxide are often used as electrodes.

Figure 6.3 summarizes the materials described earlier that are used in silicon device technology.

6.3 Compound Semiconductor Devices

The main driver of integrated circuits based on compound semiconductors, principally GaAs and InP, has historically been high speed. This was enabled by higher carrier mobilities and saturation velocities compared to those of silicon, and low parasitic capacitances due to semi-insulating substrates. GaAs ICs in particular played a critical role in the high-speed multiplexing circuitry required in the early fiber optical communications networks. However, as Si technology kept moving to smaller dimensions and higher speed, III-V ICs were largely displaced. Nevertheless, epitaxy did become a well-developed fabrication technology. Today, GaAs- and InP-based devices are widely used in the analog sections of all cellular telephone and smartphone handsets, not because of their high speed advantages, but because of low noise and power-added efficiency that lead to better battery life (see Chapter 14). Good reviews of the evolution of compound semiconductor electronic device technology can be found in references [4–6].

> **III−Vs**
>
> Compound semiconductor device technology has usually relied on advanced material structures for progress in performance.

6.3.1 MEtal-Semiconductor Field Effect Transistors (MESFETs)

Because of the lack of a well-behaved semiconductor/insulator interface, field-effect devices based on gallium arsenide have traditionally been depletion mode devices with a doped channel (Figure 6.4a).

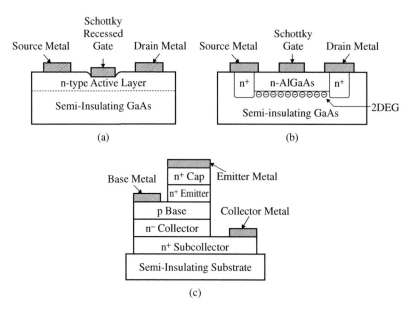

Figure 6.4 Schematic structures of typical compound semiconductor electronic devices: (a) MESFET, (b) HEMT, and (c) HBT. Metal contacts usually involve multilayer structures

The current is controlled by a metal gate which forms a Schottky barrier to the semiconductor, and reverse-bias on the gate depletes the channel to control the device current. Such MESFETs were first manufactured around 1970. The doped channel could be formed by epitaxy or by ion implantation, with Si, Se, or S as dopants. Because the gate was a Schottky barrier, the gate metallurgy was critical. Sometimes p-type layers were used near the surface for the gate (Be, Mg). Typical gate metallizations contained stacks of Ti/Cr/Ni as interfacial layers. Au was often on top for conductivity, and Pd, Pt, W (or silicides) served as barriers to prevent Au diffusion into the semiconductor during annealing. For example, a Ti/Pt/Au stack was common. For the source-drain contacts, early devices typically employed an alloyed contact based on Au/Ge or Au/Ge/Ni, although these can penetrate hundreds of nanometers into the semiconductor. Later devices (especially MODFETs and HBTs) used non-alloyed contacts to doped layers (often doped by ion implantation in FETs) and similar metals to those used for the gate. Self-aligned processes required an all-refractory gate to withstand the source-drain annealing. The substrate isolation usually relied on a semi-insulating (SI) substrate, unlike the junction isolation in silicon devices. Early SI substrates utilized chromium doping to create a deep level that insured a Fermi level near midgap despite background dopants. Chromium doping was not stable under implant and annealing conditions, so virtually all SI substrate technologies switched to using a combination of EL-2, a deep level donor created by As antisites, and carbon, the native p-type dopant that comes from the crucibles to produce stable SI substrates. SiO_2 and Si_3N_4 are used as insulating or passivating layers.

1951   Initiation of a Solid-State   ... in the Institute of   1951
         Devices Committee ...   Radio Engineers (IRE)

6.3.2 High Electron Mobility Transistors (HEMTs)

The doping of the channel region in MESFETs limits the carrier mobility due to ionized impurity scattering. According to the modulation doping principle, when a high-bandgap region is doped adjacent to a channel layer with a lower bandgap, the electrons from the dopants will transfer to the channel region to form a two-dimensional electron gas (2DEG) (Figure 6.4b). Thus, a high concentration of electrons is achieved without ionized impurities, enabling high electron mobility in the channel. The structure of the channel, doping supply layer, and capping layers is grown by heteroepitaxy techniques. The overall device architecture looks like that of a MESFET, with a Schottky-barrier gate modulating the current. These devices are known as high electron mobility transistors (HEMTs) or equivalently modulation-doped field effect devices (MODFETs). While the early emphasis was on higher channel mobility, the eventual widespread use of HEMT devices came because of their lower noise properties due to reduced scattering in the channel. They are now in virtually all cellular telephone handsets.

On gallium arsenide substrates, the widegap doped region is typically made of $Al_xGa_{1-x}As$ or $In_xAl_{1-x}As$, usually with silicon doping. The channel region can be GaAs or $In_xGa_{1-x}As$. Such devices entered production in the 1980s. On indium phosphide substrates, the channel region is also typically $In_xGa_{1-x}As$, with a widegap region of $In_xAl_{1-x}As$ or $In_xAl_yGa_{1-x-y}As$. Pseudomorphic growth and lattice-adjusting buffer layers between the channel and substrate enable many variations; with InSb or InAs channels as examples.

For high power/high voltage applications, within the past 10 years, HEMTs have also emerged based on gallium aluminum nitride ($Al_xGa_{1-x}N$), a wide bandgap semiconductor system grown typically on sapphire (Al_2O_3) or SiC substrates. Channels are typically GaN, gates are formed from Ni/Au, and source-drain Ohmic contacts from Ti/Al/Ti/Au and related materials.

6.3.3 Heterojunction Bipolar Transistors (HBTs)

A classical homojunction bipolar transistor is limited by the trade-off between a narrow low-doped base for a short transit time and high current gain versus a wide heavily-doped base for low lateral base resistance to reduce RC delays and prevent punchthrough. Increasing the bandgap of the emitter with respect to that of the base enables high current gain even with heavy base doping, which in turn allows narrow bases. Grading the bandgap of the base provides a built-in electric field to further lower the base transit time. Such devices with multiple bandgaps are known as heterojunction bipolar transistors (HBTs) (Figure 6.4c). Their low transit times and low parasitic resistances enable high performance.

While HBT technology was originally touted for its high-speed potential, the real application came due to its higher power-added efficiency in power amplifiers, and today there are two HBTs in virtually every cellular phone handset. On GaAs substrates, the wide gap material can be $Al_xGa_{1-x}As$ or $In_xGa_{1-x}P$. Doped $In_xGa_{1-x}As$ can be used to contact wide gap regions, and $Al_xGa_yIn_{1-x-y}As$ can be used as a narrow gap base. On InP substrates, the narrow gap base is typically p^+ $In_xGa_yAl_{1-x-y}As$ or $GaAs_xSb_{1-x}$ (with grading), the emitter is n-type $In_xAl_{1-x}As$, and a n^+ $In_xGa_{1-x}As$ layer may be used for the emitter contact. Many variations are possible, and some interfaces may be graded to avoid undesirable abrupt energy band discontinuities. Si is a typical n-type dopant and C, Zn, or Be are p-type dopants. Non-alloyed contacts as described earlier are preferred.

6.3.4 Optoelectronic Devices

Optoelectronic devices, such as lasers and detectors, are usually made of compound semiconductors because of their direct bandgap. Devices designed for fiber optics at 1.5 μm wavelength are usually made using an InP-based technology, with $In_xGa_{1-x}As_yP_{1-y}$ alloys being tailored for the appropriate bandgap. Shorter wavelength lasers (e.g., 980 nm) are often based on GaAs technology. Metals include those listed for HBTs and HEMTs. To modulate the light to encode data, an electro-optic modulator (traditionally based on a doped waveguide in a lithium niobate substrate) may be used. On the LED side, again many choices are possible, with emission from the IR, through the visible into the UV. These may be based on GaP or InGaN or other semiconductors. White LEDs require multiple wavelengths to be emitted. This can be achieved by a blue-emitting LED (using InGaN/GaN for the semiconductor structure, usually on a sapphire substrate) encapsulated in a package coated with a yellow-emitting phosphor, such as cerium-doped yttrium aluminum garnet (YAG), $Y_3Al_5O_{12}$:Ce.

On the detector side, yet more combinations of materials are available, as bandgaps or other energy barriers can be altered to detect specific wavelength ranges and to control the spatial region of absorption, transit times, avalanche multiplication, and so forth. One material set used for infrared (IR) focal plane imaging not covered above is HgCdTe alloys. The readout of detector arrays may include bonding on a pixel-by-pixel level to a silicon electronic readout chip. Due to the wide range of semiconductor alloys available, different device principles, and different application wavelengths, a complete listing of materials used for emitters, detectors, and modulators is too lengthy to cover here.

In the photovoltaic field, the leading thin-film technology uses absorption in a p-type cadmium telluride (CdTe) layer, with a transparent wide bandgap n-type cadmium sulfide (CdS) heterojunction. An emerging technology is based on a p-type copper indium gallium selenide absorber ($CuIn_xGa_{1-x}Se_2$, "CIGS"), which also has a wide bandgap CdS heterojunction. Transparent contacts in either technology can use tin oxide, indium tin oxide and zinc oxide; metal contacts include Al, Ni, Au, and Mo; and MgF_2 or TiO_2 may serve as an antireflection coating.

Figure 6.5 summarizes the wide range of materials used in compound semiconductor device technology.

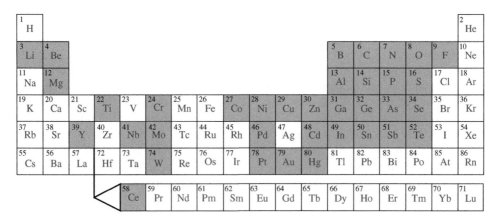

Figure 6.5 Summary of elements (shaded) used in compound semiconductor device technology

6.4 Electronic Displays

The era of high-information-content electronic displays began in the 1920s–1930s when television was invented (see also Chapter 17). Electronic displays have since gone through several generations of evolution: from cathode ray tubes (CRTs) to plasma display panels (PDPs) to active-matrix liquid crystal displays (AMLCDs) and electrophoretic ink (E-ink) to active-matrix organic light emitting diode displays (AMOLEDs).

6.4.1 1930s–2000s: Cathode Ray Tube (CRT) Displays

CRTs enabled the first receivers for the broadcasting of television [7]. In CRTs (Figure 6.6a), streams of electrons are emitted from a heated cathode (typically a tungsten filament) in a vacuum tube, are focused and accelerated by the high voltage applied between the cathode and anode (e.g., Al), and are deflected by a magnetic field to scan an image on the screen. The screen is coated with phosphors (e.g., $Y_2O_2S:Eu/Fe_2O_3$, $ZnS:Cu,Al$, $ZnS:Ag/Co$, etc.) that emit visible light when excited by energetic electrons. The CR tubes are typically made of glass, and sometimes special dopants are added (e.g., lead in the face glass for blocking radiation). CRT displays continued to dominate all applications of electronic displays for decades until ~1990–2000, when more compact flat-panel displays became popular.

6.4.2 1990s–2000s: Plasma Display Panels (PDPs)

Since the first PDP demonstration in 1964 by University of Illinois at Urbana-Champaign, the operation of PDPs fundamentally has mimicked that of fluorescent tubes [8, 9]. A PDP (Figure 6.6b) typically contains an array of tiny, luminous cells positioned between two glass plates. Each cell is filled with noble gases like Xe or Ne. A gas discharge induced by applying AC voltage on electrodes buried in insulating protective layers (e.g., MgO) creates a plasma. Energetic particles in the plasma excite gas molecules to emit UV light, which in turn excites phosphors coated in cells to emit visible light through the transparent electrode (e.g., indium tin oxide ITO) on the front plate. A partial list of phosphors includes $Y_2O_3:Eu$, $(Y,Gd)BO_3:Eu$, $Zn_2SiO_4:Mn$, $BaO\text{-}6Al_2O_3:Mn$, and $BaMg_2Al_{16}O_{27}:Eu$. In addition to their rather good image quality, PDPs can be made in large areas with printing methods. PDP panels thus gained a significant share of the large-screen

Displays

Display technology typically requires a complex integration of heterogeneous materials.

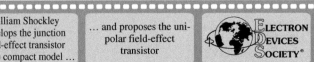

| William Shockley develops the junction field-effect transistor (JFET) compact model … | … and proposes the unipolar field-effect transistor | ELECTRON DEVICES SOCIETY® | Adjustable resistors (trimmer) come up | Introduction of PIN rectifier |

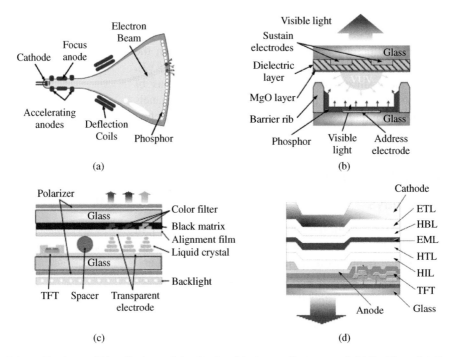

Figure 6.6 Schematic views of (a) cathode ray tube display, (b) plasma display panel, (c) liquid crystal display pixel, and (d) active matrix OLED display showing OLED and TFT circuitry

TV market (e.g., beyond 40-inch) from the 1990s to the 2000s. However, with the maturity and perfection of large-area LCD displays in the first decade after 2000, PDPs lost market share to LCD TVs.

6.4.3 1990–present: Active-Matrix Liquid Crystal Displays (AMLCDs) and Electrophoretic Ink (E—Ink)

While the use of the unique electro-optical properties of liquid crystals (LCs) for displays dates back to demonstrations at RCA in the 1960s and the invention of AMLCDs at Westinghouse in the 1970s, it took over another 20 years before AMLCDs came to prevail in all kinds of electronic devices, from handheld smart phones/tablets/notebooks to desktop computers to large high-definition TVs. An AMLCD (Figure 6.6c) typically contains an array of liquid crystal cells between glass substrates as electric-field-controlled light modulators, with each LC cell (i.e., pixel) being individually controlled by a TFT (thin film transistor) circuit. The LC cells usually consist of transparent electrodes (e.g., indium tin oxide, ITO), organic alignment layers (polymers), color filters (composed of photoresists and pigments), spacers (small plastic balls to define the cell gap), and polarizers (oriented polymers). The LC materials used in displays usually consist of one or more types of complex organic molecules.

The TFT circuit for each pixel usually contains one switch TFT and one storage capacitor (as in DRAMs). The TFT is typically a bottom-gate field-effect transistor using amorphous Si (a–Si) as the channel semiconductor, silicon nitride as the gate insulator, and Mo/Ti/Al as electrodes and conductors. To withstand the processing conditions of TFTs, alkaline earth borosilicate glasses (e.g., a mixture of barium oxide, boron oxide, and silica) that exhibit low thermal expansion and high strain points have been widely adopted in the display industry [10]. An emissive LCD requires a backlight source, such as a Cold Cathode Fluorescent Lamp (CCFL). CCFLs contain materials similar to those in PDPs, and usually a mercury vapor. Currently, they are being replaced by white LEDs.

Along with the recent rapid growth of AMLCD applications, since 2000 non-light-emitting "electronic ink" (E-ink) displays have become popular [11]. Architecturally, an active-matrix E-ink display is much like an AMLCD, except that the LC is replaced with a laminated E-ink film, and there is no backlight. The E-ink (electronic ink) film is composed of microcapsules (made of polymer membranes) suspended in a host polymer. The microcapsules contain charged pigment particles (e.g., TiO_2) in a suspended solvent that are electrophoretic - that is they can be induced by an electric field to move up or down within the capsules. This position of the particles changes the reflectance of incident light, and thus creates a "display" effect. Each pixel is addressed with an a–Si transistor as in an AMLCD. Since E-ink displays are reflective displays and are usually bistable (the particles do not move once the field is removed), they need neither backlights nor image refresh, making them very low-power displays for e-books or e-readers.

6.4.4 2000–present: Active Matrix Organic Light Emitting Diode Displays (AMOLEDs)

In the middle of the 1980s, a new type of display – organic light emitting diodes (OLEDs) was invented [12]. The operation of OLEDs is similar to that of conventional LEDs, except that they use organic non-crystalline semiconductors (see also Chapter 17). Large and dense OLED pixel arrays can easily be deposited on convenient substrates (like glasses and plastics). An efficient OLED (Figure 6.6d) typically has a multilayer structure consisting of several organic semiconducting materials, such as hole-transport materials, electron-transport materials, and emitting host and dopant materials, all between a transparent anode electrode (e.g., ITO) and a metal cathode (e.g., Al, Mg, Ag, etc.). Modern highly efficient OLEDs also require phosphorescent dopants containing transition metals (such as Ir and Pt, etc.) and carrier-injection materials (such as LiF) [13]. For high-resolution OLED displays, TFT pixel-driving circuits are used to implement active matrices. Due to higher requirements for the stability of TFTs for AMOLEDs compared to that for AMLCDs, low-temperature polycrystalline silicon (LTPS) TFTs instead of a–Si TFTs are used for AMOLEDs. The recent development of TFTs based on oxide semiconductors (e.g., amorphous InGaZnO), which provide better performance than a–Si TFTs but are compatible with a-Si TFT processing (thus less expensive than LTPS TFTs), has made them a promising alternative for pixel circuitry for AMOLEDs.

Due to their merits compared to AMLCDs – simpler structures, fewer processing steps, being thinner and lighter, more ideal viewing characteristics, faster response, higher power efficiency, no backlights, and the capability to be flexible and even rollable – AMOLEDs have recently begun to replace AMLCDs in some high-end and small-to-medium-sized applications, and are being considered for larger displays (like TVs).

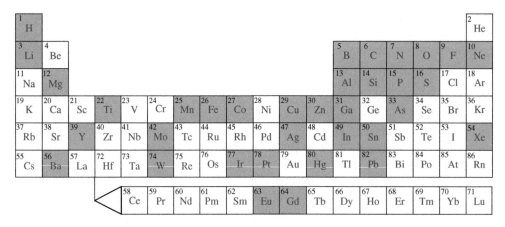

Figure 6.7 Summary of elements (shaded) used for electronic displays

Figure 6.7 summarizes elements used in the manufacturing of high information content displays.

6.5 Closing Remarks

Over some 80+ years, the range of elements used in materials for the manufacturing of electronic devices has grown to cover the better part of the periodic table (Figure 6.8). Future advances in device technology will no doubt require the development of new materials, continuing the trend of past history.

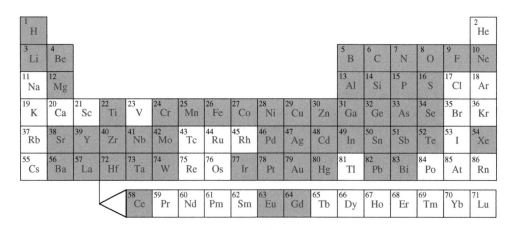

Figure 6.8 Summary of elements (shaded) used for manufacturing of various electronic devices

| Hillary and Tensing Norgay climb Mount Everest | D. Chapin, C. Fuller and G. Pearson of AT&T Bell Labs … | | … demonstrate the first silicon solar cell with 4% efficiency | Invention of cross-field amplifier by William Brown |

References

[1] T. Ghani, M. Armstrong, C. Auth, *et al.*, "A 90nm high volume manufacturing logic technology featuring novel 45nm gate length strained silicon CMOS transistors", *Tech. Dig.-IEEE International. Electron Devices Meeting (IEDM)*, pp. 978–981, 2003.

[2] E. Leobandung, H. Nayakama, D. Mocuta, *et al.*, "High performance 65 nm SOI technology with dual stress liner, and low capacitance SRAM cell", *Symposium on VLSI Technology*, pp. 126–127, 2005.

[3] R. Sbiaa, H. Meng, and S.N. Piramanayagam, "Materials with perpendicular magnetic anisotropy for magnetic random access memory", *Phys. Status Solidi RRL* Vol. 5, No. 12, pp. 413–419, 2011.

[4] S. K. Ghandi, *VLSI Fabrication Principles*, John Wiley & Sons, Inc., New York, 1994.

[5] S. Tiwari (ed.) *Compound Semiconductor Transistors*, Physics and Technology, IEEE Press, New York, 1993.

[6] W. Lu, *Fundamentals of III-V Devices: HBTs, MESFETs, and HFETs/HEMTs*, John Wiley & Sons, Inc., New York, 1999.

[7] L. E. Tannas, *Flat-Panel Displays and CRTs*, Van Nostrand Reinhold Co., New York, 1985.

[8] L. F. Weber, "The promise of plasma displays for HDTV", *SID Symposium Digest of Technical Papers*, Vol. 31, pp. 402–405, 2000.

[9] J. G. Eden, "Information display early in the 21st century: overview of selected emissive display technologies", *Proceedings of the IEEE*, Vol. 94, No. 4, pp. 567–574, 2006.

[10] A. Ellison and I. A. Cornejo, "Glass substrates for liquid crystal displays", *Inter. Journal of Applied Glass Science*, Vol. 1, No. 1, pp. 87–103, John Wiley & Sons, Inc., New York, 2010.

[11] K. Amundson, J. Ewing, P. Kazlas, R. McCarthy, J. D. Albert, R. Zehner, P. Drzaic, J. Rogers, Z. Bao, and K. Baldwin, "Flexible, active-matrix display constructed using a microencapsulated electrophoretic material and an organic-semiconductor-based backplane", *SID Symposium Digest of Technical Papers*, Vol. 32, pp. 160–163, 2001.

[12] C. W. Tang, S. A. Van Slyke, and C. H. Chen, "Electroluminescence of doped organic thin films", *Journal of Applied Physics*, Vol. 65, No. 9, 3610–3616, 1989.

[13] C. Adachi, M. A. Baldo, M. E. Thompson, and S. R. Forrest, "Nearly 100% internal phosphorescence efficiency in an organic light emitting device", *Journal of Applied Physics*, Vol. 90, No. 10, pp. 5048–5051, 2001.

1954   First Kidney transplant   ELECTRON DEVICES SOCIETY   Hendricus Lemmens files a patent application on the dispenser cathode

Chapter 7

Compact Modeling

Colin C. McAndrew and Laurence W. Nagel

7.1 The Role of Compact Models

Models have been used to guide the design of electrical systems since at least the 1880s, when Oliver Heaviside developed the "telegrapher's equations" to understand the transmission of electrical signals along wires; and the Child–Langmuir Law, dating to 1911, enabled analysis and design of vacuum tube electronic circuits. Before the advent of the digital computer the tools of the trade for electronic circuit design were engineering experience, pencil-and-paper analysis, and bread-boarding. With the rise of integrated semiconductor electronics bread-boarding was no longer feasible for design, and the ensuing increase in the number of components in, and the complexity of, circuits made hand analysis arduous. This led to the development through the 1960s and 1970s of computer programs to analyze the behavior of electronic circuits, culminating in SPICE [1].

The dc, large-signal, small-signal, and more complex analysis algorithms in SPICE (and other circuit simulators) work on the simple, but realistic, assumption that electronic circuits can be represented as an interconnected network of lumped element components; Kirchhoff's voltage and current laws applied to that network yields a system of nonlinear ordinary differential equations that SPICE solves. The models of the components used for circuit simulation are called *compact models*, and they consist of a sub-network of

Guide to State-of-the-Art Electron Devices, First Edition. Edited by Joachim N. Burghartz.
© 2013 John Wiley & Sons, Ltd. Published 2013 by John Wiley & Sons, Ltd.

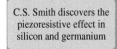

C.S. Smith discovers the piezoresistive effect in silicon and germanium

ELECTRON DEVICES SOCIETY®

Introduction of the Ebers-Moll model for bipolar transistors (J.J. Ebers and J. Moll)

current and voltage sources, charge (capacitive) current sources, flux (inductive) voltage sources, and noise current (power spectral density) sources, with the currents, voltages, charges, fluxes, and noise specified as mathematical functions of branch voltages and currents. Compact models and circuit simulators are co-dependent: without accurate compact models the most sophisticated circuit analysis algorithms will give useless results; and without being implemented in a circuit simulator compact models are useless for analysis of all but the simplest circuits.

The macroscopic electrical behavior of semiconductor devices is described by three-dimensional coupled partial differential equations (Poisson's equation and the carrier continuity and drift-diffusion transport equations for electrons and holes). Direct solution of these equations is useful for device and technology design and to help detailed understanding of device behavior (Chapter 8), but is too computationally demanding for circuit design. The art of compact modeling is to reduce the general, accurate, yet complex semiconductor differential equations into algebraic equations that are efficient to compute yet sufficiently accurate for the type of component being modeled (and meet other stringent conditions; see the sidebar *Compact Model Requirements*).

A good example of a compact model is for the current I in a *pn*-junction diode as a function of the voltage V applied across the junction

$$I = qA \left(\frac{D_p p_{no}}{L_p} + \frac{D_n n_{po}}{L_n} \right) \left(e^{V/\phi_t} - 1 \right) = I_0 \left(e^{V/\phi_t} - 1 \right) \qquad (7.1)$$

where $\phi_t = kT/q$ and the physical quantities in the first form, from Shockley's ground-breaking physics, are distilled into the parameter I_0 in the second form, while retaining the physical exponential dependence. The model can then be used for any *pn*-junction diode, with I_0 extracted from experimental $I(V)$ data from that diode. Physical analysis based on the depletion approximation and Poisson's equation gives the junction capacitance as

$$C = A \frac{\sqrt{0.5 q \varepsilon_s N_A N_D / (N_A + N_D)}}{\sqrt{\phi_t \ln (N_A N_D / n_i^2) - V}} = \frac{C_0}{(1 - V/p)^m} \qquad (7.2)$$

where again the physical quantities in the first form are replaced by model parameters (C_0, p, and m) in the second form, which enable the mathematical form of the model to be characterized for any specific device. The theory, based on an assumed abrupt junction

Compact Model Requirements

Compact models have to meet many stringent and often conflicting requirements. They need to be:

- simple yet accurate
- complete (I(V), charges, noise)
- smooth (continuous up to at least third order derivatives for distortion modeling)
- physically based
- scalable over geometry and temperature
- numerically stable and well behaved for all biases
- simple to extract parameters for
- computationally efficient.

| A.M. Clogston and H. Heffner develop the clad ... | ... periodic permanent magnet for electron beam guidance |  ELECTRON DEVICES SOCIETY® | The first issue of "IEEE Transactions on Electron Devices" is published | Transactions of the I.R.E. ELECTRON DEVICES |

between uniformly doped regions, yields a square-root form for the denominator; the parameter m is allowed to deviate from $^1/_2$ to better fit the measured $C(V)$ characteristics of real devices. This exemplifies the characteristics of a good compact model: use semiconductor physics to guide the development of an appropriate model form for a specific type of device, then judiciously adapt that form so that, via the model parameters, in can accurately represent the electrical behavior of a broad range of realizations of that device type. Note that in practice compact models are formulated in terms of charge (Q), not capacitance, but are often described in terms of capacitance because that is what most people are familiar with and what is measured for characterization. Compact model development has been driven in the past decades by elucidation of improved fundamental formulations for specific types of devices, by technology advances that have made devices sensitive to previously unimportant physical effects, and by the stringent requirements of the design of precision analog and RF circuits.

7.2 Bipolar Transistor Compact Modeling

Perhaps the most simple and elegant compact model yet developed is the Gummel-Poon bipolar transistor model [2], which is based on Gummel's integral charge control relation (ICCR, see the sidebar *Integral Charge Control Relation*)

$$I_{CE} = qA\mu_n n_i^2 \phi_t \frac{e^{V_{BE}/\phi_t} - e^{V_{BC}/\phi_t}}{Q_b'} = I_0 \frac{e^{V_{BE}/\phi_t} - e^{V_{BC}/\phi_t}}{q_b} \quad (7.3)$$

where it is assumed that mobility μ_n and the intrinsic carrier concentration n_i are constant throughout the base; here $Q_b' = \int p \, dy$ is the hole sheet density in the base, in units of cm^{-2}, seen looking perpendicular to the plane of the base.

The reduction of the ICCR to a practical compact model follows from the second form in Equation (7.3); the physical quantities in the detailed theoretical model are mapped into the compact model parameter I_0 and the Gummel number is mapped into the normalized base charge q_b (which has value 1 when the applied biases are zero).

When non-zero biases are applied to the transistor the change in q_b has two components, from the movement of the edges of the

Integral Charge Control Relation

The Gummel Integral Charge Control Relation (Ref.[3])

$$I_{CE} = qA\phi_t \frac{e^{V_{BE}/\phi_t} - e^{V_{BC}/\phi_t}}{\int\limits_{base} \frac{p(y)}{\mu_n(y)\, n_i^2(y)} dy}$$

distills the core physics of bipolar behavior into a single, coherent formula that captures the exponential dependence on applied biases, the inverse dependence on the so-called base Gummel number, the integral in the denominator, and the proportionality to emitter area A.

Gordon Teal and Willis Adcock of Texas Instruments build the first bipolar transistor in silicon

ELECTRON DEVICES SOCIETY®

Charles A. Lee, G.C. Dacey and P.W. Foy invent the mesa transistor structure

First demonstration of silicon solar cell with 6% efficiency by Calvin Fuller, Gerald Pearson and Darryl Chapin

base–emitter and base–collector depletion regions and from the injection of minority carriers into the base from the emitter and, in reverse operation, the collector. The former is proportional to the change in normalized depletion charges and the latter is proportional to the transport current (7.3)

$$q_b = 1 + \frac{C'_{j,E0}}{qQ'_{b0}} \delta q_{j,E} + \frac{C'_{j,C0}}{qQ'_{b0}} \delta q_{j,C} + \frac{\tau_b}{qAQ'_{b0}} I_{CE} \qquad (7.4)$$

where the added 0 to a subscript indicates an equilibrium (zero-bias) value, τ_b is the base transit time, and C'_j denotes zero-bias junction (depletion) capacitance per unit area. Rather than strictly enforcing the physical relationships in Equation (7.4), which are only approximate, it is recast as

$$q_b = 1 + \frac{\delta q_{j,E}}{V_{ER}} + \frac{\delta q_{j,C}}{V_{EF}} + \frac{I_0 \left(e^{V_{BE}/\phi_t} - 1\right)}{I_{KF}q_b} + \frac{I_0 \left(e^{V_{BC}/\phi_t} - 1\right)}{I_{KR}q_b} = q_1 + \frac{q_2}{q_b} \qquad (7.5)$$

where V_{EF}, V_{ER}, I_{KF}, and I_{KR} are forward and reverse Early voltages and knee currents, respectively. This has the solution

$$q_b = \frac{q_1}{2} + \sqrt{\frac{q_1^2}{4} + q_2} \approx \frac{q_1}{2} + q_1\sqrt{\frac{1}{4} + q_2} \qquad (7.6)$$

where the second form is used in SPICE. The q_1 term embodies the Early effect, giving a finite output conductance, and as the base–emitter voltage V_{BE} increases in the forward active region eventually the q_2 term in (7.6) becomes dominant, causing q_2 to vary as $e^{V_{BE}/(2\phi_t)}$ thereby, from (7.3), causing I_{CE} to transition from depending exponentially on V_{BE}/ϕ_t to varying exponentially as $V_{BE}/(2\phi_t)$; the formulation therefore also naturally captures the effects of high-level injection (see Figure 7.1).

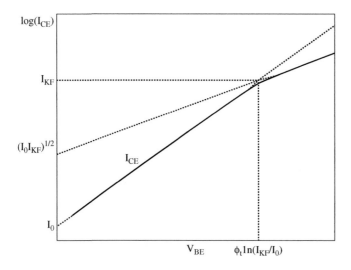

Figure 7.1   Gummel-Poon modeling of collector current including high-level injection effects

Although theoretically

$$C'_{0j,E} V_{ER} = C'_{0j,C} V_{EF} = \tau_b I_{KF}/A = \tau_b I_{KR}/A = q Q'_{b0} \tag{7.7}$$

this is not enforced. Rather, the form of the model is maintained and V_{EF}, V_{ER}, I_{KF}, and I_{KR} are considered to be separate model parameters that are determined from experimental data. This has the added benefit of (apart from the "shape" parameters for the depletion capacitance model) allowing flexibility in independently fitting dc and capacitance data.

Additional elements and capabilities must be added to Equation (7.3) to enable accurate modeling of practical devices, such as charge and noise models, parasitic resistances and capacitances, the substrate transistor for four-terminal devices, and self-heating. But these are refinements; the core physics of bipolar transistor operation is captured in Equation (7.3) in a computationally efficient, simple, and elegant form, which exemplifies what a compact model is supposed to achieve.

7.3 MOS Transistor Compact Modeling

In principle the operation of a MOS transistor is simple: the field transverse to the gate dielectric-to-silicon interface induces an inversion layer charge (the "channel") under the gate, and the drain-source voltage V_{DS} applied across that inversion layer causes current to flow. In some respects, a MOS transistor is like a resistor whose conductance is modulated by the gate and body (or "substrate" or "bulk") potentials. Despite that apparent simplicity, developing good compact models for MOS transistors has proven quite difficult, especially for modeling charges and for avoiding undesirable and unphysical behavior such as: asymmetry about and singularity at $V_{DS} = 0$; violating passivity; discontinuities or glitches at boundaries between operating regions; unphysical negative capacitances; and inconsistency between large-signal and small-signal models.

MOS transistor compact models fall into one of three general categories. The most common, the basis for analyses in most text books on CMOS circuit design, is the source-referenced, threshold voltage based formulation. Hallmarks of this approach are that the inversion charge is modeled as

$$Q_I = -WLC'_{ox}\left[V_{GS} - V_T\left(V_{SB}\right)\right], \tag{7.8}$$

where V_T is the body-bias dependent threshold voltage, and the drain-source current in strong inversion nonsaturation is of the form

$$I_{DS} = \mu C'_{ox} \frac{W}{L}\left(V_{GS} - V_T - \alpha V_{DS}\right) V_{DS} \tag{7.9}$$

where α is the bulk charge linearization coefficient. This is the form of the first MOS transistor model implemented in SPICE, the Shichman-Hodges or Level 1 model [4], and the BSIM3 and BSIM4 models [5] represent the pinnacle of development of this type of model. Mathematical techniques are used to make Equation (7.9) approximately asymptotically reasonable in saturation and weak inversion, however this type of model has no physical basis in moderate inversion which, because of reduced supply voltages, is the major region of operation in modern CMOS processes. All widely used source-referenced threshold voltage based MOS transistor models also have singularities at $V_{DS} = 0$ which lead to an inability to model

precision analog circuits such as passive mixers and switched capacitor circuits, for example, as used in some data converters.

A second type of MOS transistor model comes from the inversion charge based approach; this was first developed in [6] and has become most popular as the EKV model [7]. The hallmarks of this approach are a normalized inversion charge q_I computed from an expression like (the exact form is model specific)

$$\frac{q_I}{n} + \phi_t \ln\left(\frac{q_I}{n\phi_t}\right) = \frac{V_{GB} - V_{FB}}{n} - \phi_n - n_1, \qquad (7.10)$$

where n is the reciprocal of the slope of surface potential ψ_s versus gate voltage in weak inversion, and a drain current of the form

$$I_{DS} = \mu \frac{W}{L}\left(\frac{Q_{I0}'^2}{2nC_{ox}'} - \frac{Q_{IL}'^2}{2nC_{ox}'}\right) - \mu \frac{W}{L}\phi_t\left(Q_{I0}' - Q_{IL}'\right) \qquad (7.11)$$

where the first term represents the drift component of current and the second term the diffusion component; note the symmetry inherent in Equation (7.11). Iterative solution is required for the inversion charge density at source and drain ends of the channel, and clearly the basis of this approach is not valid in accumulation and depletion where there is no inversion charge. Surface potential based models do not suffer from this problem; it can be shown that there is a close relationship between inversion charge based and surface potential based MOSFET models, and the latter turn out to be more amenable as a basis for implementing physical effects important to transistor operation, such as velocity saturation (which is more naturally modeled via field, that is, potential difference per unit length, than via inversion charge density).

Surface potential based models are the third major category of MOS transistor models. The first practical model of this type was based on the charge-sheet approximation [8], and forms the basis of present day models.

The drain-source current in a MOS transistor consists of drift and diffusion components. At any point x along the channel

$$I_{DS} = W\mu\left(-Q_I'\right)\frac{d\psi_s}{dx} + W\mu\phi_t\frac{dQ_I'}{dx} \qquad (7.12)$$

where the inversion charge density is

$$Q_I' = -C_{ox}'\left(V_{GB} - V_{FB} - \psi_s\right) - Q_B'$$

$$\approx -C_{ox}'\left(V_{GB} - V_{FB} - \psi_s - \gamma\sqrt{\psi_s}\right) \qquad (7.13)$$

Symmetric Linearization

Although threshold voltage based and inversion charge based MOSFET models have proven popular, surface potential based MOSFET models are closest to the essential physics of MOS transistor operation. In 1978 Brews derived an ideal charge-sheet model for I_{ds} for long channel MOSFETs [8]. However, that model form is somewhat complex, especially when extended to model charges, and difficult to add short channel effects to. Symmetric linearization [9] simplified the I_{DS} and, especially, charge model expressions greatly and led to a simple, accurate, and numerically well-behaved surface potential based MOSFET model.

(here $\gamma = \sqrt{2q\varepsilon_s N}/C'_{ox}$ is the body effect coefficient). The square root term in Equation (7) introduces complexities into modeling expressions, especially for charge, and as the bulk charge density Q'_B varies only weakly with the surface potential along the channel it can be linearized with respect to the mid-point surface potential $\psi_{sm} = (\psi_{s0} + \psi_{sL})/2$ [9] (see sidebar *Symmetric Linearization*)

$$Q'_B = -\gamma C'_{ox}\psi_{sm} - \frac{\gamma C'_{ox}}{2\sqrt{\psi_{sm}}}\left(\psi_s - \psi_{sm}\right) = -C'_{ox}\left[\gamma\psi_{sm} + \left(\alpha_m - 1\right)\left(\psi_s - \psi_{sm}\right)\right], \qquad (7.14)$$

where $\alpha_m = 1 + \gamma/\left(2\sqrt{\psi_{sm}}\right)$, with essentially no loss in accuracy but significant reduction in complexity, especially for charge expressions. Integrating Equation (7.12) from source to drain gives

$$I_{DS} = \mu C'_{ox}\frac{W}{L}\left(V_{GB} - V_{FB} - \psi_{sm} - \gamma\sqrt{\psi_{sm}}\right)\left(\psi_{sL} - \psi_{s0}\right) + \mu C'_{ox}\frac{W}{L}\alpha_m\phi_t\left(\psi_{sL} - \psi_{s0}\right) \qquad (7.15)$$

where the first term is the drift component and the second term the diffusion component.

The charge and capacitance modeling expressions are too complex to provide details of here; however, the simplification symmetric linearization brings to charge modeling expressions is astounding with essentially no loss in modeling accuracy, see Figure 7.2 (details are provided in [9]).

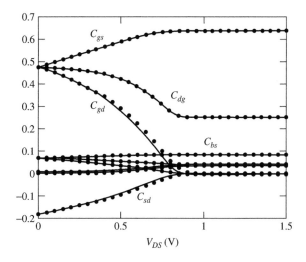

Figure 7.2   Capacitances versus V_{DS} in strong inversion (normalized to LWC'_{ox}). Symbols are symmetric linearization model, lines are from a model that does not linearize Q'_B

7.4 Compact Modeling of Passive Components

Passive components (Chapter 4) are important in many circuits. Although "passives" are often considered to be simple linear resistors, capacitors, and inductors, modeled via

$$I = G \cdot V, I = C \frac{dV}{dt}, V = L \frac{dI}{dt} \tag{7.16}$$

respectively, real passives in semiconductor technologies are significantly more complex. Not only do they have parasitic series resistance and inductance from interconnect, and parasitic capacitance to adjacent elements, that must be taken into account to properly model their behavior over frequency, but they can be nonlinear. This can be intentional, for example, in MOS varactors (variable capacitors) used for frequency tuning, or unintentional, for example, in both diffused and polysilicon resistors. Modern communications systems can have stringent requirements for distortion and harmonic content, and as these are controlled by nonlinearities it is a requirement for compact models of passives to accurately represent those nonlinearities.

Integrated resistors, despite Ohm's law, are not linear, and as their resistance is modulated by the silicon region ("substrate") they sit in, or above, and are not even two-terminal. Depletion region movement at the bottom of a resistor, from *pn*-junction physics for diffused resistors and MOS physics for polysilicon resistors, modulates the conducting depth, and thereby the resistance, of the resistor body. The average conducting region depth depends on both the substrate bias and the voltage across a resistor, which requires a three-terminal nonlinear model. Velocity saturation causes additional nonlinearity in diffused (but not polysilicon) resistors, and although this is generally considered to be a "sub-micron" effect, of importance only for lengths below about 1 μm, observable nonlinearity, which is critical for distortion modeling, is apparent in experimental data from devices of up to 50 μm, so it must be taken into account. Self-heating can also cause significant nonlinearities in resistors, and is usually the dominant cause of nonlinearity in poly resistors. The time constants associated with self-heating are of order 0.1 to 10.0 μs, hence the small-signal ac resistance varies with frequency; at low frequencies the effect of self-heating tracks with variations in applied bias, at high frequencies it has no effect (except on the quiescent operating point). An accurate model, R3, that incorporates all of these effects, and their geometry and temperature dependence, has been developed [10], and Figure 7.3 shows a fit of the R3 model to measured data from a highly nonlinear resistor.

MOS varactors are commonly used in *LC*-tanks for frequency tuning of voltage-controlled oscillators (VCOs) in wireless communication ICs. The surface potential approach for MOSFET modeling is generally used for MOS varactor modeling, but in a greatly simplified form as there is no V_{DS} dependence. Although the basic physics of the MOS system has been well-known for several decades the performance of VCOs depends heavily on parasitics, which control the quality factor Q of oscillators; VCO design therefore entails accurate modeling of parasitics (see Figure 7.4). MOS capacitors are made from thin oxides, so quantum mechanical and polysilicon doping effects are important, and because the top (polysilicon) and bottom (silicon) plates can be of the same or different doping types (to optimize the $C(V)$ dependence of the varactor for design) handling of polysilicon doping effects is more complex than in MOS transistors. There is no direct source of inversion charge carriers in MOS varactors, so there is a frequency dependence to inversion capacitance, from thermal recombination and generation of the inversion charge. Although VCOs generally operate around flatband, where dC/dV is largest and independent of frequency, they can

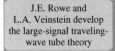

J.E. Rowe and L.A. Veinstein develop the large-signal traveling-wave tube theory

1957

ELECTRON DEVICES SOCIETY®

Launch of Sputnik, the first earth satellite

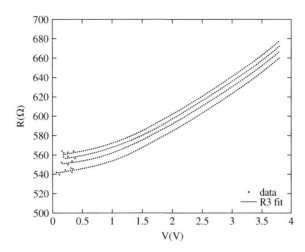

Figure 7.3 R3 model fit to measured data from a diffused resistor significantly affected by depletion pinching, velocity saturation, and self-heating. Well bias is 0, 5, 10, and 15 V bottom-to-top curves

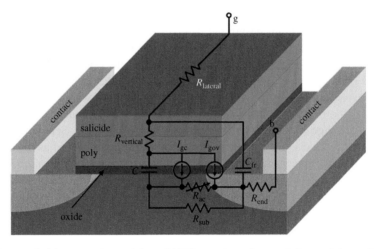

Figure 7.4 Parasitics needed for accurate modeling of MOS varactors. Accumulation resistance R_{ac} is bias-dependent

be biased towards strong inversion at times and if the time-dependent $C(V)$ is not modeled properly then the locking behavior of VCOs is not modeled properly. The relaxation time approximation (RTA) is used to model this behavior,

$$\frac{dQ_I}{dt} = -\frac{Q_I - Q_{I0}}{\tau} \tag{7.17}$$

where the added superscript 0 indicates an equilibrium value.

Integrated inductors do not generally require bias dependent models, and so are commonly modeled via subcircuits constructed from linear two-terminal inductors, resistors, and capacitors. The sub-circuits can be complex because inductor performance is significantly limited by parasitics (inter-winding capacitance, capacitance to substrate, resistance, eddy current losses in the substrate) and the skin effect, all of which must be taken into account to accurately model the frequency dependence of the electrical behavior of inductors.

Overall, accurate compact modeling of passive components entails a lot more than the simple linear relations (Equation 7.16), and the complexity of models for passives can exceed the complexity of simple models for active devices.

7.5 Benchmarking and Implementation

With so many constraints on what they must do, it is very easy to inadvertently overlook one or more requirements when developing a compact model. The best way to ensure the quality of a compact model is to define benchmark tests that expose common problems, and then rigorously apply those tests to verify whether or not a model passes them. MOS transistor models have proven to be the most susceptible to problems, and although the first MOSFET model benchmarks were reported 30 years ago [11], and have been significantly expanded since [12], many common MOSFET models still fail to pass some tests; this may not be important for digital circuit design but can be catastrophic for precision analog and RF design.

Implementation of compact models can be a major task, and historically was a significant barrier to the migration of good developments in modeling from theory into practical use. A complex model can be more than 20 000 lines of C code, which often contains a significant number of errors in hand-coded derivatives (model developers usually have a device physics, not software, background); the errors can impair convergence, but more important cause inaccuracy in small-signal analysis. In the late 1990s Verilog-A [13] emerged as a language for behavioral modeling of analog circuits. Serendipitously, it was almost ideal for defining compact models, and despite initial resistance it has emerged as the *de facto* standard for compact modeling. A model defined in Verilog-A is roughly 10% of the size of a model defined in C, and because it automatically computes derivatives it completely removes the most common source of errors in compact model codes. Using Verilog-A also enforces the requirement that the large-signal model and the small-signal model be consistent (the latter the derivative of the former).

References

[1] L. W. Nagel, "SPICE2: A computer program to simulate semiconductor circuits", Memo. ERL-M520, University of California, Berkeley, May 1975.

[2] H. K. Gummel and H. C. Poon, "An integral charge control model of bipolar transistors", *Bell Syst. Tech. J.*, vol. 49, no. 5, pp. 827–852, May–Jun. 1970.

[3] H. K. Gummel, "A charge control relation for bipolar transistors", *Bell Syst. Tech. J.*, vol. 49, no. 1, pp. 115–120, Jan. 1970.

[4] H. Shichman and D. A. Hodges, "Modeling and simulation of insulated-gate field-effect transistor switching circuits", *IEEE J. Solid-State Circuits*, vol. 3, no. 2, pp. 285–298, Sep. 1968.

Carl Frosch and Link Derick receive the patent on the silicon oxidation furnace

Invention of tunnel diode by Leo Esaki (Esaki Diode)

[5]   W. Liu, *MOSFET Models for SPICE Simulation, Incuding BSIM3v3 and BSIM4*, New York: John Wiley & Sons, Inc., 2001.

[6]   M. A. Maher and C. A. Mead, "A physical charge-controlled model for MOS transistors", in *Advanced Research in VLSI*, P. Losleben (ed.), Cambridge, MA: MIT Press, 1987.

[7]   C. C. Enz and E. A Vittoz, *Charge-Based MOS Transistor Modeling: The EKV Model for Low-Power and RF IC Design*, New York: John Wiley & Sons, Inc., 2006.

[8]   J. R. Brews, "A charge-sheet model of the MOSFET", *Solid-State Electron.*, vol. 21, no. 2, pp. 345–355, Feb. 1978.

[9]   T.-L. Chen and G. Gildenblat, "Symmetric bulk charge linearisation in charge-sheet MOSFET model", *Electron. Lett.*, vol. 37, pp. 791–793, 2001.

[10] C. C. McAndrew, "Integrated resistor modeling", in *Compact Modeling: Principles, Techniques and Applications*, G. Gildenblat (ed.), Springer, pp. 271–297, 2010.

[11] Y. Tsividis, "Problems with precision modeling of MOS LSI", *Techn. Dig. IEEE International Electron Devices Meeting (IEDM)*, pp. 274–277, 1982.

[12] X. Li, W. Wu, A. Jha, G. Gildenblat, R. van Langevelde, G. D. J. Smit, A. J. Scholten, D. B. M. Klaassen, C. C. McAndrew, J. Watts, C. M. Olsen, G. J. Coram, S. Chaudhry, and J. Victory, "Benchmarks tests for MOSFET compact models with application to the PSP model", *IEEE Trans. Electron Devices*, vol. 56, no. 2, pp. 243–251, Feb. 2009.

[13] Accellera, *Verilog-AMS Language Reference Manuals*. Available at http://www.eda.org/verilog-ams/htmlpages/lit.html (accessed March 30, 2012).

1958   Strain gauges became commercially available   ELECTRON DEVICES SOCIETY®   First solar powered satellite is built at the U.S. Naval Research Laboratory   Manufacturing of mesa transistors starts

Chapter 8

Technology Computer Aided Design

David Esseni, Christoph Jungemann, Jürgen Lorenz, Pierpaolo Palestri,
Enrico Sangiorgi and Luca Selmi

8.1 Introduction

Shortly after the invention of the bipolar junction transistor, William Shockley published the corresponding theory in 1949 based on approximate solutions of the semiconductor equations which form the drift-diffusion model (DD) [1]. These analytical expressions, which were subsequently refined by many researchers, led to the development of the compact models described in Chapter 7. With the improvement in semiconductor processing technology (drift transistor, narrow base, etc.) the need for exact solutions of the DD model became apparent. Since analytical solutions of this set of nonlinear equations exist only under very special conditions, numerical approaches were required giving rise to Technology Computer Aided Design (TCAD). At the same time the results from semiconductor processing have increasingly differed from simple approximations that can be expressed by analytical equations (see also Chapter 10). Consequently, since the 1970s process simulation has been developed to predict the geometry and dopant profiles of devices, resulting from the various manufacturing steps. Device simulation is based on this information, whether it is obtained by process simulation or by other means. For a given geometry and doping profiles the flow of electrons and holes is calculated in one, two or three dimensions, depending on the problem at hand, as a function of the bias conditions. Besides the terminal characteristics, device simulation provides insight into the electrostatics and the transport processes within the devices, which cannot be measured, and, thus, improves the

Guide to State-of-the-Art Electron Devices, First Edition. Edited by Joachim N. Burghartz.
© 2013 John Wiley & Sons, Ltd. Published 2013 by John Wiley & Sons, Ltd.

understanding of the device operation. As a consequence TCAD has played a major role in understanding the physical phenomena underlying fundamental effects like reliability, variability and scaling that have been identified as the main road-blocks threatening the still valid assumptions of Moore's law. Today, TCAD is an integral part of process technology development.

The workhorse of device simulation TCAD is the DD model, which is described in Section 8.2, but it became soon clear that the DD model does not capture many important effects such as carrier confinement and far from equilibrium transport effects. This led to the development of quantum corrections to the basic DD equations (e.g., the density gradient one [2]) and of microscopic transport models based directly on the semi-classical Boltzmann Transport Equation (BTE) which are discussed in Section 8.3. In recent years it has become possible to build devices which are so small that quantum transport effects like, for instance, direct source/drain tunneling occur and full quantum transport models have been developed (Section 8.4). An outline of the developments in process modeling, starting from the simulation of single process steps in one dimension to a full three-dimensional simulation is given in Section 8.5.

8.2 Drift-Diffusion Model

The introduction of the digital computer made numerical solutions of the partial differential equations of the DD model possible and in 1964 H.K. Gummel presented the first self-consistent solution of the DD model for a one-dimensional pnp-BJT [3]. In this paper stationary solutions were presented together with small-signal results for low frequencies (Figure 8.1).

Within a few years two-dimensional solutions of the DD model for FETs and BJTs (e.g., [4]) were presented. Dimensional splitting together with box integration (finite volume method) resulted in efficient and stable formulations of the multidimensional problem. A serious impediment was the weak numerical robustness of the DD model when the constitutive equations for the current densities were discretized with finite differences. If the electrostatic potential changed by more than the thermal voltage between two grid nodes, instability occurred in the concentrations. D.L. Scharfetter and H.K. Gummel solved this problem by introducing a new type of stabilization scheme [5] which made it possible to use rather coarse grids. This, in turn, improved the efficiency of the simulation, and the Scharfetter-Gummel stabilization scheme is at the core of the success of device simulation.

> **Drift-Diffusion Model**
>
> The DD model describes the flow of electrons and holes in semiconductors due to an electric field (drift) and due to their finite thermal velocity (diffusion). These constitutive equations for the current densities in conjunction with charge conservation and the Poisson equation for the electrostatic potential form the drift-diffusion model, which is the basis of most compact models and today's TCAD.

1959   Richard Feynman gives his famous talk "There's plenty of room at the bottom" … **E**LECTRON **D**EVICES **S**OCIETY®   … at a meeting of the American Physical Society.

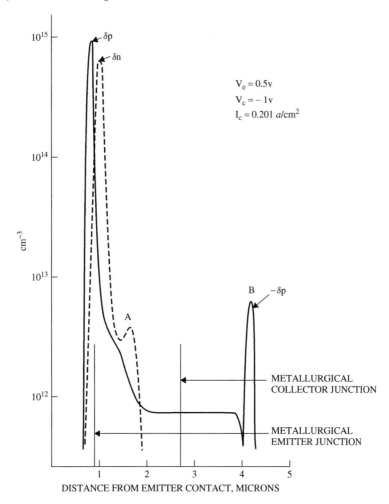

Figure 8.1   Change of the electron and hole concentration in a pnp-BJT due to a change of the emitter voltage equal to the thermal voltage [3]. © 1964 IEEE. Reprinted with permission Gummel, H.K.; "A self-consistent iterative scheme for one-dimensional steady state transistor calculations," *IEEE Trans. Electron Devices*, vol. 11, no. 10, pp. 455–465, October 1964

It is also used in other fields than semiconductor device modeling (e.g., ion channels in cells). This paper also contained the first transient large-signal solution of the DD model for an oscillator. In 1966 W. Shockley *et al.* published a paper on how to calculate electronic noise in the framework of the DD model [6]. Later, general-purpose device simulators were developed, which could handle all kinds of semiconductor devices.

In small devices or more complex semiconductors (e.g., GaAs) higher order transport equations beyond the DD approximation are often used [7]. The DD model is based on the first two moments of the particle distribution function (particle density and current density). The energy transport and hydrodynamic

models are based in addition on the moments corresponding to the energy density and energy current density. With these models velocity overshoot and impact ionization in small devices are modeled more accurately than with the DD model.

8.3 Microscopic Transport Models

The DD model, as well as the energy balance and hydrodynamic models are obtained from the BTE taking assumptions that make them inaccurate when applied to scaled devices, such as ultra thin base BJTs, MESFETs, HFETs and MOSFETs with channel length below approximately 100 nm, or when hot carrier phenomena need to be examined.

Numerical solutions of the BTE using statistical methods (e.g., the Monte Carlo (MC) approach) have been pursued since the 1960s and the 1970s to analyze uniform transport in semiconductors, mainly with the aim to interpret the experimental velocity-field curves for electrons and holes in various materials. An excellent review of this era is provided in [8]. Until the mid-1980s the computational burden was too heavy to allow for the simulation of realistic devices and thus limited to uniform cases or drastically simplified structures. Investigation of realistic MOSFETs dates back to the late 1980s [9, 10] when MC transport was coupled with 2D Poisson equation providing a so-called "self-consistent" solution (Figure 8.2 which is taken from Ref. [10]).

During the 1980s, models have become more and more sophisticated; for example the simple parabolic models for the conduction band has been replaced by a full-band description taken from accurate non-local pseudo-potential calculations, first in uniform cases [11] and then in realistic devices [9].

During the 1990s, full-band MC simulators have been widely employed to study the current drive of short channel MOSFETs, the performance of high-frequency BJTs, as well as hot-carrier phenomena in MOSFETs (bulk and gate currents, device degradation) and non-volatile-memories (injection efficiency during write in Flash devices, mechanisms for cell erase, . . .). At the same time, models have evolved including more scattering mechanisms and improving the numerical efficiency and convergence properties. Many groups worldwide have been active in this field. An important landmark has been the mutual benchmarking of the developed models and simulators, as documented in [12].

In MOSFETs quantization in the inversion layer is very strong and cannot be neglected. This led to the development of the so-called multi-subband MC technique where the Schrödinger equation is solved perpendicular to the transport direction, resulting in subbands. Within

The Monte Carlo Method for Carrier Transport

Monte Carlo simulators describe the motion of electrons and holes in semiconductors as a sequence of "free-flights" (i.e., collision-less motion according to Newton's law) interrupted by scattering events which modify the particle's momentum. The simulation is statistical in the sense that parameters such as the duration of the free-flight, the selection of the scattering mechanism and of the state-after-scattering are computed using random numbers (with suitable probability density functions).

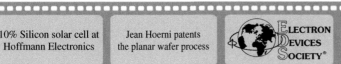

| 10% Silicon solar cell at Hoffmann Electronics | Jean Hoerni patents the planar wafer process | ELECTRON DEVICES SOCIETY® | Oberservation of drift effects in transistors due to Sodium |

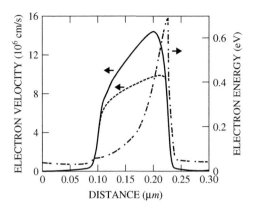

Figure 8.2 Average electron velocity in the MC (solid) and DD (dashed) simulations and surface electron energy in the MC solution (dashed-dotted). L_G = 150 nm [10]. © 1989 IEEE. Reprinted with permission Venturi, F.; Smith, R.K.; Sangiorgi, E.C.; Pinto, M.R.; Ricco, B.; "A general purpose device simulator coupling Poisson and Monte Carlo transport with applications to deep submicron MOSFETs"; 1989; *IEEE Trans. on Computer-Aided Design*; 8:4

these subbands transport is still described with the semi-classical BTE [13]. The multi-subband MC technique has been employed to study inversion layers and advanced MOSFET architectures, such as ultra-thin-body SOI.

The stochastic MC method is not the only technique to solve the BTE. Some of the first numerical solutions of the BTE were obtained by a spherical harmonics expansion (SHE) for the 3D k-space. Later, this method was used to simulate whole devices [14]. Particularly in low-dimensional systems, the direct numerical solution is much faster than the MC (e.g., nanowires). In inversion layers, the distribution function in the 2D k-space is expanded with Fourier harmonics [15]. With the direct numerical approach exact solutions of the BTE for stationary systems and small-signal solutions in the frequency domain for RF simulations, including noise, can be obtained.

> **The multi-subband Monte Carlo**
>
> Multi-subband Monte Carlo simulators provide the coupled solution of the following equations: the 1D Schrödinger equation in each section of the device, multi-subband BTE describing transport along the channel and the 2D Poisson equation.

8.4 Quantum Transport Models

The experimental observation of quantum mechanical effects in the terminal currents of the electron devices is not a recent discovery, in fact the tunneling current through the dielectrics and the negative differential resistance in single barrier or resonant tunneling diodes have been known since about the 1950s. Nevertheless, the development of quantum transport models able to describe complete devices is a relatively recent development.

Robert Noyce of Fairchild Co. invents the planar integrated circuit (IC)

ELECTRON DEVICES SOCIETY®

The Non-Equilibrium Green's Function (NEGF) formalism is today the most mature framework for the simulation of a complete device by using a quantum transport model. Some of the seminal works for the NEGF formalism can be identified in the contributions that introduced the inelastic phonon scattering, first at phenomenological level [16], and then, more rigorously, by demonstrating that the effect of phonon scattering resembles that of the device contacts, and, thus, can be accounted for in the NEGF formalism according to appropriate self-energies [17].

Besides the inclusion of the scattering, the NEGF approach has the distinct advantage that the formalism does not depend on the specific form of the Hamiltonian operator, so that the NEGF has been successfully used for a simple effective mass Hamiltonian, but also for $\mathbf{k \cdot p}$ as well as atomistic, tight-binding Hamiltonians.

The NEGF formalism is computationally very demanding and, in practice, the modeling of complete devices would not have been possible without the use of recursive algorithms based on the Dyson equation [18]. Even with recursive algorithms, the calculations of the necessary blocks of the Green's function for 3D devices is computationally very challenging if one embraces the standard finite differences discretization schemes for the effective mass or $\mathbf{k \cdot p}$ Hamiltonian and, in particular, it is so for a tight-binding Hamiltonian. A remarkable improvement in the computational efficiency has been obtained by resorting to the so-called mode-space approach [19]; the fully coupled-mode approach is theoretically equivalent to the problem in real space but it allows for huge savings in memory and CPU requirements. Finally, a key enabler for the simulation of real devices with the NEGF formalism is the development of efficient algorithms for the calculation of the self-energies of the contacts [20].

Thanks to the improvements in the theoretical aspects and numerical algorithms the NEGF formalism has been successfully used for the simulation of resonant tunneling diodes [21], and more recently for the simulation of MOS transistors [19, 22] as well as Tunnel FETs. The quantum transport approach is also very well suited for the simulation of those sources of variability (e.g., the surface roughness or line edge roughness) that are due to random variations of the device morphological features.

> **Non Equilibrium Green's Function (NEGF)**
>
> The NEGF formalism allows for the simulation of electron devices according to a full quantum transport model. The NEGF approach can be used for both coherent and dissipative transport by accounting for the phonon scattering through appropriate self-energies. The formalism can be used for a simple effective mass, but also for a k·p or an atomistic, tight-binding Hamiltonian.

8.5 Process and Equipment Simulation

Similar to the need to replace analytical device models by more accurate physical ones like DD, it became also apparent that analytical approximations for the geometries and dopant profiles of devices

1960s   The Decade of Integration   **1960s**

were no longer sufficient. This gave rise to the development of the first programs for the one-dimensional simulation of ion implantation, diffusion and oxidation, namely SUPREM [23] and ICECREM [24], and for the two-dimensional simulation of deposition, etching, and especially optical lithography, SAMPLE [25]. Here, various physics-based analytical models were used for the simulation of ion implantation, whereas dopant diffusion and activation was modeled by rather simple diffusion–reaction equations, and oxide growth by simple analytical expressions. For these programs meshing was not yet a problem. Physical mechanisms for dopant diffusion and activation have since then been a focal point of research. Whereas in the beginning the question whether dopant diffusion and activation would be mediated by vacancies or interstitials in the silicon crystal lattice was the central point of debate, in the meantime physical models became much more sophisticated, describing the diffusion and reaction of dopants, various point defects, and their complexes and agglomerates, including the amount of electrical activation, by rather complex systems of partial differential equations. A detailed review of this field was published in 2004 [26]. This complexity is among others due to the shift to short-term processes (e.g., flash or laser annealing), which result in the need to accurately describe various transient effects, for example, transiently enhanced diffusion and reduced activation (Figure 8.3). Generally, the design and conduction of experiments that allow for separating between the various physical effects occurring is a key challenge for the development of process models, and especially for diffusion and activation. Whereas continuum simulations using systems of partial differential equations continue to be the main method for the simulation of dopant diffusion and activation, since several years also atomistic simulation methods have been developed, for example [27], and are in the meantime available as options in leading-edge simulation programs. Similarly, for ion implantation, analytical models are mostly used, whereas Monte-Carlo simulations provide additional benefits for special situations.

For the two-dimensional simulation of topography steps, in SAMPLE the surfaces were discretized by polygons which were shifted due to some physical rates calculated. Various kinds of analytical models were used for the processes in question, including transfer functions in optical lithography and angular characteristics of sources in sputter deposition.

At the beginning of the 1980s it became apparent that not only vertical but also two-dimensional dopant profiles need to be known. This led to two-dimensional versions of both SUPREM [28] and ICECREM. Since then, meshing has been, and will always be, a key issue in process simulation, due to the need to deal with moving, highly nonplanar surfaces and interfaces. Furthermore, in order to provide useful

Process and Equipment Simulation

In equipment simulation physical quantities directly affecting the wafer, such as flux distributions of reactive ions across the wafer, are calculated from basic equipment data and settings. Frequently starting from the results of equipment simulation, process simulation deals with the prediction of the development of device geometries, dopant distributions and other physical quantities caused by the various process steps used in semiconductor fabrication. Due to the large diversity of equipments, materials and processes to be described, there are many different classes of models and algorithms used in these areas of TCAD.

| MOS quality SiO$_2$ becomes available | Bipolar power transistors come up |  ELECTRON DEVICES SOCIETY® | Aluminum interconnects are introduced | Diffused source/drain contacts are being used in MOSFETs |

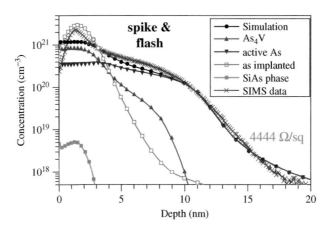

Figure 8.3 Simulation of the diffusion and activation of arsenic implanted with 1 keV, 10^{15} cm^{-2}, and annealed with a combination of spike and flash processes (Simulations: Fraunhofer IISB, experiments: Mattson Thermal Products). Courtesy of Fraunhofer

input for two-dimensional device simulation, for example of a MOS transistor, it became necessary to study the interaction of the various process steps employed, which means to simulate the development of device geometries and dopant distributions throughout all process steps. Since then, the diversity of processes, materials and effects to be dealt with in process simulation has continuously grown, giving rise to the need to develop and implement a large set of physical models.

The obvious next step in the development of process simulation tools was the transition to three-dimensional simulation. A first presentation of worldwide activities in this area was the "Workshop on 3-Dimensional Process Simulation" [29] held 1995 in conjunction with the SISDEP 1995 conference. From the beginning it was clear that integration between the different process steps to be considered is mandatory, and that besides the computational effort needed especially the meshing problems resulting from three-dimensional non-planar time-dependent geometries would be the key challenge. Although in the following years much effort was spent on this problem, meshing continues to be a key requirement for improvement in simulation, among others identified in the simulation chapter of the International Technology Roadmap for Semiconductors (ITRS). Nevertheless, the timely start of the developments in three-dimensional process simulation, both by academia and software houses, before research on true three-dimensional device architectures like FinFETs started, made it possible that now simulation strongly contributes to the development and optimization of such advanced devices.

With physical limits of semiconductor technology development approaching and/or to be circumvented, various requirements and challenges are currently being addressed by process simulation. Among others, further device scaling depends on the extension of optical lithography beyond its traditional resolution limits and/or the economically viable introduction of Extreme Ultraviolet Lithography (EUV). Here, lithography simulation is extensively being used to assess and optimize emerging technological options such as Multiple Patterning or Source Mask Optimization. Lithography is a good example where the simulation of equipment and process cannot be separated. Etching and deposition equipment is being simulated to extract etching

and deposition rates as well as homogeneities across the wafer or between wafers, and effects of pattern density. This includes aspects like plasma simulation or computational fluid dynamics.

Moreover, besides the functionality of devices and circuits, also their variability and reliability have become crucial. Because most sources of variability (e.g., focus variations in lithography, iso-dense bias in deposition/etching) and several effects influencing reliability are at the level of the equipment and/or processes, coupled equipment/process/device and partly circuit simulation is mandatory to address these issues.

References

[1] W. Shockley, "The Theory of p-n Junctions in Semiconductors and p-n Junction transistors", *Bell Syst. Tech. J.*, vol. 28, pp.435–489, July 1949.

[2] M.G. Ancona, Z. Yu, W.-C. Lee, R.W. Dutton, P.V. Voorde, "Density-Gradient Simulations of Quantum Effects in Ultra-Thin-Oxide MOS Structures", *Techn. Dig. IEEE International Electron Devices Meeting (IEDM)*, 1997, pp. 97–100, 1997.

[3] H.K. Gummel, "Self-Consistent Iterative Scheme for One-Dimensional Steady State Transistor Calculations", *IEEE Trans. Electron Devices*, vol. 11, pp. 455–465, October 1964.

[4] J.W. Slotboom, "Iterative Scheme for 1- and 2-Dimensional D.C.-Transistor Simulation", *Electronics Lett.*, vol. 5, pp. 677–678, 1969.

[5] D.L Scharfetter and H.K. Gummel, "Large-Signal Analysis of a Silicon Read Diode Oscillator", *IEEE Trans. Electron Devices*, vol. 16, pp. 64–77, January 1969.

[6] W. Shockley, J.A. Copeland, R. P. James, "The Impedance Field Method of Noise Calculation in Active Semiconductor Devices", in *Quantum Theory of Atoms, Molecules and Solid State*, P.O. Lowdin (ed.), Academic Press, 1966.

[7] K. Blotekjaer "Transport equations for electrons in two-valley semiconductors", *IEEE Trans. Electron Devices*, vol. 17, no.1, pp. 38–47, January 1970.

[8] C. Jacoboni and L. Reggiani, "The Monte Carlo method for the solution of charge transport in semiconductors with applications to covalent materials", *Rev. Mod. Phys.*, vol. 55, pp. 645–705, 1983.

[9] M.V. Fischetti and S.E. Laux, "Monte Carlo analysis of electron transport in in small semiconductor devices including band-structure and space-charge effects", *Phys. Rev. B*, vol. 38, pp. 9721–9745, 1988.

[10] F. Venturi, R.K. Smith, E. Sangiorgi, M.R. Pinto, and B. Riccó, "A General Purpose Device Simulator Coupling Poisson and Monte Carlo Transport with Applications to Deep Submicron MOSFETs", *IEEE Trans. on Computer-Aided Design*, vol. CAD-8, p. 360, 1989.

[11] H. Shichijo and K. Hess, "Band-structure-dependent transport and impact ionization in GaAs", *Phys. Rev. B*, vol. 33, p.4197, 1981.

[12] A. Abramo, L. Baudry, R. Brunetti, *et al.*, "A comparison of numerical solutions of the Boltzmann transport equation for high-energy electron transport in silicon", *IEEE Trans. Electron Devices*, vol. 41(9), p. 1646, 1994.

[13] M.V. Fischetti and S.E. Laux, "Monte Carlo study of electron transport in silicon inversion layers", *Phys. Rev. B*, vol. 48, pp. 2244–2274, 1993.

[14] A. Gnudi, D. Ventura, G. Baccarani, and F. Odeh, "Two-dimensional MOSFET Simulation by Means of a Multidimensional Spherical Harmonics Expansion of the Boltzmann Transport Equation", *Solid State Electronics*, vol. 36, pp. 575–581, 1993.

[15] A.T. Pham, C. Jungemann, B. Meinerzhagen, "On the numerical aspects of deterministic multisubband device simulations for strained double gate PMOSFETs", *Journal of Computational Electronics*, vol. 8, p. 242, 2009.

[16] M. Buttiker, "Symmetry of electrical conduction", *IBM Journal Res. Dev.*, vol. 32, pp. 317, 1988.

[17] S. Datta, "Steady-state quantum kinetic equation", *Phys. Rev. B*, vol. 40, pp. 5830, 1989.

[18] H.U. Baranger, D.P. DiVincenzo, R.A. Jalabert, A.D. Stone, "Classical and quantum ballistic-transport anomalies in microjunctions", *Phys. Rev. B*, vol. 44, pp. 10637–10675, 1991.

[19] R. Venugopal, S. Goasguen, S. Datta, M.S. Lundstrom, "Quantum mechanical analysis of channel access geometry and series resistance in nanoscaletransistors", *J. Appl. Phys*, vol. 95, p. 292, 2004.

[20] M.P. Lopez Sancho, J.M. Lopez Sancho, J. Rubio, "Quick iterative scheme for the calculation of transfer matrices: application to Mo (100)", *Journal of Physics F: Metal Physics*, vol. 14, p. 1205, 1984.

[21] R.C. Bowen, G. Klimeck, R. K. Lake, W. R. Frensley, T. Moise, "Quantitative simulation of a resonant tunneling diode", *J. Appl. Phys. vol.* 81, p. 3207, 1997.

[22] A. Svizhenko and M.P. Anantram, "Role of scattering in nanotransistors", *IEEE Trans. Electron Devices*, vol. 50, pp. 1459, 2003.

[23] D.A. Antoniades, R.W. Dutton, "Models for Computer Simulation of Complete IC Fabrication Processes", *IEEE Trans. Electron Devices*, vol. 26, pp. 490–500, Apr. 1979.

[24] H. Ryssel, K. Haberger, K. Hoffmann, G. Prinke, R. Dümcke, A. Sachs, "Simulation of Doping Processes", *IEEE Trans. Electron Devices*, vol. 27, pp. 1484–1492, Aug. 1980.

[25] W. Oldham, S. Nandgaonkar, A. Neureuther and M. O'Toole, "A general simulator for VLSI lithography and etching processes: Part I – Applications to projection lithography", *IEEE Trans. Electron Devices*, vol. 26, pp. 717–722, Apr. 1979.

[26] P. Pichler, *Intrinsic Point Defects, Impurities, and Their Diffusion in Silicon*, Wien: Springer, 2004.

[27] M. Jaraiz, L. Pelaz, E. Rubio, J. Barbolla, G.H. Gilmer, D.J. Eagelsham, H.J. Gossmann, J.M. Poate, "Atomistic Modeling of Point and Extended Defects in Crystalline Materials", *Mat. Res. Soc. Symp. Proc.*, vol. 532, 43–53, 1998.

[28] M.E. Law and R.W. Dutton, "Verification of Analytic Point Defect Models using SUPREM-IV", *IEEE Transactions on Computer Aided Design*, 7(2), p. 181–190, 1988.

[29] J. Lorenz (ed.), *3-Dimensional Process Simulation*, Wien: Springer, 1995.

Emannuel Rashba proposes spin orbit interaction that can be modulated by an electric field | Landauer explores energy dissipation limits in computing | ELECTRON DEVICES SOCIETY® | Robert Brian and Gary Pittman discover infrared radition from GaAs ... | ... and file a patent on infrared LED

Chapter 9

Reliability of Electron Devices, Interconnects and Circuits

Anthony S. Oates, Richard C. Blish, Gennadi Bersuker and Lu Kasprzak

9.1 Introduction and Background

During World War II military field radios and other electronic equipment which relied on electron devices, were less than highly reliable. In many cases field radios and spare tubes were too frequently not functioning. Furthermore, as avionic control systems became pervasive in military aircraft, there emerged a demand for highly reliable transistorized equipment. What ensued was a revolution of sorts in both commercial and military applications: dies that contained a single transistor were soon replaced with dies that contained several transistors and provided a circuit function. As process defects were reduced the transistors could be made smaller, further processing improvements allowed the die to grow in size. Therefore, while more and more transistors of a smaller size were packed into die, many more circuit functions could be provided on a single die. Small Scale Integrated (SSI) circuits were born. Medium Scale Integrated (MSI) circuits followed, with even larger die and many more transistors. Bipolar transistors and metal oxide silicon field effect transistors (MOSFETs) carved out applications where their characteristics were beneficial. This integrated circuit revolution continues to this day and Large Scale Integrated (LSI) circuit dies are pervasive, from TVs to cruise ships, from satellites to road signal lights. Today, there can be billions of transistors on a single die.

Guide to State-of-the-Art Electron Devices, First Edition. Edited by Joachim N. Burghartz.
© 2013 John Wiley & Sons, Ltd. Published 2013 by John Wiley & Sons, Ltd.

1962   M.R. Boyt invents the multiple beam klystron   **E**LECTRON **D**EVICES **S**OCIETY®   Nick Holonyak demonstrates the red LED

This achievement took about 50 years. As each successive generation of smaller transistors was fabricated, unique reliability problems were identified, which had to be overcome. This was accomplished by finding the root cause of the failure and changing the process and, or, the material, as well as the design, if necessary, to eliminate the cause of failure. This relentless pursuit of decreasing transistor size, increasing die and package size, and increasing circuit functions on a die, ultimately improved reliability, such that the denser die failed not more frequently, and often less frequently, than its predecessor. This is the key to the success of the integrated circuit industry. As the maturity of a process improved, defects were reduced and the reliability and yield of the die increased.

It was well established that accelerated testing (at high temperature, voltage, current, temperature cycling or THB: Temperature – Humidity – Bias) of a large number of presumably identical devices resulted in a distribution of failures in time, such that some would fail early in the test and some would fail throughout the test, while a significant number would last an extremely long time. The characteristic distribution of failures in time, which is shown schematically in Figure 9.1, has become known as the bathtub curve.

Early and intermediate failures were invariably due to some sort of defect introduced during device processing, such as a significant deviation in horizontal dimension or layer thickness. Some failures were caused by how the device was used or powered, necessitating rethinking of the model for the device behavior at the fundamental level of the physics of operation, DFR (design-for-reliability). Elimination of the early failures can be accomplished by a process known as "burn in", whereby the devices are exercised and tested for a few tens to a hundred hours to weed out the early failures. This whole set of activities has been called building-in-reliability (BIR).

This approach has revealed many causes of failure, which have been overcome by material selection, processing, and understanding the details of the transition in physics from the hundreds of micrometers to atomic dimensions, some 6 orders of magnitude in dimensions. Some of the more important reliability problems solved were: (i) purple plague – a brittle, high resistivity Au−Al intermetallic compound causing open contacts at wire bonding pads on the surface of die and its mitigation by low temperature thermosonic Au−Al bonds (see sidebar *The Long History of the Purple Plague*) and by replacing wire bonds with C4 (Controlled Collapse Chip Connection) for high pin count packages; ionic contamination – (ii) mobile ions like Na^+ which move in an electric field and drift through the gate dielectric, thereby changing the threshold voltage of a MOSFET, (iii) corrosion – loss of contact to a metal surface because of an insulating film,

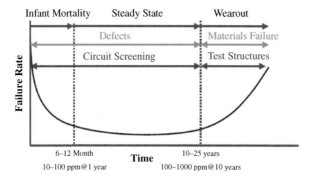

Figure 9.1   The bathtub curve for IC failure rates

(iv) electromigration – opening of cracks in metal lines at grain boundaries due to the electron wind force, (v) stress – induced voiding due to a competition between diffusion in thin metal films and CTE (Coefficient of Thermal Expansion) mismatch between the film and silicon substrate, (vi) CPI (Chip Package Interaction) wherein package stresses or materials drive failure mechanisms on the chip such as metallization movement, fatigue or cracking, or interaction of chip metallization with encapsulant impurities (ionics and flame retardants), (vii) time-dependent dielectric breakdown (TDDB) - dielectric shorting in a gate dielectric or inter-level dielectric used to separate two levels of metal wiring, (viii) leakage of charge from isolated gates (flash memories), (ix) hot carrier injection - high energy electrons scattered into the gate dielectric from the MOSFET channel current, (x) soft errors – loss of memory cell data integrity due to interaction with the Si substrate of energetic particles produced either by radioactive decay of package materials (alpha particles) or from cosmic rays (high energy neutrons), and (xi) NBTI and PBTI – shifts in MOSFET threshold and trans-conductance due to hole and electron trapping and de-trapping in the gate dielectric. As each problem was solved, guidelines and ground-rules were produced as the level of integration became greater and greater. This paradigm is well-established and will continue into the future. New problems will be discovered and solved as electron devices continue to change both the world and the way we live in it. In the following sections we highlight those failure mechanisms that are pertinent to current and future Si IC technologies.

The Long History of the Purple Plague

One of the earliest failure mechanisms to be identified involved Al wires that connected the chip to Au-plated TO-5 headers. Premature failure when $AuAl_2$ (Al-rich InterMetallic Compound (IMC)) was formed. The problem was solved partially by using Dual In Line (DIP) packages (Ag-plated leadframes, with Au wire) which supplanted Al wire/TO-5 headers. IMC formation also occurred during Au wire bonding by thermocompression (TC), which required both high temperature and pressure. Thermosonic (TS) bonders supplanted TC in 1984, significantly reducing the problem. The final chapter in the wire bond saga involved the addition of impurities in the Au wire, which eliminated IMC formation.

9.2 Device Reliability Issues

9.2.1 Transistor Degradation Mechanisms

Transistors are susceptible to multiple degradation mechanisms, the relative importance of which changes depending on the technology node and device operational mode. With aggressive scaling of transistor dimensions, reliability characteristics such as TDDB, both NBTI and PBTI and Random Telegraph Noise (RTN) remain at the forefront. With the increasing emphasis on low power technologies other well-known mechanisms, such as Hot Carrier Injection, remain important, but of lesser concern.

TDDB describes the increase in the gate dielectric leakage leading to eventual breakdown and loss of transistor functionality. This occurs because a gate voltage stress leads to the generation of electrically active defects that form an irreversible, conductive percolation path between the cathode and anode. A major challenge for reliability

Demonstration of visible laser light at 77K from GaAs-GaP alloys …   ... by Nick Holonyak and a team at General Electric (GE)    ELECTRON DEVICES SOCIETY®   50th anniversary of IRE

assessment is to determine the mechanism of the defect generation process; several models have been proposed [1]. The anode hole injection ("1/E") model [2], which is applicable at larger electric fields (although a low voltage modification was developed [3]), assumes that electrons are injected from the cathode and generate holes at the anode, which then diffuse into the dielectric film causing defects. The electrochemical ("E") model relies on the process of bond polarization and breakage driven by the oxide electric field (it may be assisted by hole trapping in the oxide) [4, 5]. The hydrogen release model explores widely observed strong correlations between the hydrogen content in SiO_2 and gate leakage/breakdown. This model suggests that the hydrogen is released from the anode/oxide interface by the injected electrons, and generates electrically active defects [6, 7]. Each model has limitations in ability to describe the entire range of breakdown characteristics observed over the range of materials and structures that are of current technological importance. Further progress will require establishing links between the atomic structure of the defect involved in failure and the measured electrical characteristics. This will be particularly important for novel channel materials (e.g., III–V channels on Si) and multi-gate transistors, which are now being developed for low power applications.

Shifts of the threshold voltage of pFETs with temperature and inversion gate voltage stress (NBTI) were first observed in the late 1970s [8]. Failure times due to NBTI have decreased rapidly with technology scaling, while their variability has increased. Enormous efforts have been expended to understand the process factors that control NBTI, so that transistors can be designed to ameliorate potential circuit reliability problems. Despite these efforts it is clear that mitigation of NBTI with continued technology scaling will require both process and circuit design approaches.

The mechanisms responsible for NBTI have been of intense interest for over 40 years and yet they are still hotly debated. NBTI is proposed to involve positive charge generation at the gate dielectric interface with the Si channel, and is observed irrespective of the choice of SiO_2 or high-k gate dielectric, since the latter usually incorporates a thin SiO_2 layer in contact with the Si substrate. The mechanism has been successfully described by the Reaction–Diffusion (R–D) model [9, 10], which suggests that interface states are formed by hydrogen release from passivated Si dangling bonds at the SiO_2/Si interface, following capture of a hole from the Si channel. However, the R–D model experiences difficulties in explaining the extremely wide range of threshold voltage (V_{th}) relaxation times observed when the voltage stress is removed. An alternative model has recently been proposed where the injected holes

Gate Dielectric Reliability

Reliability of the gate dielectric of MOS transistors has been an on-going concern throughout the history of the IC industry. Initial problems involved contamination of the dielectric by Na ions, which resulted in threshold voltage drifts. By the 1980s the long-term reliability of the dielectric became a major issue as thickness approached 10 nm. This necessitated development of physics based breakdown models; a vigorous and highly productive debate ensued that still continues to some extent to this day. In the late 1990s, there was concern that IC scaling might be limited by gate dielectric reliability as SiO_2 dielectrics reached 2 nm. Improved physics based models provided confidence in continuing scaling, and materials innovations, such as the introduction of SiON and more recently high-k materials, have progressed without major reliability problems.

tunnel directly to and are trapped at defects in the bulk oxide [11]. Therefore, establishing accurate microscopic models of the defects is a key issue in identifying manufacturing processes that most directly impact NBTI.

The relationship between NBTI degradation of transistors and its impact on circuits is important. Recent studies have shown that SRAMs are particularly vulnerable to NBTI [12], which can lead to unacceptably large increases in the minimum operational voltage of the bit-cell (V_{min}). Careful design of the SRAM bit-cell is required to prevent premature failure. Digital logic circuits are less vulnerable since NBTI-induced change in the timing delay of logic stages is largely obscured by manufacturing process-related timing variability [13].

In the case of nFETs, PBTI has become an issue (V_{th} drift with inversion stress) with the introduction of high-k gate stacks. PBTI is believed to occur by trapping of injected electrons in the high-k film [14–16]. Although this trapping can be suppressed by decreasing the high-k layer thickness, it is still not negligible, especially in the case of devices for low power applications.

In small area RTN (see sidebar, *Random Telegraph Noise*) in the drain current becomes a concern since fluctuations of the current amplitude might interfere with the digital circuit operations. SRAM is especially sensitive to these fluctuations which manifest themselves as shifts in the minimum operational voltage (V_{min}) [17]. The RTN is caused by a stochastic process of the electron exchange between the channel and a single trap (or several traps in the case of the multi-level RTN) in the dielectric [18].

Random Telegraph Noise (RTN)

While transistor current fluctuation with a certain stable amplitude seems to be characterized as "noisy" it can be described by two characteristic times, corresponding to the system being either in the low current (when the defect is charged) and high current (defect is empty) state. The charge exchange rate defining these times is determined by the distance separating the trap and substrate and the defect structural properties. Analysis of the temperature and voltage dependencies of RTN times allows extracting the defect structural characteristics.

9.2.2 Interconnect Degradation Mechanisms

The earliest interconnect issues involved intermetallic formation between Au wire bonds and the Al wires on the IC as a result of high temperatures used during chip packaging [19]. It was eliminated with the introduction of low temperature thermo-sonic bonding techniques and Au alloy wires. Corrosion of Al during THB was also an early issue resulting from the penetration of ionic contaminants from the package environment [20]. The introduction of silicon nitride layers to encapsulate ICs effectively solved this issue [21]. There are, however, several failure mechanisms that have defied comprehensive process solutions, remaining as important technological issues.

Thin film IC interconnects carry relatively high current densities $\sim 10^5 - 10^6$ A/cm^2, which leads to a large flux of atoms in the direction

Syncom II, the first geo-synchronous telecommunications satellite (TWT), …

… developed at Hughes Co.

ELECTRON DEVICES SOCIETY®

CMOS patent application

of electron flow – this is termed electromigration (see sidebar, *Electromigration*). In regions of a conductor where this flux experiences a divergence, in particular at grain boundary triple point intersections and at the interfaces between different materials in contacts and vias, open and short failures can occur. Following the discovery of electromigration failures in Al interconnects in the mid-1960s [22], it was realized that the only feasible means of prevention was to limit the current density. Since smaller currents negatively impacted circuit performance, process solutions were vigorously sought. The initial process solution to the problem involved reducing the rate of transport along grain boundaries by doping the Al layer with Cu, which segregated at grain boundaries [23]. Subsequently, systematic control of the grain structure was implemented to further improve lifetimes [24]. With increasing circuit integration and density, multiple levels of metal began to be employed (below the 1 μm node), requiring W-plugs as contacts between the metal levels, which exacerbated the electromigration problem since the plug interface is an inherent site of flux divergence [25]. This led to the introduction of refractory cladding layers (e.g., TiN above and below the AlCu layer) to prevent open circuit at vias connecting metal levels.

Al interconnects were industry standard until the 0.13 μm node, which saw the introduction of damascene Cu/SiO$_2$ and ultimately Cu/low-k structures for cost and performance reasons. From a reliability perspective, Cu offered larger current carrying capability > 10^6 A/cm^2 due to the slower self-diffusion of Cu compared to Al [26]. Cu exhibits several failure modes due to electromigration, as shown in Figure 9.2 [27], which complicates the optimization of interconnect processes for robust performance.

Electromigration

Failure of circuit interconnects has been an issue of IC technologies for almost 50 years. In 1969 Jim Black produced his now famous model for the failure time. In 1976 Ilan Blech discovered the "short-length" effect whereby electromigration may be eliminated by a mechanical gradient induced back flow. These discoveries form the basis on which the engineering of electromigration stands. Despite the enormous amount of effort that has been expended to understand the fundamentals of this destructive process, and the manufacturing processes that influence it, electromigration remains as one of the most important, and intractable issues in IC technology development.

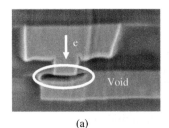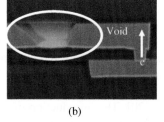

(a)   (b)

Figure 9.2   Examples of electromigration voids in dual damascene Cu/low-k interconnects for electron flow from the upper level metal to the lower (a) and vice versa (b)

However, Cu failure times have decreased significantly with technology scaling [28], posing a serious challenge to attaining the increasing current densities required for continued performance enhancement and density increase. This has necessitated the introduction of analogues of the techniques used to improve Al electromigration performance: namely Cu alloys to limit grain boundary diffusion [29, 30], and metallic cap layers at the Cu surface to impede diffusion along the top surface of the Cu [31].

Sress-Induced voiding became an important technological issue as feature sizes approached 1 µm in the late 1980s [32]. The growth of stress-voids is driven by two competing mechanisms: thermal stresses arising from CTE mismatch between the interconnect and the Si substrate, and vacancy diffusion that allows void growth [33]. The interplay of these mechanisms leads to the rate of void growth showing a peak typically in the temperature range 150–250°C. Al interconnects typically exhibit stress-voids at grain boundary intersections with the film surface, and so refractory metal cladding (as described previously for electromigration improvement) also serves to provide electrical redundancy to limit the electrical impact of the voids (see sidebar *Stress-Induced Voiding*). For Cu damascene interconnects, voiding occurs preferentially under vias attached to large Cu plates, where stress gradients are coupled with a large atomic sink [34]. Multiple via placement is required to prevent open circuit failure, necessitating the use of via placement design rules. However, at the most advanced technology nodes (20 nm and below), voids can also form where narrow trenches are attached to wider plates due to large stress gradients [35]. Voids in these regions can only be controlled by limiting stresses in the interconnect materials during processing.

With the introduction of Cu interconnects, breakdown of inter-level dielectric (ILD) was observed [36]. Porous low-k materials are particularly vulnerable to breakdown due to their relatively low electrical breakdown strength [37]. While the details of the breakdown mechanism still remain to be clarified, it is evident that it is accelerated by Cu impurities in the dielectric that are either residual from processing, or are introduced from the trench during voltage stress [38]. The reliability of porous low-k materials is limited by the presence of the pores that enable lower k values to be attained: pores are equivalent to shorting defects. The lower limit of porous low-k reliability due to this reliability constraint is k ~2.3 [39]. To attain lower effective k, structures such as air-gaps are required.

Stress-Induced Voiding (SIV)

SIV was characterized in 1986 as a competition between CTE (Coefficient of Thermal Expansion) mismatch and thermally activated stress relaxation. This failure mechanism was particularly insidious since it produced what amounted to a shelf-life for circuits: no power was needed for it to cause failure. Indeed, it has been suggested that the failure of a large DRAM manufacturer at the time was at least partially due to this issue. Elimination of the issue required a careful management of thermal cycles during process, and eventually the introduction of refractory cladding layers for Al metallizations. Stress-voiding re-surfaced as a major issue again with the introduction of Cu, and especially low-k materials. This time via placement design rules were needed for mitigation. It continues to be a major reliability issue.

| First issue of IEEE Transactions on Electron Devices | | ELECTRON DEVICES SOCIETY® | IRE Professional Group on Electron Devices becomes … | … IEEE Professional Group on Electron Devices |

9.3 Circuit-Level Reliability Issues

There are several reliability issues that are manifest at the circuit level. In this section we discuss Soft Errors in memory and logic circuits and the impact of mechanical stresses on circuit reliability (see sidebar *Soft Errors*).

9.3.1 Soft Errors Caused by Single Event Upsets

Single event upsets (SEU) in circuits [40] encompass a range of phenomena associated with the interaction of energetic particles (α particles, cosmic ray neutrons, energetic ions, X- and γ rays) and Si. Charge generation in the active areas of transistors causes changes in internal voltages, which can lead to corruption of stored or transmitted data. Electrically, the most important consequences of SEU include temporary data loss in memories and flip-flops, transistor latch-up and corruption of logic waveforms. The errors associated with SEU are commonly referred to as soft-errors (SER) since they usually do not cause system destruction, and they can be recovered by re-writing the data.

Alpha particles are emitted from trace amounts of U^{238} impurities in package materials and solder bumps [41]. High-energy neutrons originate in the cosmic ray background [42]. The passage of a α particle leads to the generation of electron–hole pairs. Neutrons interact via nuclear collisions to produce energetic charged species in the Si that then produce electron–hole pairs. Additional SER upset mechanisms include the capture of slow thermal neutrons by B^{10} isotopes [43], a reaction that produces a α particle and an energetic Li^7 positive ion. The impact of α – particle SER on circuits is mitigated by control of purity of packaging materials. Neutron SER requires specialized design techniques to eliminate or correct errors. Memory protection is commonly achieved using error detection and correction techniques (ECC) for single bit upsets; bit interleaving, where bits in the same logical word are physically separated, is required to minimize multi-bit errors in the same word [44]. NAND memories with very long words often experience multibit errors within their long words, which can be addressed economically with additional ECC bits. Protection of logic storage (flip-flop) elements requires design techniques such as triple modular redundancy (TMR) [45], DICE design [46], or layout modification to reduce charge collection (LEAP) [47]. Scaling trends for SER are shown in Figure 9.3 covering the last 10 years of technology progression. These trends can be understood by considering that SER is determined by the critical charge to change a node voltage and the area over which charge collection occurs. SRAM shows a decreasing trend with scaling as area reduction has outpaced

Soft Errors

The year 1978 was the opening chapter of SER in DRAMs in which alpha particles from Pb-rich glass in ceramic DIP (Dual-in-Line) packages caused transient errors. Early mitigation was to remove the zircon additive (i.e., the trace U and Th impurities) from the glass while later efforts focused on impurities in molding compound. More recently efforts have concentrated on maximizing the critical capacitance to upset a cell and on realizing that thermal neutrons can cause ^{10}B fission, thus releasing an alpha particle and a Li ion. Another effective mitigation is ECC (Error Correction Code) which relies on the probability that multiple errors within a single word are very unlikely under alpha irradiation or cosmic ray neutrons.

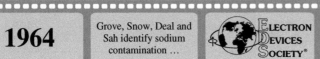

| 1964 | Grove, Snow, Deal and Sah identify sodium contamination … | ELECTRON DEVICES SOCIETY® | … as reason for instability of MOS transistors | Donald L. Bitzer and H. Gene Slottow develop the first plasma screen display |

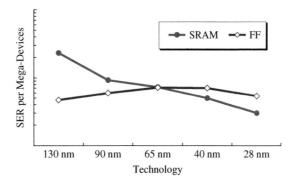

Figure 9.3 SER failure rates as a function of technology progression from the 0.13 μm to 28 nm technology nodes. The figure presents experimental data for SRAM and logic flip-flops

reduction in the critical charge. Flip-flop SER has remained approximately constant over the same technology range. With the availability of ECC for memories, flip-flop SER is now a dominant concern. SER of random logic is also of increasing interest as failure rates can approach that of flip-flops for GHz frequencies [48]. SER will continue to be an important issue for the foreseeable future: challenges exist in determining the SER susceptibility of new transistor structures such as FinFETs [49], and in the characterization of new SEU mechanisms, such as from muons [50] present in the cosmic radiation background.

9.3.2 Mechanical Stress Effects on ICs

The most commonly used stress test to actuate mechanical stress issues is temperature cycling (alternating between hot and cold in air or nitrogen) to accelerate failure mechanisms such as cracking and delamination [51]. Temperature shock and drop tests have been de-emphasized since they are more suited to hermetic packages, which have been largely displaced by many different "plastic" packages.

The Coffin–Manson formulation [52] is commonly used to model fatigue lifetime and characterize cumulative reliability lognormal failure as a function of the logarithm of the number of temp cycles. The slope and intercept of the curve provide estimates for the variance in failure time (sigma) and the time to 50% failure, respectively. Weibull modeling is an equivalent analysis technique (based on power law rather than Gaussian), for which the slope and intercept provide an index of defects versus wearout (variance in failure time) and the time to 63%

Package Reliability Testing

The Panama Test (THB: Temperature Humidity and Bias) originated in 1955 and subjected electronic systems to jungle conditions, roughly 40°C and 80% Relative Humidity (RH). Gradually, device performance improved, especially when epoxy molding compound supplanted silicone in 1972, and the standard THB became 85/85. However, a second major breakthrough was needed, namely the replacement of PSG (PhosphoSilicate Glass, which is permeable to and soluble in water, liberating destructive ions) by plasma silicon oxynitride for final passivation. After this transformation HAST (Highly Accelerated Stress Test, 130/85, in 1981) replaced 85/85. Current devices are substantially immune to THB when failure rate is measured as FITs.

H. K. J. Ihantola and J. L. Moll introduce the first DC MOSFET model

ELECTRON DEVICES SOCIETY®

Harvey Nathanson demonstrates the first batch fabricated MEMS device at Westinghouse

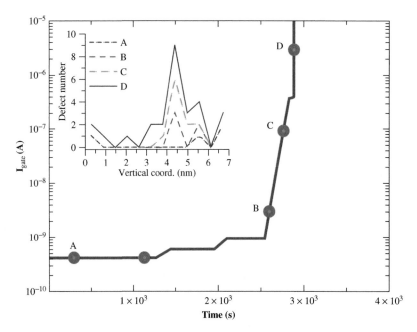

Figure 9.4 An example of the simulation of the gate current through a single conductive path associated with the grain boundary in a 7 nm TiN/HfO$_2$/TiN stack during constant voltage stress at V$_g$ = 2.8 V. The inset shows the number of defects generated along the conduction path responsible for the current at different stress times A, B, C, D. © 2011 IEEE. Reproduced with permission from L. Vandelli *et al.*, *Proc. IEEE Rel. Phys. Symp.*, 807, 2011

(1−1/e) failure, respectively. Practical equivalence of lognormal versus Weibull is shown in JEDEC JEP122.

Chip Package Interaction (CPI) is a relatively new term to cover the "boundary" between chip fabrication and package assembly, producing a class of failure mechanisms that rely upon the interaction of chip topology and package materials. This area is now garnering a considerable amount of attention as new, mechanically weak low-k dielectric materials are introduced for improved device performance (see Figure 9.4). Recent efforts to understand these issues have used multiple dimensional tiered Finite Element Analysis and Moiré patterns to calculate mechanical stresses over the vastly different distance scales for chips versus packages. Another productive approach is to study the fundamentals of interfacial adhesion. Yet another approach is purely empirical utilizing test chips with multi-level ladder networks near the periphery where the stresses driving delamination and cracking are largest. It is also likely that step stress techniques will find increase use to find the threshold for failure, which is subjected to lognormal analysis. This approach is also described in JEDEC JEP122.

9.4 Microscopic Approaches to Assuring Reliability of ICs

The continued scaling of IC technologies has resulted in the reduction of failure times, and an increase in variability of most major IC reliability failure mechanisms. In the name of improved performance, new gate and channel materials are being rapidly introduced, and transistor architectures are moving toward multi-gates. The need to reduce RC of interconnects necessitates the introduction novel barrier metal and dielectric materials. To continue to improve reliability at the rate required by Moore's law, there is a need to develop "microscopic" physics-based degradation models to complement the widely used statistical approach to reliability. This should enable a connection to be made between the electrical characteristics of the defects involved in degradation and the specific processes used in fabrication. Valuable examples of such an approach are recent the attempts to understand RTN [18], NBTI [53], high-k dielectric (Figure 9.4) [54], and high-k gate stack [55] breakdown at the microscopic level using multi-phonon assisted (i.e., non-elastic) charge transport processes. Moreover, with improved physics- based models it will be possible to develop accurate simulation tools to more closely determine the link between each transistor failure mechanism and the degradation of circuit performance. In-depth knowledge of circuit degradation characteristics will allow reliability specifications for individual transistors and interconnects to be more accurately set. Finally, increased variability in failure times presents a challenge to collection of accelerated testing data, upon which the development of reliability physics is built. New experimental procedures are required to efficiently test the larger sample sizes required to incorporate larger variability.

References

[1] S. Lombardo, J. H. Stathis, B. P. Linder, K. L. Pey, F. Palumbo, and C. H. Tung, "Dielectric breakdown mechanisms in gate oxides", *J. Appl. Phys.*, 98, 121301, 2005.

[2] C. Schuegraf and C. Hu, *J. Appl. Phys.*, "Scanning tunneling microscopy and spectroscopy characterization of ion-beam-induced dielectric degradation in ultrathin SiO_2 films and its thermal recovery process", 76, 3695, 1994.

[3] M. Alam, B. Weir, J. Bude, P. Silverman, and A. Ghetti, "A computational model for oxide breakdown: theory and experiments", *Microelectron. Eng.*, 59, 137, 2001.

[4] J. W. McPherson and H. C. Mogul, "Underlying physics of the thermochemical E model in describing low-field time-dependent dielectric breakdown in SiO_2 thin films", *J. Appl. Phys.*, 84, 1513, 1998.

[5] J. McPherson, J.-Y. Kim, A. Shanware, and H. Mogul, "Thermochemical description of dielectric breakdown in high dielectric constant materials", *Appl. Phys. Lett.*, 82, 2121, 2003.

[6] D. J. DiMaria and J. W. Stasiak, "Trap creation in silicon dioxide produced by hot electrons", *J. Appl. Phys.*, 65, 2342, 1989.

[7] J. H. Stathis, *IBM J. Res. Dev.*, "Reliability limits for the gate insulator in CMOS technology", 46, 3242, 2004.

[8] K. O. Jeppson, C. M. Svensson, "Negative bias stress of MOS devices at high electric fields and degradation of MNOS devices", *J. Appl. Phys.*, 48, 2004, 1977.

[9] M. A. Alam and S. Mahapatra, "A comprehensive model of PMOS NBTI degradation", *Microelectron. Reliab.*, 45, 71, 2005.

[10] S. Mahapatra, V. Maheta, A. Islam, M. Alam, "Isolation of NBTI stress generated interface trap and hole trapping components in PNO p-MOSFETs", *IEEE Trans. Elec Dev*, 56, 236, 2009.

[11] T. Grasser, B. Kaczer, W. Goes, H. Reisinger, T. Aichinger, P. Hehenberger, P.-Jürgen Wagner, F. Schanovsky, J. Franco, M. T. Luque, and M. Nelhiebel, "The paradigm shift in understanding the bias temperature instability: From reaction–diffusion to switching oxide_traps", *IEEE Trans. Elec Dev*, 58, 3652, 2011.

Gordon Moore formulates Moore's Law: ... ... doubling of transistor density every two years

IEEE Professional Group on Electron Devices becomes IEEE Electron Devices Group

[12] J. C. Lin, A. S. Oates, H. C. Tseng, Y. P. Liao, T. H. Chung, K. C. Huang, P. Y. Tong, S. H. Yau, Y. F. Wang, "Prediction and Control of NBTI–Induced SRAM V$_{ccmin}$ Drift", *Techn. Dig. IEEE Int. El. Dev. Meeting (IEDM)*, p. 1–4, 2006.

[13] B. Vaidyanathan and A. S. Oates, *IEEE Trans. Dev. Mat. Rel.*, 12, 428, 2012.

[14] A. Kerber and E. A. Cartier, "Reliability challenges for CMOS technology qualifications with hafnium oxide/titanium nitride gate stacks", *IEEE Trans. Dev. Mat. Rel.*, 9, 147, 2009.

[15] D. P. Ioannou, S. Mittl, and G. La Rosa, "Positive bias temperature instability effects in nMOSFETs with HfO$_2$/Ti gate stacks", *IEEE Trans. Dev. Mat. Rel.*, 9, 128, 2009.

[16] G. Bersuker, J. H. Sim, C. S. Park, C. D. Young, S. Nadkarni, R. Choi, and B. H. Lee, "Mechanism of electron trapping and characteristics of traps in HfO$_2$ gate stacks", *IEEE Trans. Dev. Mat. Rel.*, 7, 138, 2007.

[17] M. J. Kirton, and M. J. Uren, "Noise in solid-state microstructures: A new perspective on individual defects, interface states and low-frequency (1/f) noise", *Advances in Physics*, 38, 367, 1989.

[18] L. D. Veksler, G. Bersuker, S. Rumyantsev, M. Shur, H. Park, C. Young, K. Y. Lim,W. Taylor, and R. Jammy, "Understanding noise measurements in MOSFETs: the role of traps structural relaxation", *Proc. IEEE Int. Rel. Phys. Symp.*, pp. 73, 2010.

[19] E. Philofsky, "Purple plague revisited", *Proc. Int. Rel. Phys. Symp.*, pp. 177, 1970.

[20] G.L. Schnable and R.S. Keen, "Metalization and bonds-a review of failure mechanisms", *Int. Rel. Phys. Symp.*, pp. 170, 1967.

[21] M. M. Atalla, A. R. Bray, and R. Lindner, "Stability of thermally oxidized silicon junctions in wet atmospheres", *Proc. IEEE*, 106, 1130, 1959.

[22] I. A. Blech in *Rome Air Develop. Center , Contr. No.* AF30(602)-3776, December 1965.

[23] I. Ames, F. M. D'Heurle, and R. E. Horstmann, "Reduction of electromigration in aluminium films by copper doping", *IBM J. Res. Dev.*, 14, 461, 1970.

[24] S. Vaidya and A. Sinha, "Effect of texture and grain structure on electromigration in Al-. 0.5% Cu thin films", *Thin. Sol. Films*, 75(3), 253, 1981.

[25] A. S. Oates, E. P. Martin, D. Alugbin, and F. Nkansah, "Comparison of Al electromigration in conventional Al alloy and W-plug contacts to silicon", *Appl. Phys. Lett.*, 62, 3273, 1993.

[26] C. K. Hu, R. Rosenberg, H. S. Rathore, D. B. Nguyen, and B. Agarwala, "Scaling effect on electromigration in on-chip Cu wiring", *IEEE Int. Interconnect Tech. Conf.*, pp. 267, 1999.

[27] S. C. Lee and A. S. Oates, "Identification and analysis of dominant electromigration failure modes in Copper/Low-k Dual-Damascene interconnects", *IEEE Int. Rel. Phys. Symp.*, pp. 107, 2006.

[28] A. S. Oates and M. H. Lin, *IEEE Int. Rel. Phys. Symp.*, pp. 6B.2.1, 2012.

[29] C. K. Hu and B. Luther, "Electromigration in two-level interconnects of Cu and Al alloys", *Mat., Chem., Phys.*, 41, pp. 1, 1995.

[30] M. H. Lin and A. S. Oates, "The effects of Al doping and metallic-cap layers on electromigration transport mechanisms in Copper nanowires", IEEE Trans. *Dev. Mat. Rel.*, 11, pp. 540, 2011.

[31] C.-K. Hu, D. Canaperi, S. T. Chen, L. M. Gignac, B. Herbst, S. Kaldor, M. Krishnan, E. Liniger, D. L. Rath, D. Restaino, R. Rosenberg, J. Rubino, S.-C. Seo, A. Simon, S. Smith, and W.-T. Tseng, "Effects of overlayers on electromigration reliability improvement for Cu/low K interconnects", *IEEE. Int. Rel. Phys. Symp.*, 2004.

[32] J. Klema, R. Pyle, and E. Domangue, "Implication of nitrogen contaminated during deposition of sputtered aluminum/silicon metal films", *IEEE Int. Rel. Phys. Symp.*, pp. 1, 1984.

[33] J. W. McPherson and C. F. Dunn, "A model for stress-induced metal notching and voiding in very large-scale-integrated Al-Si", *J. Vac. Sci. Tech. B*, 5, pp. 1321, 1987.

[34] E. T. Ogawa, J. W. McPherson, J . A. Rosal, K. J. Dickerson, T.-C. Chiu, L. Y. Tsung, M. K. Jain, T. D. Bonifield, J. C. Ondrusek, and W. R. McKee, "Stress-induced voiding under vias connected to wide Cu metal leads", *IEEE Int. Rel. Phys. Symp.*, pp. 312, 2003.

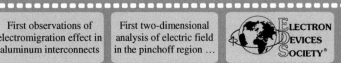

First observations of electromigration effect in aluminum interconnects   First two-dimensional analysis of electric field in the pinchoff region ...   ELECTRON DEVICES SOCIETY®   ... of a MOSFET by S.R. Hofstein and G. Warfield

[35] C. C. Lee and A. S. Oates, "A new stress migration failure mode in highly scaled Cu/low-k interconnects", *IEEE Trans. Dev. Mat. Rel.*, 12(2), pp. 529. 2012.

[36] A. L. S. Loke, J. T. Wetzel, R. Changsup, W. J. Lee, and S. S. Wong, "Copper drift in low-K polymer dielectrics for ULSI metallization", *IEEE VLSI Tech. Symp.*, pp. 26, 1998.

[37] E. Ogawa, K. Jinyoung, G. S. Hasse, H. C. Mogul, and J. W. McPherson, "Leakage, breakdown, and TDDB characteristics of porous low-k silica-based interconnect dielectrics", *IEEE Int. Rel. Phys. Symp.*, pp. 166, 2003.

[38] F. Chen, O. Bravo, K. Chanda, P. McLaughlin, T. Sullivan, J. Gill, J. R. Lloyd, R. Kontra, and J. Aitken, "A comprehensive study of low-k SiCOH TDDB phenomena and its reliability lifetime model development", *IEEE Int. Rel. Phys.*, 46, 2006.

[39] S. C. Lee and A. S. Oates, "Reliability limitations to the scaling of porous low-k dielectrics", *Int. Rel. Phys., Symp.*, pp. 1541, 2011.

[40] R. C. Baumann, "Radiation-induced soft errors in advanced semiconductor technologies", *IEEE Trans. Dev. Mat. Rel.*, 5, 305, 2005.

[41] T. C. May and M. H. Woods, "A new physical mechanism for soft errors in dynamic memories", *IEEE Int. Rel. Phys. Symp.*, pp. 33, 1978.

[42] T. J. O'Gorman, "The effect of cosmic rays on the soft error rate of a DRAM at ground level", *IEEE Trans. Elec. Dev.*, 41, 553, 1994.

[43] R. C. Baumann, and E. B. Smith, "Neutron-induced boron fission as a major source of soft errors in deep submicron SRAM devices", *IEEE Int. Rel. Phys., Symp.*, pp. 152, 2000.

[44] J. Maiz, S. Hareland, K. Zhang, and P. Armstrong, "Characterization of multi-bit soft error events in advanced SRAMs", *IEEE. Int. Elec. Dev. Meeting*, pp. 21.4.1, 2003.

[45] R. Hentschke, F. Marques, F. Lima, L. Carro, A. Susin, and R. Reis, "Analyzing area and performance penalty of protecting different digital modules with hamming code and triple modular redundancy", *IEEE Int. Rel. Phys. Symp.*, pp. 95, 2002.

[46] D. G. Mavis and P. H. Eaton, "Soft error rate mitigation techniques for modern microcircuits", *IEEE Int. Rel. Phys. Symp.*, pp. 216, 2002.

[47] H. H. K. Lee, K. Lilja, M. Bounasser, P. Relangi, I. R. Linscott, U. S. Inan, and S. Mitra, "LEAP: Layout design through error-aware transistor positioning for soft-error resilient sequential cell design", *IEEE Int. Rel. Phys. Symp.*, pp. 203, 2010.

[48] N. N. Mahatme, S. Jagannathan, T. D. Loveless, L. W. Massengill, B. L. Bhuva, S.-J. Wen, and R. Wong, "Comparison of combinational and sequential error rates for a deep submicron process", *IEEE Trans. Nuclear Science*, 58(6), 2719, 2011.

[49] Y. P Fang and A. S. Oates, "Neutron-induced Ccarge collection simulation of bulk FinFET SRAMs compared with conventional planar SRAMs", *IEEE Trans. Dev. Mat. Rel.*, 11, 551, 2011.

[50] B. Sierawski, R. A. Reed, M. H. Mendenhall, R. A. Weller, R. D. Schrimpf, S. Wen, R. Wong, N. Tam, and R. C. Baumann, "Effects of scaling on muon-induced soft errors", *Int. Rel. Phys. Symp.*, pp. 3C.3.1, 2011.

[51] P. C. Paris and F. Erdogan, "A Critical Analysis of Crack Propagation Laws", *Trans. ASME, Series D*, 85, pp. 528–517, 1963.

[52] S. S. Manson, "Predictive Analysis of Metal Fatigue in the High Cyclic Life Range", *ASME, Winter Meeting in New York, Dec. 1979*, pp. 145–183, 1979.

[53] T. Grasser, *Microelectronics Reliability*, "Stochastic Charge Trapping in Oxides: From Random Telegraph Noise to Bias Temperature Instabilities", 52, 39, 2012.

[54] L. Vandelli, G. Bersuker, A. Padovani, J. H. Yum, L. Larcher, and P. Pavan, "A physics-based model of the dielectric breakdown in HfO_2 for statistical reliability prediction", *Proc. IEEE Int. Rel. Phys. Symp.*, pp. 807, 2011.

[55] G. Bersuker, D. Heh, C. D. Young, L. Morassi, A. Padovani, L. Larcher, K. S. Yew, Y. C. Ong, D. S. Ang, K. L. Pey, and W. Taylor, "Mechanism of high-k dielectric-induced breakdown of interfacial SiO_2 layer", *Proc. IEEE Int. Rel. Phys. Symp.*, pp. 373, 2010.

| 1966 | R.W. Bower and H.D.Dill present the first self-aligned gate structure at IEDM 1966 … | ELECTRON DEVICES SOCIETY® | … the so-called SAGFET | Introduction of the MeSFET by C.A. Mead |

Chapter 10

Semiconductor Manufacturing

Rajendra Singh, Luigi Colombo, Klaus Schuegraf, Robert Doering and Alain Diebold

10.1 Introduction

Since the invention of integrated circuits (ICs) in 1958, semiconductor manufacturing, popularly known as "chip manufacturing," has played a vital role in enabling the information revolution that started in the last half of the twentieth-century and is continuing to shape the world of tomorrow. Due to advancements in semiconductor manufacturing, the critical dimension of ICs has been reduced from about 20 μm in 1960 to about 20 nm in 2012 (see also Chapter 2). Reduced cost, higher integration density and improved reliability have been largely responsible for providing a wide variety of semiconductor products with increased functionality. The basic sequence in IC manufacturing is transistor formation (front-end processing), interconnect formation (back-end processing), and assembly and test. Single wafer processing (SWP) dominates both front- and back-end processing, and SWP steps are also being introduced into assembly [1]. Starting with planar processing (two-dimensional film growth or deposition), silicon IC processing has evolved to the use of the third dimension for transistor channels (e.g., FinFET transistors), device isolation (e.g., trench isolation), DRAM cells (e.g., trench capacitors), and chip assembly (e.g., stacked packaging). The basic steps of semiconductor manufacturing are fundamentally the same for silicon and compound semiconductor based ICs. More than 90% of the world's US$1.5 trillion electronics business is based on silicon. The objective of this chapter is to provide an overview of the fundamental aspects of semiconductor manufacturing, with the main focus on silicon ICs.

Guide to State-of-the-Art Electron Devices, First Edition. Edited by Joachim N. Burghartz.
© 2013 John Wiley & Sons, Ltd. Published 2013 by John Wiley & Sons, Ltd.

| Double-integral MOSFET model by C.-T. Sah and H.-C. Pao | | ELECTRON DEVICES SOCIETY® | IEEE Electron Devices Group with 5 technical committees: Electron Tubes, Solid-State Devices, ... | ... Energy Source Devices, Integrated Electronics, and Quantum Electronics |

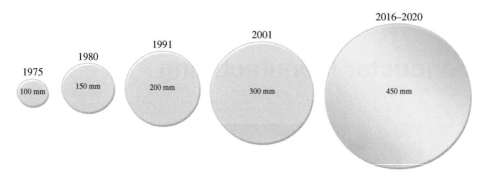

Figure 10.1   Increase in size of silicon wafers over the last about 40 years

10.2 Substrates

Today, the most advanced silicon IC manufacturing facilities for logic, memory and analog devices use 300-mm wafers, with plans to introduce 450-mm wafers. Silicon crystals with a metal purity of at least 99.9999999% are used to generate wafers for IC manufacturing. During the last 50 years, an increase in the size of silicon wafers (Figure 10.1) and reduced defect densities (Figure 10.2) have been primarily responsible for providing low-cost and high-yield ICs. In addition to bulk silicon, silicon-on-insulator (SOI) wafers have also been developed and are used for some high-performance ICs.

10.3 Lithography and Etching

Lithography in the semiconductor manufacturing context is typically the transfer of a pattern to a photosensitive material ("photoresist") by selective exposure to light. Several revolutions have driven lithography systems into new frontiers including the introduction of novel exposure paradigms, illumination sources, mask technology, resist technology, and wafer ambient. The state-of-art has progressed from full-wafer exposure, contact printing with broad-band mercury lamps to extremely sophisticated systems using laser illumination, step-and-repeat exposure with synchronized mask and wafer scanning using phase-shift masks, and chemically-amplified resists in an aqueous ambient to reach high numerical apertures. To top off these tremendous technological gains, wafer throughput, system uptime and registration accuracy have advanced significantly. The complexity of this solution is perhaps best reflected in the dramatic escalation of lithography system cost, where next-generation systems proposing to employ extreme ultra-violet (EUV) illumination in a vacuum ambient are expected to cost upwards of US$100 million.

Despite the repeated revolutions in lithography, the latest revolution towards EUV technology is proving to be quite challenging. As a result, first with 30 nm NAND memory production, and from 20 nm on for logic technologies, alternative techniques using double-patterning approaches have been developed to extend "Moore's Law" in the interim. Double patterning comes in two forms. First, memory is using a spacer-based technique for pitch division to create features at half the resolution pitch of 193 nm immersion lithography systems [2]. This opened the path to the manufacture of structures with half-pitch below 40 nm, now

1967   Harvey Nathanson invents surface micromachining   ELECTRON DEVICES SOCIETY®   Amana Co. fabricates the countertop microwave oven

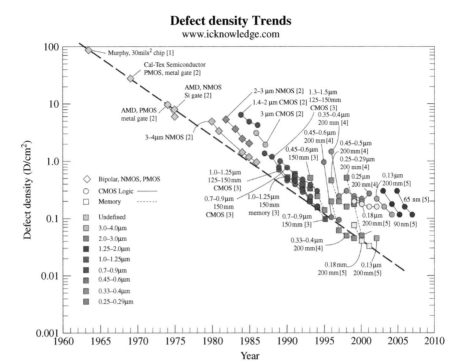

Defect density Trends
www.icknowledge.com

Figure 10.2 Reduction of defect density over the last 50 years. Reproduced with permission from IC Knowledge

extended to around 20 nm half-pitch. This technique requires an additional "non-critical" lithography step to trim features from array edges, hence "double" patterning. For logic, a second technique that is better suited for random layouts and their interconnection requirements decomposes a layout into two separate patterns for sequential patterning following a sequence of lithography-etch-lithography-etch or LELE [3]. In addition, the LELE technique requires stringent overlay tolerance to enable robust production-worthy patterning solutions. Both techniques add process complexity and cost with additional deposition, lithography and etch steps. Nonetheless, these schemes are now being extended to self-aligned quad patterning and sequences such as LELELE and perhaps even LELELELE. The competition for the future of lithography is divided into two camps. In the first, EUV delivers on its promise to resolve features less than 20 nm; in the other, multi-patterning techniques are adopted to augment the most cost-effective and productive lithography capability available, whether immersion 193 nm or EUV.

Etch technology, which transfers features from resists defined by photolithography, has also seen its share of revolution and innovation – though perhaps not of the same profound scale as for lithography. Etch technology has progressed from batch, wet chemical process, to single-wafer, plasma-based systems that operate with excellent throughput, uptime and low defectivity. Innovations in etch technology continue to

the two-year cadence of Moore's Law by delivering regular improvements to critical dimension uniformity, pattern-loading effects, etch rate and defect levels by optimizing the uniformity and density of plasmas and gas flows. Further, for today's ever more three-dimensional structures, etch technology addresses increasing aspect ratios while expanding the range of materials etched.

With the rise of multi-patterning techniques to address the challenges of advanced lithography, etch plays a key role where additional steps need tight dimensional control for precision manufacturability. Since overlay tolerance remains a critical challenge for multi-patterning, self-aligned etches will be increasingly employed, not just for memory, but also for logic technologies. For new materials, like copper, cobalt, high-k dielectrics, and magnetic materials, conventional subtractive etch is no longer the only path to film definition. Increasingly, new materials are deposited in molds etched in conventional dielectric and conductive films, but removed by chemical–mechanical planarization (CMP). As such, CMP has become an increasingly important process technology to work in harmony with etch to define new materials.

10.4 Front-End Processing

Front-end processing refers to the sequence of steps that construct the basic devices (e.g., transistors and capacitors) that comprise the integrated circuit. After bipolar, PMOS and NMOS device technologies, in the early 1980s, CMOS device technology evolved to the mainstream by enabling great power reduction and has been the workhorse technology for driving Moore's Law scaling [4] from transistor gate lengths of about 2 μm to today's 20 nm (see also Chapters 2 and 11). As the primary circuit technology, CMOS forms the foundation of the major IC product families, logic, analog, DRAM, SRAM, non-volatile memory, and so on. In the past decade, feature-size scaling has led CMOS technology to significant structural changes to attempt to meet the power-performance-cost requirements. Figures 10.3(a) and 10.3(b) illustrate the changes made in shrinking the process technology from 180 nm to 32 nm, respectively, – with more radical changes in memories [5].

Front-end processing, in particular for logic technologies, has advanced from a technology based mostly on batch, thermal furnace processes into one that now relies almost exclusively on a diverse set of unique, highly-specialized single-wafer-process sequences. The key modules for transistor formation comprise of isolation, gate-stack, junction and contact. In the early 1980s, furnaces and batch processing were

Rayleigh Criteria

Lord Rayleigh derived some basic optical formulas, known as "Rayleigh criteria," back in the 1800s. The rst of them is the proportionality between resolution and the ratio of wavelength to numerical aperture of an optical system. Depending on the geometry being imaged, the proportionality factor typically ranges from about 0.5–1.25. Thus, in many optical systems, we regard resolving power as being roughly comparable to the wavelength. However, due to a variety of innovations, as described in this section, we are presently using optical lithography to print IC features that are about an order of magnitude smaller than the wavelength of the light being used (193 nm)! The second Rayleigh criterion is another proportionality, in this case, between depth-of-focus and the ratio of wavelength to the square of numerical aperture. Numerical aperture is just the sine of the half-angle subtended by the lens from the focal point. A large lens captures more diffracted light from the object and creates a more precise image although in a shallower focus, which requires atter wafers.

| G. Chen describes "Purple Plague" in gold wire bonding on aluminum pads | Rare-earth permanent magnets shown by K.J. Strnat and G. Hoffer | ELECTRON DEVICES SOCIETY® | 1968 | Yamagami patents the insulated gate bipolar transistor (IGBT) |

pervasively used in IC manufacturing, as shown in Figure 10.3(c). Isolation was achieved by thermal oxidation in a steam ambient while polysilicon, silicon-nitride, and silicon-dioxide films were deposited by the newly introduced thermal low-pressure-chemical-vapor-deposition (LPCVD) technology. Many etches were performed with batch chemical processes, while junctions were formed by combining solid-phase deposition for dopant introduction with furnace-based thermal annealing. Single-wafer or mini-batch processes were introduced to improve control over dimensions, materials properties and interfaces. Later, ion implantation enabled precision control of impurity doping. Reactive ion-etch was also introduced to improve dimensional control. Sputter technology was developed and introduced into manufacturing flows to provide high-quality, high-purity elemental-metal and metal-nitride thin films.

As CMOS technology advanced further, with ever greater requirements for dimensional scaling, wafer scaling from 150–300 mm, new materials, shallow junctions and interface control, single-wafer process technologies were developed to address process control, flexibility, process clustering, and high throughput. Today's advanced CMOS transistors rely on new process technologies, such as atomic-layer deposition, selective epitaxy, plasma nitridation, CMP, advanced dielectric gap-filling materials, and laser annealing. As shown in Figure 10.3(d), these technologies have advanced to the level where clustered systems, enabling process steps to be sequenced without breaking vacuum, are becoming increasingly prevalent to maintain precise control over interfaces to avoid or eliminate unwelcome oxidation and surface adsorbates. As an example, consider the evolution of the gate stack forming the heart of the MOS transistor. Whereas, previously, the silicon-oxide dielectric was thermally grown in a furnace, now, a complicated sequence, encompassing processes to first clean the silicon interface and, then, form a near atomically precise oxide interface layer, followed by high-k gate-dielectric deposition and anneal, represents the state-of-the-art.

Gate Dielectrics

Silicon dioxide is one of most critical materials that have enabled CMOS to achieve its level of success. At the end of the 1990s, however, the leakage current of SiO_2 gate dielectric at a thickness of ~ 2 nm reached its limits. At that time, nitrogen was introduced in the SiO_2 matrix to form $Si_xO_yN_z$ in order to reduce the leakage and improve reliability. The introduction of nitrogen enabled the industry to continue scaling equivalent oxide thickness (EOT) to about 1.2 nm, however, as the industry was nearing the 45-nm technology node, it became clear that high-k and metal gates had to be introduced for high-speed devices to further reduce gate leakage. Today, while high-k and metal gates can be found in high-speed devices, $Si_xO_yN_z$ gate dielectrics are still pervasive in the most other product lines (see also Chapter 6).

10.5 Back-End Processing

An essential part of the integrated circuit is the ability to connect the transistors, capacitors, resistors, and so on with wiring to form a desired circuit. This was the integrated circuit that Kilby and Noyce envisioned over 50 years ago. From that time interconnect technology has advanced from the simplest implementation, wires over the transistors used by Kilby in 1958, Figure 10.4(a) to a fully-planar large-scale integrated

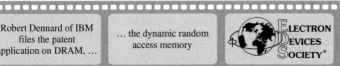

Robert Dennard of IBM files the patent application on DRAM, …

… the dynamic random access memory

ELECTRON **D**EVICES **S**OCIETY®

R. Kerwin, D. Klein and J. Sarace invent silicon gate technology

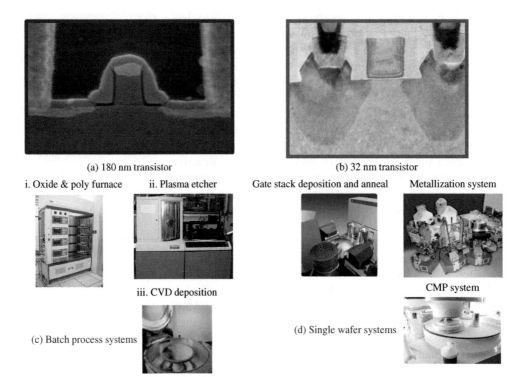

(a) 180 nm transistor   (b) 32 nm transistor

i. Oxide & poly furnace   ii. Plasma etcher   Gate stack deposition and anneal   Metallization system

iii. CVD deposition   CMP system

(c) Batch process systems   (d) Single wafer systems

Figure 10.3 Front-End processing. (a), (c-ii, c-iii), (d) Courtesy of Applied Materials; (b) Reproduced from [6], © 2009 IEEE (c-i) Courtesy of Bruce Technologies

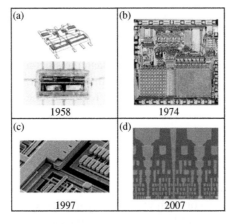

Figure 10.4 (a) Jack Kilby's first integrated circuit [7]; (b) fully integrated planar TMS-1000 by Texas Instruments in 1974 [7]; (c) Cu wiring revealed after dielectric removal by IBM [8] and (d) 45-nm device cross-section showing multilevel metals by Intel [9]. Courtesy of Texas Instruments, Inc.

H. Shichman and D.A. Hodges introduce a MOSFET model later used in SPICE Level 1

ELECTRON **D**EVICES **S**OCIETY®

George H. Heilmeier at RCA demonstrates the first liquid crystal display (LCD)

Early mass production of red LED's

circuits using Al metal interconnects in the 1970s, such as the Texas Instruments TMS-1000, Figure 10.4(b) [7], to today's nanoelectronic integrated circuits shown in Figure 10.4(c) [8] using copper wiring (now with features less than 30 nm) within a low-k dielectric medium and CMP as a key process [9] to achieve flat surfaces between the various metal levels. CMP was a major innovation that has enabled ultra-scaling of ICs.

Numerous materials have been studied and used to scale interconnects in order to achieve the level of integration of today's typical devices as shown in Figure 10.4(d). In order to integrate wiring to form circuits, it was necessary to develop metal-deposition processes, suitable dielectric materials and deposition processes for electrical isolation, the associated cleaning, patterning and etching technologies and the aforementioned CMP. In parallel, it was necessary for metallurgists and chemists to produce the high-purity metal targets for physical vapor deposition and the precursors for the various types of chemical vapor deposition (CVD) techniques, such as thermal and plasma, and, later, for atomic-layer deposition (ALD). The development of high-quality materials was not only necessary to achieve the high levels of integration and device requirements but, also, to achieve reliable circuits that could be operated over a wide range of voltages and ambient conditions. At first, Al metal was used for wiring, and, as it was scaled into the sub-micrometer regime, it became necessary to deposit very high-purity, principally oxygen-free metal, in order to meet the increasing demands on low-resistivity wiring and device reliability. This need led to the development of ultra-high-vacuum sputtering systems. At the same time, high-quality dielectrics and processes, principally CVD silicon dioxide, were used. Later, lower-k dielectrics deposited by various techniques were introduced. As the devices were scaled, the RC delay of circuits tended to increase as a result of increased resistance due to smaller wires and increasing capacitance due to smaller dielectric gaps. This led to the introduction of copper metallization with a much lower resistivity than aluminum, and to lower-k dielectrics to decrease the metal-to-metal capacitance. The introduction of copper and low-k dielectrics into the CMOS process flow introduced a host of new challenges associated with etching, low-k dielectric breakdown and integrity, and Cu diffusion. Up to this point, Al was dry etched and then covered with the dielectric, but it is very difficult to dry etch Cu, so it was necessary to modify the flow to deposit the dielectric first, etch it, and then fill the vias and trenches with Cu. In addition, because of the low density and poor mechanical integrity of low-k materials in comparison to the traditional SiO_2, Cu-barrier and Cu-seed technology had to be developed. This meant developing conformal processes

Micro-Electromechanical Systems (MEMS) Manufacturing

In many respects, MEMS processing resembles silicon integrated-circuit manufacturing. Usually starting with silicon, materials are added or removed by many techniques that are mostly shared with IC manufacturing, such as optical lithography, dry etching (RIE and DIE), wet etching, and plating. However, processes such as molding and electro-discharge machining are also used for MEMS manufacturing. Due to the complex nature of MEMS systems (moving parts in many cases), assembly is typically more challenging in the manufacturing of reliable MEMS (see also Chapter 18).

1969   First man on the moon (Apollo 11)   **E**LECTRON **D**EVICES **S**OCIETY®   S.E. Miller proposes optoelectronics integration with several components on the same substrate

that could deposit materials like TaN, Ta, TiN or Ru with nanometer thickness control. To accomplish this, atomic-layer-deposition technology was developed and implemented. The totality of integrated-circuit and individual-device performance requirements has led to a myriad of manufacturing-technology inventions, including materials integration schemes, thin-film deposition processes, and the associated patterning, etching and cleaning technologies that, together, have enabled the fabrication of devices that power the ubiquitous electronic devices now in the hands of most people on the planet.

10.6 Process Control

Process control is critical for the manufacturing of integrated circuits. Process control requires both the measurement equipment and an automated data management system. Measurements are now done in situ (inside the process equipment), in-line (inside the clean room), and off-line (in a laboratory). In this section, many of the key advances in process control are highlighted.

In the early days, process engineers could estimate the thickness of silicon dioxide or silicon nitride film thickness from the color of the dielectric layer. As feature dimensions decreased, optical microscopes were used to measure the width of key features such as the line width of the transistor gate (CD: critical dimension). With further decreases in feature dimensions, scanning electron microscopy became necessary as highlighted in Figure 10.5. Reflectometers were introduced to measure film thickness with greater accuracy. The correct positioning of one layer on top the layer below is a key measurement need. This is still done using an optical microscope based system and is known as an overlay measurement. By the mid-1980s, in-line physical and electrical metrologies were widely used for process control [10, 11]. Some

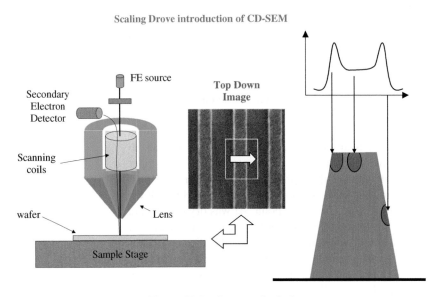

Figure 10.5 Process Control

of the key classes of measurement were and still are defect detection and review, critical dimensions, film (thickness), and electrical properties such as sheet resistance measurements became available [10, 11].

The need to achieve improved yield and device performance drove the introduction of Statistical Process Control (SPC). Film thickness is a good example of the application of SPC. Based on the electrical properties of the device, an upper and lower limit to the thickness is established. The film thickness is measured across the wafer for many or all of the wafers in a lot. When the measurement is rapid, this is done for every wafer in a lot of usually 24 wafers. The frequency of other more time consuming measurements were determined by an engineering team based mostly on the maturity of the process. The key to SPC is having a measurement method (metrology) that can reproducibly measure a needed parameter, for example, the thickness, over a long period of time (precision) with good accuracy. Accuracy refers to the ability to measure the true value of the variable. Over the past 15 years, this concept has evolved and the total measurement uncertainty is determined in a manner that includes the ability of measurement systems to determine the same value on the same control sample, which is known as tool matching. Process control charts are kept for most of process equipment in the clean room using an automated factory control system.

The next step in process control was the use of sensors on the process equipment. The in situ approach is coupled to automated data analysis systems. This 15-plus-year trend is part of the Advanced Equipment Control and Advanced Process Control (AEC/APC) approach to improving process control. By the mid-1990s, the need toward more equipment based process control was widely recognized [12].

The need for improvements in yield, especially for memory, resulted in the introduction of a variety of defect detection and review systems through the years. The first systems were based on optical microscopes. This evolved into a system that linked automated image processing software with optical microscopy and is now known as Bright Field Optical Inspection. Later, light scattering systems known as dark field systems were introduced. When defects and particles became too small to detect with optical systems, scanning electron microscopes with image analysis capability were introduced.

10.7 Assembly and Test

After back-end processing, semiconductor wafers are sent to an assembly and test facility for sawing the wafers into individual ICs and creating "packages" which allow them to be conveniently interconnected as components of larger electronic systems [13]. The sawing operation is usually performed with a diamond-saw, and the bulk of the package is most commonly a protective plastic body.

The larger systems are typically organized into printed-circuit "boards" (PCBs) that have printed and etched metal traces serving as the wiring between IC attachment structures and passive electrical components on the board.

The most common IC package-to-board connection technique is still based on using a metal "lead frame," as shown in Figure 10.6. The leads extend from the package and conduct individual input and output signals as well as ground and power between the IC and a "socket" on the PCB into which the package is inserted. At appropriate circuit nodes, the top layer of interconnect on the IC terminates at "bond pads" which are connected inside the package to individual leads. Historically, this connection has mostly been made via fine gold wires which are welded by electrically melting the ends of the wire into balls shaped by surface tension. Thus, this process is called *ball bonding*. To reduce cost, copper has recently been replacing gold for this connection.

William Boyle and George E. Smith invent the Charge-Coupled-Device (CCD)

ELECTRON DEVICES SOCIETY®

Invention of molecular beam epitaxy by J.R. Arthur and Alfred Y. Cho

Figure 10.6   IC in lead-frame-based package. Courtesy of Texas Instruments, Inc.

In the past few years, a newer assembly technology called *flip-chip* has been replacing lead-frame-based packaging for an increasing number of applications. In the flip-chip technique, solder "bumps" are created on the IC bond pads and the chip is "flipped-over" during assembly so that bond pads match up with pads on the PC board. The solder is then heated to form the electrical bonds. A principal advantage of flip-chip technology is increased density of I/O, for example, via the *ball-grid-array* (BGA).

In both the lead-frame and flip-chip technologies, the IC is encapsulated in a protective package, which is typically a plastic envelope in common cost-sensitive applications. For ICs operating in harsher environments, the package may have a hermetically-sealed ceramic body. In either case, one of the primary challenges for packaging high-performance ICs is heat removal, since they can generate as much as 100 Watts/cm^2. Packaging for such applications is designed with finite-element modeling of heat flow from hot spots on the chip to heat sinks. Similarly-detailed electrical modeling is also important for managing I/O inductance, etc., especially for ICs handling high-frequency signals.

The most advanced IC assembly processes use a myriad of approaches to fabricate multi-chip packages (MCPs). Lead-frames, bumps, and even through-silicon vias (TSVs) are now being mixed and matched into various types of "systems-in-package" (SIP).

After the assembly is completed, the IC undergoes "final test," which determines the functionality of the product unit [14]. The chip has already been tested in wafer form, called "multiprobe," which determines wafer-fabrication yield. Of course, the testers, except for their interface hardware to the device-under-test

| Jim Black at Motorola describes a mathematical model of mean time to failure (MTTF) … | … due to electromigration | **E**LECTRON **D**EVICES **S**OCIETY® | Device Research Conferences on "solid-state devices" and on … | … "vacuum tubes and other devices" merged back into one conference |

(DUT) are essentially identical. For logic testing, test programs are written to sequentially apply "test-vector" inputs and sort any incorrect responses into various "fault bins" for failure-analysis feedback to the circuit design and manufacturing processes. Today, it's not unusual to have circuits with so many potential logic states that exhaustively testing them is impractical. In fact, the cost of testing is now its largest issue. Several growing-complexity-driven IC cost challenges are partly being addressed through "design for X." Thus, we now have "design for test" as another part of overall "design for manufacturing." For example, design for test includes ICs with internal test circuitry; that is, *built in self-test* (BIST).

10.8 Future Directions

In 2012, 32-nm IC manufacturing is in mainstream production and, one manufacturer has started shipping products based on manufacturing at 22 nm. For improved CMOS performance, high-mobility Ge and compound semiconductors (e.g., InGaAs) are being considered as channel-region semiconductor materials. It is possible that future generations of lithography may be based on extreme-ultra-violet (EUV) light sources of wavelength of 13.5 nm. However, the cost of EUV is unacceptably high at this time, which could make way for alternative techniques. Future advancements in 3D packaging have the potential of integrating logic, analog, new types of memory, optics, and so on into "systems-in-package." The introduction of 450-mm wafers is expected to take place in the timeframe of 2016–2020, at least for development. Process variability is one of the key challenges to continue semiconductor manufacturing below critical dimensions of roughly 10 nm. Fundamental research on new materials, their interfaces, new processes and new devices is certainly required if we are ever to achieve practical ICs with sufficiently low power consumption for products manufactured with critical dimensions approaching 5 nm or less.

References

[1] R. Singh and R. Thakur, "Chip Making's Singular Future", *IEEE Spectrum*, pp. 40–45, February 2005.
[2] W. Y. Jung, S. M. Kim, C. D. Kim, G. H. Sim, S. M. Jeon, S. W. Park, B. S. Lee, S. K. Park, J. S. Kim, and L. S. Heon, "Patterning with amorphous carbon spacer for expanding the resolution limit of current lithography tool", *Proc. SPIE*, vol. 6520, 65201C (9 pp.), 2007.
[3] P. Zimmerman, "Double patterning lithography: double trouble or double the fun?", July 20, 2009, *SPIE Newsroom*. DOI: 10.1117/2.1200906.1691 http://spie.org/x35993.xml.
[4] C. A. Mack, "Fifty Years of Moore's Law", *IEEE Transaction on Semiconductor Manufacturing*, vol. 24, pp. 202–207, 2011.
[5] C. Edwards, "The Big 22 nm Gamble", *Engineering and Technology*, vol. 7, no. 5, pp. 76–79, June 2012.
[6] P. Packan, *et al.*, "High performance 32nm logic technology featuring 2nd generation high-k + metal gate transistors", *Electron Devices Meeting (IEDM)*, 2009 IEEE International. 659–662, 2009.
[7] K. J. Kuhn, "Considerations for Ultimate CMOS Scaling", *IEEE Transactions on Electron Devices*, vol. 59, pp. 1813–1828, 2012.
[8] R. Iassac, J. Heidenreich, D. Edelstein, *et al.*, *Copper Interconnects, The Evolution of Microprocessors*, IBM Icons. Available at http://www.ibm.com/ibm100/us/en/icons/copperchip/ (accessed October 15, 2012) July 2011.
[9] P. Moon, V. Chikarmane, K. Fischer, R. Grover, T. A. Ibrahim, D. Ingerly, K. J. Lee, C. Litteken, T. Mule, and S. Williams, "Process and Electrical Results for the On-die Interconnect Stack for Intel's 45nm Process Generation", *Intel Technology Journal*, 2, pp. 87–92, 2008.

1970   New light on
                                        electron devices   1970

[10] *Twenty Years at Tencor, the Chip History Center*, VLSI Research, 2010. WWW.chiphistory.org/exhibits/ (accessed October 15, 2012).

[11] *Process Diagnostics, the Chip History Center*, VLSI Standards, 1990. WWW.chiphistory.org/exhibits/ (accessed October 15, 2012).

[12] *Changes in Chip Making and How it is Driving Process Diagnostics*, VLSI Research, June 1996. WWW.chiphistory.org/exhibits/ (accessed October 15, 2012).

[13] M. Lamson, "Integrated-Circuit Packaging", in *Handbook of Semiconductor Manufacturing Technology*", 2nd edition, R. Doering and Y. Nishi (eds.), Boca Raton, FL, CRC Press, pp. 100–200, 2008.

[14] "Test and Test Equipment", 2011 *International Technology Roadmap for Semiconductors*, www.itrs.net (accessed October 15, 2012), SIA, 2011.

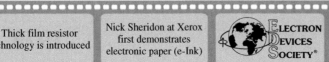

Part III

APPLICATIONS BASED ON ELECTRON DEVICES

| A micromachined pressure sensor using a silicon membrane … | … is developed at IBM Research by Kurt Peterson | **E**LECTRON **D**EVICES **S**OCIETY® | Ion-implanted source/drain contacts for MOSFETs come up | Polysilicon gate structure for MOSFETs is introduced |

Chapter 11

VLSI Technology and Circuits

Kaustav Banerjee and Shuji Ikeda

11.1 Introduction

Very-Large-Scale Integration (VLSI) is the process of creating integrated circuits and systems by combining billions of transistors into a single chip. VLSI circuits constitute one of the most pervasive and indispensable technologies of the modern world with applications in nearly every domain of human advancement including consumer electronics, computing, telecommunications, energy, space, automobile, aerospace, biomedicine, entertainment, military, and so on. CMOS devices (PMOS and NMOS) have been the workhorse of VLSI circuits for over three decades. Aggressive scaling of the CMOS device size makes it possible to integrate a large number of elements including CPU core, bus architecture, memories and other logic components on a single chip. This integrated architecture scheme improved system performance drastically and is commonly known as system LSI or SoC (system-on-a-chip). In SoC, the chip design is customized and the intellectual properties are included in the chip design. Since the 1960s, the number of transistors per unit area has doubled

First Integrated Circuit

Jack Kilby (Texas Instruments) and Robert Noyce (Fairchild) separately demonstrated the monolithically integrated circuit or microchip (1958).

Kilby received the Nobel Prize in 2000.

Guide to State-of-the-Art Electron Devices, First Edition. Edited by Joachim N. Burghartz.
© 2013 John Wiley & Sons, Ltd. Published 2013 by John Wiley & Sons, Ltd.

1970 — Kulite Co. presents the first silicon accelerometer

ELECTRON DEVICES SOCIETY®

Introduction of Gummel-Poon model for bipolar transistors (H. Gummel and H.C. Poon)

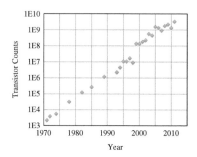

Figure 11.1   Trend of transistor counts (Data source: ISSCC)

every 1.5 years (18 months). This observation is known as Moore's law, predicted in the landmark article in the April 19, 1965 issue of the *Electronics Magazine* [1]. Today, more than 40 years since the advent of Moore's law, it is still providing guidelines for technology development but has slowed down to transistors doubling every 18–24 months. (Figure 11.1) [2].

11.2 MOSFET Scaling Trends

The rapid increase in the number of transistors integrated on a chip has been supported by the overall size reduction of transistors (see Chapter 2). With regard to the miniaturization of Metal-Oxide-Semiconductor Field-Effect-Transistor (MOSFET) fabrication processes, Robert Dennard and co-workers of IBM, presented in 1974 a prediction of device performance when related parameters (density of impurities, power-supply voltage, etc.) together with the physical dimension of devices were changed in a certain scaling ratio [3]. This scaling principle became a basic guideline for the miniaturization of MOSFET devices for more than 30 years.

To reduce power, supply voltage was reduced from 5 V to 3.3 V in 1993. Since then, the supply voltage has continuously reduced until around 2000. (Figure 11.2) [4]. As can be observed from this figure, scaling of supply voltage has significantly slowed down since the early 2000s.

To achieve higher drive current with lower supply voltage, the gate length was reduced (drain current is proportional to 1/L (L: gate length)) more than the conventional trend (more aggressively than DRAM half pitch shrink rate). In 2001, the DRAM half pitch was 130 nm and the LOGIC gate length (printed) was 90 nm. In 2009, they were reduced to 52 nm and 34 nm respectively (Figure 11.3) [4].

Early History of Integrated Circuits

- Invention of BJT (1948)
- First silicon transistor (1954)
- First integrated circuit (1958)
- MOS transistor (1960)
- MOS integrated circuit (1962)
- DRAM cell (1968)
- Intel formed (1968)
- AMD formed (1969)
- Microprocessor invented (1971)
- 32-bit microprocessors (1980)

Note: although NMOS was conceived in the early 1960s and found application in wrist watches and portable electronics, but device quality was poor, hence PMOS was used until 1974.

- The Intel 4004 Microprocessor (1971) was PMOS only (2300 transistors, 10 um process).
- The Intel 8080 (8-bit) Microprocessor (1974) was NMOS only 6000 transistors, 6 um process, 2 MHz).

| DRAM patent is issued to Robert Dennard of IBM | Intel builds the first DRAM Chip (1kb pMOS dynamic RAM) |  | Else Kooi and co-workers at Philips Co. invent the LOCOS process | |

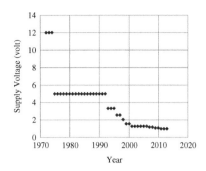

Figure 11.2 Trend of supply voltage scaling (Data source: ITRS Roadmap)

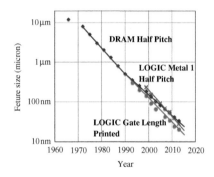

Figure 11.3 Trend of feature size (Data source: ITRS Roadmap)

The threshold voltage was also reduced to increase drive current. Higher drive current is essential for higher performance of the system and increases the operation power for a given supply voltage (power when system is ON). However, lower threshold gives higher leakage (off-current) and increases stand-by current and power (stand-by power: power when the system is OFF). Reducing power both during stand-by mode and operation mode is becoming a major issue for system LSI. The design for power is called power management, which seeks to achieve a balance between power reduction and high performance.

11.3 Low-Power and High-Speed Logic Design

Requirements for higher performance and more functionality in a LSI system increase both operation and standby power. Hence, power management has been one of the major issues in VLSI systems [5, 6].

Transistor Performance

Delay time, τ
  Operation Power, P

$\tau = CV/I$
$P = VI$

V: supply voltage
C: capacitance
I: transistor current

Z. Alferov and V. Andreev demonstrate the first GaAs heterostructure solar cell

Zhores Alferov presents the first laser diode achieving continuous wave operation

ELECTRON DEVICES SOCIETY®

D. Frohman-Bentchkowsky demonstrates the first charge-trap device (MNOS)

In fact, designing low-power and energy-efficient VLSI systems constitutes a key area for sustaining the irreversible growth of the global information technology industry [7]. Achieving energy-efficiency is also of critical importance for all VLSI circuits used in mobile applications (such as cell phones, PDAs and laptop computers) for increasing the battery life.

To achieve higher drive current with lower supply voltage, threshold voltage of the device must be reduced and this increases off-current (or leakage power) of the device drastically, thereby reducing the energy-efficiency of the devices and circuits [7]. Moreover, leakage power increases at higher chip operating temperatures, due to the exponential dependence of subthreshold current on temperature. This has a profound impact on the power-performance optimization schemes [8].

Energy-efficiency can be achieved by lowering both dynamic (switching) and leakage power consumption. Lower power also leads to lower operating temperatures of electronic devices resulting in improved performance and reliability [9], since most reliability mechanisms tend to degrade with increasing temperature. However, lowering of power using traditional techniques becomes increasingly difficult beyond the 22-nm technology node. This is due to the fact that in such nanoscale devices, the most effective "knob" used for lowering power, namely the power supply voltage, cannot be scaled as rapidly as in earlier technology generations without incurring significant performance penalty arising from the inability to simultaneously reduce the threshold voltage. Simultaneous scaling of threshold voltage, which is essential for maintaining a certain ON to OFF ratio of the device currents (that is essential in digital circuits where the transistors are used as switches), leads to a substantial increase in the subthreshold leakage (OFF state) current (Figure 11.4a), owing to the "non-abrupt" nature of the switching characteristics of MOSFETs (Figure 11.4b), thereby making the devices very energy inefficient.

The metric that is used to capture the abruptness of this switching behavior is known as the subthreshold swing (S) (inverse of the subthreshold slope = *dlogId/ dVgs* in Figure 11.4(b), which has a fundamental lower limit of *2.3 kT/q = 60 mV/decade* for MOSFETs, indicating that in order to reduce the drain-to-source current by one decade, one must reduce the gate voltage by 60 mV [10]. This value is essentially due to the "thermionic emission" of carriers from the source to the channel region in MOSFETs. Practical nanoscale CMOS devices have even higher subthreshold swings (80–90 mV/decade). Hence, although lowering of the subthreshold swing below 60 mV/decade is a non-trivial task, it is extremely critical for achieving any significant improvements in energy-efficiency of LSI systems. While some

Transistor Drive Current

$$I_{ds} = (W/L)\,\mu C$$
$$\left[(V_{gs} - V_{th})\,V_{ds} - V_{ds}{}^2/2\right]$$

I_{ds}: drive current
W: transistor width
L: transistor gate length
μ: carrier mobility
C: gate capacitance
V_{gs}: gate-to-source voltage
V_{th}: threshold voltage
V_{ds}: drain-to-source voltage

Dynamic Power Dissipation

$P_{dynamic} = \alpha\,C_{eff}\,V^2\,F$
(assuming full swing)

α: activity (switching) factor
C_{eff}: effective output load capacitance of the switching gate
V: supply voltage
F: maximum operation frequency

Leakage Power Dissipation

$P_{subthresold\ leakage} = I_S\,10^{(-V_{th}/S)}$
$(1 - 10^{(-V_{ds}/S)})\,W_{eff}\,V_{ds}$

I_S: nominal zero-threshold leakage current
V_{th}: threshold voltage
S: subthreshold swing
W_{eff}: effective transistor width
V_{ds}: drain-to-source voltage

Invention of the micro processor at Intel by Marcian Hoff

ELECTRON DEVICES SOCIETY®

Casio markets the first pocket computer

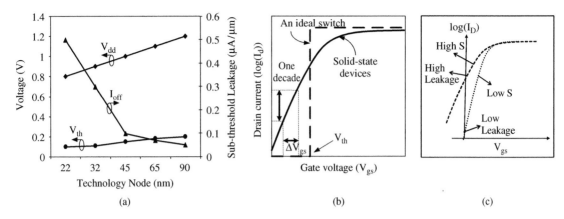

(a)   (b)   (c)

Figure 11.4   (a) CMOS technology scaling trends as predicted by ITRS; (b) comparison between switching characteristics of an ideal switch and that of solid-state devices; (c) relation of subthreshold swing to leakage current

circuit- and architecture-level techniques exist for lowering power, these efforts can produce even better results with more energy-efficient transistors as CMOS technology scales beyond the 22-nm node. It is clear from Figure 11.4(c) that reducing the S leads to a substantial reduction of the OFF-state current.

Multi-threshold voltage technique has been implemented to reduce both operation and standby power of the system while maintaining higher performance [11]. In the ITRS road map, three types of devices are listed, high performance, low power and regular (in between). These devices are provided by basically changing threshold voltage of the device through various ways. Low threshold voltage transistors are used in critical paths for high-speed logic and high threshold voltage transistors are employed elsewhere to reduce power. Gate length tuning and body biasing to control threshold voltage of transistors provide system designers further options to optimize system performance.

11.4 Scaling Driven Technology Enhancements

Continuous CMOS scaling has driven and is still driving the introduction of many new materials, as well as structure and device engineering. For example, gate insulator thickness is reduced as gate length reduces to achieve higher performance. In 1990s, gate oxide thickness was reduced to 2 nm, and direct tunneling current through the gate oxide could not to be ignored anymore and became one of the key factors limiting transistor scaling. To solve this problem, the use of an insulator with a higher dielectric constant (high-k) than the conventional silicon dioxide film (SiO_2) has been investigated since early 1990s (see Chapter 6). This reduced gate leakage drastically while maintaining high gate insulator capacitance (thinner equivalent oxide thickness-EOT). After introducing high-k gate insulator, thickness reduction of the gate insulator is getting more aggressive to further increase transistor performance. Intel led the high-k gate insulator implementation in 45 nm technology generation in 2009 as shown in Figure 11.5 [12, 13].

Reducing the gate length requires shallow junctions of source/drain. However, reducing the source/drain depth increases resistance, which can significantly hamper efforts to obtain a high drive current. Reducing

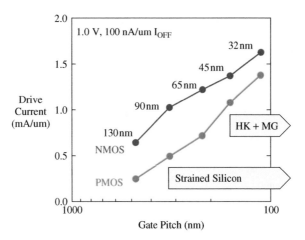

Figure 11.5 Transistor drive current versus gate pitch trend showing the introduction of strained silicon and high-k metal gate innovations. © 2011 IEEE. Reprinted with permission from M. Bohr, "The evolution of scaling from the homogeneous era to the heterogeneous era," *Technical Digest of IEDM*, 2011, Fig. 6

the gate length without an increase in resistance requires a junction formation technology that permits the use of high dopant concentrations while allowing the construction of shallow junctions. To form a shallow junction, dopant must be introduced near the surface, which requires a low-energy ion implantation technique. Furthermore, reducing resistance by increasing the dopant concentration requires a high-temperature thermal annealing treatment. Since higher temperature causes deeper junction due to increased diffusivity of dopants, the duration of the thermal treatment must be significantly limited.

The scaling challenges, including gate length and gate oxide scaling as well as reduction of the threshold voltage, has driven other technologies to enhance device performance, such as mobility enhancement via strain engineering [14].

PMOS and NMOS respond differently to different types of strain. Specifically, PMOS performance is best served by applying compressive strain to the channel, whereas NMOS transistors benefit from tensile strain. Many approaches to strain engineering induce strain locally, allowing both n-channel and p-channel strain to be modulated independently. In 90-nm node, uniaxial strained silicon technology for both N- and PMOS FET has been introduced by Intel as shown in Figure 11.5 [12, 15].

Fully depleted devices (FDD) such as fully depleted silicon-on-insulator (FD-SOI) (Figure 11.6) and FinFET have also been investigated to enhance device performance [16, 17]. These devices have advantages in terms of short-channel effects (due to better electrostatics) and switching capacitance, allow relaxed gate oxide thickness scaling as shown in Figure 11.7 [18]. Fully depleted devices also have lower body effect, which allows higher logic fan-in and higher supply voltage, thereby bringing more options for circuit designers (Figure 11.8) [13]. For FinFETs, design of high performance logic circuits is challenging due to the width quantization effect (FinFET device width = 2H, where H is the height of the fin), which restricts the width to be changed by factors of 2H [19]. Intel announced FinFET (or Tri-gate) implementation into production in 2012 (Figure 11.9) [20, 21].

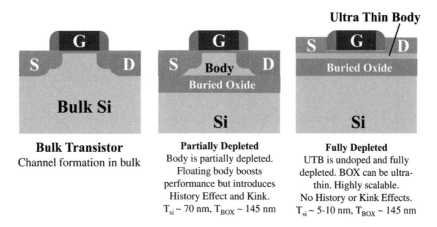

Figure 11.6 Fully Depleted (FD)-SOI Transistors versus Bulk and Partially Depleted (PD)-SOI Transistors

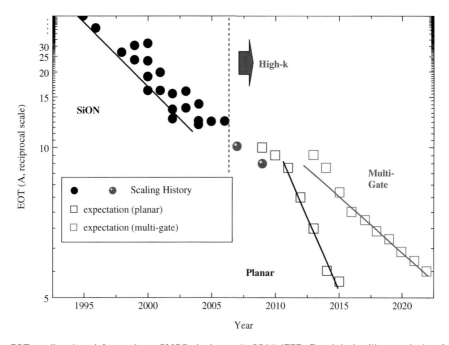

Figure 11.7 EOT scaling trend for various CMOS devices. © 2011 IEEE. Reprinted with permission from M. Bohr, "The evolution of scaling from the homogeneous era to the heterogeneous era," *Technical Digest of IEDM*, 2011, Fig. 6

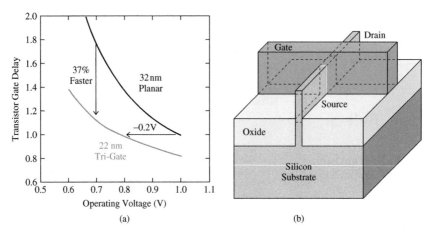

Figure 11.8 (a) Tri-gate's low voltage performance as a function of operating voltage. © 2011 IEEE. Reprinted with permission from M. Bohr, "The evolution of scaling from the homogeneous era to the heterogeneous era," *Technical Digest of IEDM*, 2011, Fig. 6. (b) Schematic of a Fin-FET or Tri-gate transistor

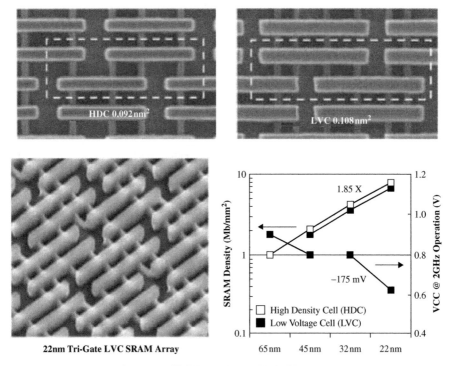

Figure 11.9 The 22-nm HDC and LVC Tri-gate SRAM bitcells. © 2012 IEEE. Reprinted, with permission from E. Karl, *et al.*, "A 4.6GHz 162Mb SRAM design in 22 nm tri-gate CMOS technology with integrated active VMIN-enhancing assist circuitry", *Digest of Technical Papers, ISSCC*, p 230, 2012

Planar Fully Depleted Silicon-on-Insulator (FD-SOI) technology relies on an ultra-thin layer of silicon over a Buried Oxide (commonly called BOX). Transistors built on this top silicon layer are ultra-thin body devices and have unique, extremely attractive characteristics including body biasing made possible due to the ultra thin BOX. Soitec/IBM reported in 2009 that their unique FD Extra-Thin SOI (ET-SOI) transistors exhibit ultra-low spatial variability of the silicon layer and provide 40% lower power when operated at current performance levels because of the drastic cut in leakage current supplied by the buried oxide layer [22].

As threshold voltage of devices is reduced to achieve higher performance, variation of the threshold voltage becomes a major problem for power reduction while maintaining higher performance [23, 24]. In general, parameter (process, temperature or voltage) variations can significantly impact leakage power since subthreshold leakage is a strong function of channel length, oxide thickness, supply voltage, and temperature. Hence, statistical estimation becomes crucial for accurately estimating leakage power [25]. Fully depleted devices reduce random dopant fluctuations and could provide less variation of the threshold voltage [26]. Reducing variations is essential for power management and for enhancing the operation margin of embedded SRAMs in a chip under continuous scaling of operation voltage [27]. For all nanoscale transistors employing metal as the gate electrode, the work-function variation of the gate due to randomness of the metal grain orientations has been identified as the most significant source of random threshold voltage variation [28, 29].

Nanowires made from different materials including silicon and SiGe provide an attractive pathway for making fully depleted gate-all-around field effect transistors (FETs) [30]. Nanowire FETs can be made both horizontally and vertically as shown in Figure 11.10. CMOS devices based on nanowires can provide even better electrostatics than planar FD-SOI or FinFETs/Tri-gate FETs.

Samsung announced in 2010 IEDM keynote speech that their LOGIC technology will use various techniques to enhance performance but still continue using silicon technology (Figure 11.11).

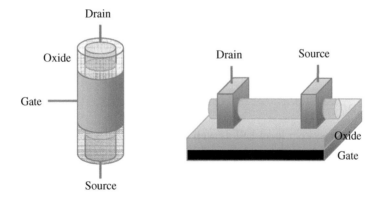

Figure 11.10   Vertical and horizontal nanowire FETs

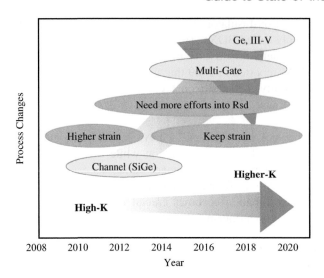

Figure 11.11   Key technology for the future Si CMOS era (Source: IMEC). © 2010 IEEE. Reprinted with permission from Kinam Kim, "From the future Si technology perspective: Challenges and opportunities," *Technical Digest of IEDM*, 2010

11.5 Ultra-Low Voltage Transistors

FD-SOI, FinFETs and NW-FETs are considered "non-classical" CMOS devices but are still limited by the 60 mV/decade subthreshold swing due to the thermionic emission of carriers in the high energy tail of the Boltzmann distribution of carriers in the source. Therefore, this effect is sometimes referred to as the "Boltzmann Tyranny" effect.

   A device that can overcome the 60 mV/decade fundamental limitation of all MOSFETs is the Tunnel-Field-Effect-Transistor (TFET). Hence, TFETs can potentially replace the CMOS for achieving future ultra-low voltage (≤0.5 V) and energy-efficient VLSI systems. TFETs are essentially a gated reverse biased diode operating based on band-to-band tunneling at the source-channel junction (Figure 11.12). Vertical Si nanowire based n-type and p-type TFETs have been demonstrated with 30 mV/decade subthreshold swing [31, 32], although the low ON current of TFETs continues to be a challenge.

11.6 Interconnects

Interconnects are crucial for system level LSI as well as transistor design. These two should be considered together to optimize system performance. A typical high performance IC employs several layers of metal interconnects separated by insulating materials, with "local" (or short wires) employed for local communication and "global" or long wires used for global communication within a chip. Interconnects play a dominant role in determining the performance (due to increase in signal delay) and power dissipation (due to interconnect capacitance that must also be driven by the logic gates) of most high-performance ICs.

Richard Martin achieves microscopic analysis of the piezo electric effect

1973

ELECTRON DEVICES SOCIETY®

Brian D. Josephson receives the Noble Prize in Physics for …

… theoretical predictions of the properties of a supercurrent through a tunnel barrier (Josephson Effect)

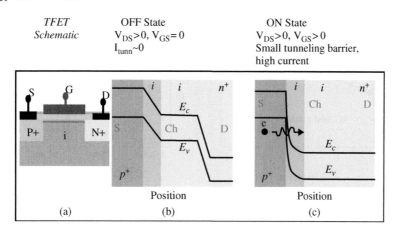

TFET Schematic	OFF State $V_{DS}>0$, $V_{GS}=0$ $I_{tunn}\sim0$	ON State $V_{DS}>0$, $V_{GS}>0$ Small tunneling barrier, high current

Figure 11.12 (a) Device schematic and (b), (c) Working principle, of an n-type Tunneling-Field-Effect-Transistor (TFET)

Delay of a wire increases quadratically with its length. Repeater (inverter) or buffer insertion results in lower delay increase. Repeater size and inter-repeater separation can be optimized to give minimal delay under a given power constraint [33]. Buffered wires are also needed for distributing the clock throughout the chip. As a result, over 50% of the power consumption in ICs can be attributed to the interconnects [34]. Interconnects are also needed for power/ground distribution in a chip.

As system functionality increases, number of metal layers increases. In 2010, 10 levels of interconnect have been used in mass production [4].

As shown in Figure 11.13(a), interconnect delay becomes important even for local wires (that connect adjacent or nearby transistors and gates and are less than few micrometers in length) [35]. This is mainly due to the interconnect capacitance. Figure 11.13(b) shows that the delay improvement in buffered global interconnects is much slower than that of the transistor delay for the global wires [35]. As frequency of signals increases, inductance of interconnects starts to impact the delay and needs careful consideration in the design optimization of interconnects [36]. For emerging communication-centric designs such as Multi-Core ICs and Network-on-Chip (NoC) designs, interconnects become the main bottleneck.

Important requirements in wiring technology are the reduction in wire resistance and capacitance so that the pulses generated by the logic gates can be propagated at high speeds. To this end, the use of copper instead

Interconnect Repeater Insertion

Interconnect delay (w/o repeater),
$$\tau = RC = rcL^2 \text{ (lumped model)}$$

Interconnect delay (with repeater inserted at L/2):
$$\tau = 2*R/2.C/2 = 1/2 \ rc \ L_2$$

  L: wire length
  R: wire resistance
  C: wire capacitance
  r: resistance per unit length
  c: capacitance per unit length

Note: lower resistivity interconnect materials reduce R, while low-k dieletrics reduce C.

Leo Esaki receives the Nobel Prize in Physics …

ELECTRON **D**EVICES **S**OCIETY®

… "for experimental discovery of tunneling phenomena in semiconductors"

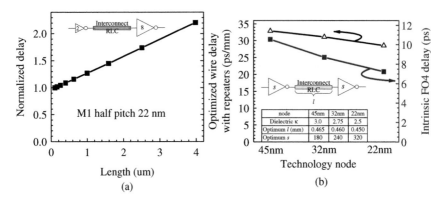

Figure 11.13 (a) Delay of a fanout-of-4 (FO4) driver as a function of local interconnect length, based on SPICE simulations and parameters in ITRS 2008. (b) Comparison between intrinsic (FO4 gate) delay and delay of optimally buffered global interconnects for three technology nodes using SPICE simulations and PTM model. The interconnect cross section is assumed to remain constant as 0.3×0.3 μm; and wire resistivity is assumed to be 2.2 μΩ cm

of aluminum has been introduced in low-resistance wiring technology in late 1990s (Figure 11.14) [37, 38]. In addition, fluorinated oxide film (SiOF) with a low dielectric constant (low-k) has come into use in place of silicon oxide film as an inter/intra-metal insulating material in late 1990s [4, 39]. Similar to transistor scaling, aggressive reduction of dielectric constant of backend insulating materials have been investigated (to lower interconnect capacitance and RC delay) and lower than 2.5 dielectric constant low-k materials has been implemented in mass production in 2011 [4].

Similar to wire resistance, contact resistance is one of the main challenges of interconnect technology. Contact resistance is about 100 Ω in 28 nm generation and increases by over 15% every two years [40]. Lowering of contact resistance between the first metal layer and silicided source/drain is more challenging than metal-to-metal contacts.

Inter-metal layer connectivity is achieved using short vertical interconnects known as *vias* (Figure 11.14). Resistance of minimum size vias becomes a concern for nanoscale technologies due to increase in effective resistivity of the metal (which results in greater Joule heating) arising from increased grain boundary and surface scattering as well as the presence of a barrier metal to encapsulate the copper [41]. Various barrier materials have been investigated to reduce the via resistance.

For sub-22 nm technologies, interconnect reliability is a major concern. For scaled interconnects (including vias), due to their increasing resistivity, the current carrying capacity of these wires becomes significantly lower due to increased Joule heating. As a result, the

Delay of a Driver-Interconnect-Load

$$t_{50\%} = 0.5rcl^2 + 0.7R_{driver}\,cl + 0.7(rl + R_{driver})C_{load}$$

$t_{50\%}$: delay w.r.t the 50% point in the input/output waveforms

l: interconnect length between driver and load

r: resistance per unit length of interconnect

c: capacitance per unit length of interconnect

R_{driver}: driver resistance

C_{load}: load capacitance of the driver.

1974   MOS scaling theory is presented by …    **E**LECTRON **D**EVICES **S**OCIETY®   … Dennard, Gaensslen, Yu, Rideout, Bassous and LeBlanc at IBM   Tak H. Ning publishes early work on hot carrier characterization in MOSFETS

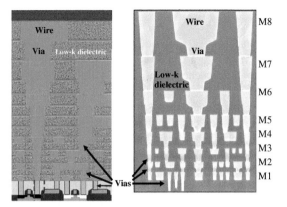

Figure 11.14 (*left*) Schematic view of the cross-section showing hierarchical scaling of interconnects in advanced ICs [4]. (*Right*) A cross-sectional micrograph of dual-damascene Cu interconnect layers used in Intel's 65 nm microprocessors

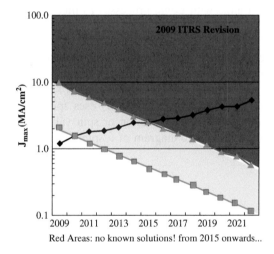

Red Areas: no known solutions! from 2015 onwards...

Figure 11.15 Required current density (black line) is increasing with scaling. Maximum allowed current density for Cu wire (blue line) is decreasing. The yellow region between the blue and green lines represents a transitional region where some evolutionary solutions still exist

performance driven current density in these wires and vias exceeds the maximum sustainable current density [41, 42], which has also been highlighted by the ITRS (Figure 11.15). In the domain of emerging interconnect materials, carbon nanotube bundles and graphene nanoribbons are being investigated [43]. It is envisioned that some type of electrical interconnects will continue to be the primary form of on-chip communication links.

11.7 Memory Design

The modern computer system consists of a series of memories from the fastest to the slowest (see also Chapters 3 and 13). Cache (SRAM) are fabricated on a chip to maintain high response between CPU and memory. Other memories, slower than cache, such as DRAM, and Flash are commonly integrated in a package or on a board (Figure 11.16).

Lower supply voltage and higher frequency of system LSI give smaller operation margins for embedded SRAMs. Since smaller geometries of devices gives larger variation, the difficulties of SRAM design are increasing as feature size and supply voltage are reducing.

Such slow down in the scaling trend is shown in Figure 11.17 [44]. High-k metal gate introduced in 45-nm node and 3D MugFET in 32 nm node reduce variability of the devices and enhance further scaling. Furthermore, circuit innovations are needed to enhance SRAM operation margin to increase system speed and enable low voltage operation. Using multiple power supplies makes it possible to realize high performance design with low power dissipation. Partial swing SRAM uses a sense amplifier to detect small voltage differences. The full swing design relies on a classic dynamic precharge and evaluate scheme and achieves better noise robustness.

Various alternative memories are being evaluated for replacing SRAM, DRAM and Flash. These emerging memories are designed to synchronize with Logic processes as much as possible for easier integration in system LSIs [45].

Ferroelectric RAM (FeRAM) is similar in construction to DRAM but uses a ferroelectric layer instead of a dielectric layer to achieve non-volatility [46]. Beyond Flash, other novel forms of non-volatile memory are also developed including magnetoresistive random-access memory (MRAM) where data is stored by magnetic storage elements formed from two ferromagnetic plates [47].

Spin-transfer torque random-access memory (STT-RAM) uses spin-transfer torque in which the orientation of a magnetic layer in a magnetic tunnel junction or spin valve is modified using a spin-polarized current [48]. Figure 11.18 shows a schematic diagram of a spin valve/magnetic tunnel junction.

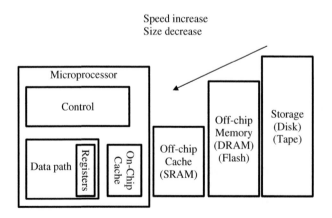

Figure 11.16   Memory hierarchy of computing systems

| David House of Intel corrects Moore's law to doubling of density every 18 months | ELECTRON DEVICES SOCIETY® | 6000 life-size pottery figures found in China | Introduction of 3-μm silicon technology node in manufacturing | First demonstration of 1-μm MOS at IBM |

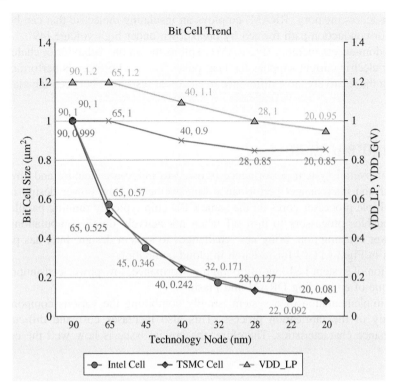

Figure 11.17 SRAM bit cell trend. Reproduced with permission from *ISSCC Trends Report*, ISSCC 2012

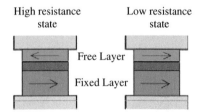

Figure 11.18 A schematic diagram of a spin valve/magnetic tunnel junction. In a spin valve the spacer layer (purple) is metallic; in a magnetic tunnel junction it is insulating

Resistive random-access memory (RRAM) employs an insulating dielectric that can be made to conduct through a filament or conduction path formed by breakdown under high voltage [49].

Phase-change random-access memory (PC-RAM) exploits the unique behavior of chalcogenide glass. For write operations, an electric current supplies the heat pulse. The read process is performed at sub-threshold voltages by utilizing the relatively large difference in electrical resistance between the glassy and crystalline states [50].

11.8 System Integration

The enhancement of overall system performance is owed to micro-architecture and circuit innovation, as well as to transistor and interconnect performance improvements. Low power design in microprocessors innovates using multiple processor cores on the same CPU chip typically running parallel threads at lower frequencies. This enables processors to turn off when not actively doing computations. These trends in integration and power consumption, bring new challenges to system design. Intel has produced a 48-core processor as shown in Figure 11.19 for research in cloud computing.

The implementation of system LSI that offers high performance, low power consumption, and low costs requires the technique of embedding DRAM and Flash.

For the on-chip implementation of a system, simply combining the various components can result in increased complexity of the fabrication process. This also increases cost and difficulties in achieving the desired performance characteristics. Therefore, the crucial issue is how well the components can be

Figure 11.19 The 48-core x86 Processor as single-chip cloud computer (Source: Intel, 2009)

1976 D. Carlson and C. Wronski present the first solar cell based on an amorphous silicon thin film (a-Si:H solar cell) ELECTRON DEVICES SOCIETY® R.S. Pengelly and J.A. Turner introduce the GaAs Monolithic Microwave IC (MMIC)

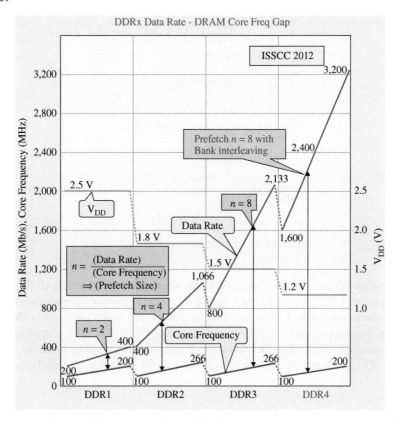

Figure 11.20 Trends for DRAM and High-Speed I/O for DDRx. Reproduced with permission from *ISSCC Trends Report*, ISSCC 2012

combined and how to manufacture devices and develop processes that are well-suited for on-chip implementation. For image processing in mobile devices, an increase in bandwidth, a reduction in the number of I/O pins, and even a decrease in power consumption can be attained by incorporating DRAM on a mix and match basis into the CMOS logic unit. However, scaling of feature size has also increased difficulties for both fabrication and circuit design. In late 1990s system-in-package (SiP) became popular and stand alone memories were mounted in one package instead of a single chip. This causes other issues in system design. Memory core frequency and external-data rate are much slower than those of system LSIs and the frequency gap between system LSI and memories is getting bigger and bigger as shown in Figure 11.20 [44].

New interface technology such as 3D IC technology with TSV (Through Silicon Via) as shown schematically in Figure 11.21, is essential to reduce this mismatch. Three-dimensional ICs can reduce the length of longest interconnects by vertically stacking multiple Si layers and thereby reduce chip area and power dissipation, and improve chip performance [51]. Three-dimensional ICs also offer the most promising platform

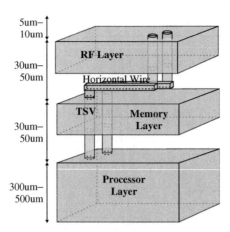

Figure 11.21 Schematic plot of a 3D IC [54]

to implement "More-than-Moore" technologies, bringing heterogeneous materials (Silicon, III–V semiconductors, carbon nanotube: CNT/graphene, etc.) and technologies (memory, logic, radio frequency, mixed signal, micro-/nano-electromechanical systems, optoelectronics, etc.) on a single chip [51]. The fabrication technology of TSVs has been demonstrated by various groups [52, 53]. The electrical characteristics of TSVs including resistance, capacitance and inductance, as well as coupling noise to active devices can strongly impact the 3D system performance and requires careful modeling and analysis [54].

References

[1] G. E. Moore, "Cramming more components onto integrated circuits", *Electronics*, Vol. 38, No. 8, April 19, 1965.
[2] Transistor counts, reported in the *International Solid State Circuits Conference* (ISSCC).
[3] R. H. Dennard, F. H. Gaensslen, V. L. Rideout, E. Bassous, A. R. LeBlanc, "Design of ion-implanted MOSFET's with very small physical dimensions", *IEEE Journal of Solid-State Circuits*, Vol. 9, No. 5, pp. 256–268, 1974.
[4] International Technology Roadmap for Semiconductors (ITRS) Available at http://www.itrs.net (accessed October 16, 2012).
[5] G. Yeric, "Technology Roadmaps and Low Power SoC Design", *Techn. Dig. IEEE International Electron Devices Meeting (IEDM)*, pp. 362, 2011.
[6] P. P. Gelsinger, "Microprocessors for the new millennium: Challenges, opportunities, and new frontiers", Digest of Technical Papers, *Int. Solid-State Circuits Conf.*, pp. 22–25 2001.
[7] V. De and S. Borkar, "Technology and design challenges for low power and high performance", in *Proc. International Symposium on Low Power Electronic Design*, pp. 163–168, 1999.
[8] S.-C. Lin and K. Banerjee, "A design-specific and thermally-aware methodology for trading-off power and performance in leakage-dominant CMOS technologies", *IEEE Transactions on Very Large Scale Integration Systems*, Vol. 16, No. 11, pp. 1488–1498, 2008.
[9] S-C. Lin and K. Banerjee, "Cool chips: Opportunities and implications for power and thermal management", *IEEE Transactions on Electron Devices, Special Issue on Device Technologies and Circuit Techniques for Power Management*, Vol. 55, No. 1, pp. 245–255, 2008.

[10] Y. Taur and T. H. Ning, *Fundamentals of Modern VLSI Devices*, 2nd edn, Cambridge University Press, 2009.

[11] S. Mutoh, T. Douseki, Y. Matsuya, T. Aoki, S. Shigematsu, and J. Yamada, "1-V power supply high-speed digital circuit technology with multithreshold-voltage CMOS", *IEEE Journal of Solid-State Circuits*, Vol. 30, No. 8, pp. 847–854, 1995.

[12] K. Mistry, C. Allen, C. Auth, *et al.*, "A 45 nm logic technology with high-k+ metal gate transistors, strained silicon, 9 Cu interconnect layers, 193 nm dry patterning, and 100% Pb-free packaging", *Techn. Di. IEEE International Electron Devices Meeting (IEDM)*, p. 274, 2007.

[13] M. Bohr, "The evolution of scaling from the homogeneous era to the heterogeneous era", *Techn. Dig. IEEE International Electron Devices Meeting (IEDM)*, p. 1.1.1, 2011.

[14] J. Welser, J. L. Hoyt and J. F. Gibbons, *et al.*, "NMOS and PMOS transistors fabricated in strained silicon/relaxed silicon-germanium structures", *Techn. Dig. IEEE International Electron Devices Meeting (IEDM)*, p. 1000, 1992.

[15] T. Ghani, M. Armstrong, C. Auth, *et al.*, "A 90 nm high volume manufacturing logic technology featuring novel 45 nm gate length strained silicon CMOS transistors", *Techn. Dig. IEEE International Electron devices Meeting (IEDM)*, p. 978, 2003.

[16] D. Hisamot, W. C. Lee, J. Kedzierski, *et al.*, "A folded-channel MOSFET for deep-sub-tenth micron era", *Techn. Dig. IEEE International Electron devices Meeting (IEDM)*, p. 1032, 1998.

[17] X. Huang, W. C. Lee, C. Kuo, D. Hisamoto, L. Chang, J. Kedzierski, E. Anderson, H. Takeuchi, Y.-K. Choi, K. Asano, V. Subramanian, T.-J. King, J. Bokor, and C. Hu, "Sub 50-nm FinFET: PMOS", *Techn. Dig. IEEE International Electron Devices Meeting (IEDM)*, pp. 67–70, 1999.

[18] K. Kim, "From the future Si technology perspective: Challenges and opportunities", *Techn. Dig. IEEE International Electron Devices Meeting (IEDM)*, p. 1.1.1, 2010.

[19] S. H. Rasouli, H. F. Dadgour, K. Endo, H. Koike, and K. Banerjee, "Design optimization of FinFET domino logic considering the width quantization property", *IEEE Transactions on Electron Devices*, Vol. 57, No. 11, pp. 2934–2943, 2010.

[20] E. Karl, Y. Wang, Y. G. Ng, *et al.*, "A 4.6GHz 162Mb SRAM design in 22 nm tri-gate CMOS technology with integrated active VMIN-enhancing assist circuitry", *Dig. of Techn. IEEE International Solid-State Circuits Conference (ISSCC)*, p. 230, 2012.

[21] C. Auth, C. Allen, A. Blattner, *et al.*, "A 22 nm high performance and low-power CMOS technology featuring fully-depleted tri-gate transistors, self-aligned contacts and high density MIM capacitors", *Dig. Techn. P., VLSI Technology Symposium*, p. 131, 2012.

[22] K. Cheng, A. Khakifirooz, P. Kulkarni, *et al.*, "*Extremely thin SOI (ETSOI) CMOS with record low variability for low power system-on-chip applications*", *Techn. Dig. IEEE International Electron Devices Meeting (IEDM)*, 2009.

[23] S. Borkar, T. Karnik, S. Narendra, J. Tschanz, A. Keshavarzi, and V. De, "Parameter variations and impact on circuits and microarchitecture", in *Proc. Design Automation Conference*, pp. 338–342, 2003.

[24] T. Tsunomura, A. Nishida, F. Yano, *et al.*, "Analyses of 5σ Vth fluctuation in 65nm-MOSFETs using Takeuchi plot", *Dig. Techn. P. Symposium on VLSI Technology*, p. 156, 2008.

[25] H. F. Dadgour, S.-C. Lin and K. Banerjee, "A statistical framework for estimation of full-chip leakage-power distribution under parameter variations", *IEEE Transactions on Electron Devices*, Vol. 54, No. 11, pp. 2930–2945, 2007.

[26] S. Mukhopadhyay, K. Kim and C.-T. Chuang, *et al.*, "Device design and optimization methodology for leakage and variability reduction in sub-45-nm FD/SOI SRAM", *IEEE Transactions on Electron Devices*, Vol. 55, No. 1, p. 152, 2008.

[27] D. Burnett, K. Erington, C. Subramanian, and K. Baker, "Implications of fundamental threshold voltage variations for high-density SRAM and logic circuits", *Dig. Techn. P. Symposium on VLSI Technology*, p. 15, 1994.

T.J. Rogers, a student at Stanford University, …   … invents the vertical MOS (VMOS) power transistor   ELECTRON DEVICES SOCIETY®   Steve Jobs and Steve Wozniak bring the Apple II computer to market

[28] H. F. Dadgour, K. Endo, V. De and K. Banerjee, "Grain-orientation induced work-function variation in nanoscale metal-gate transistors – Part I: Modeling, analysis, and experimental validation", *IEEE Transactions on Electron Devices*, Vol. 57, No. 10, pp. 2504–2514, 2010.

[29] H. F. Dadgour, K. Endo, V. De and K. Banerjee, "Grain-orientation induced work-function variation in nanoscale metal-gate transistors – Part II: Implications for process, device, and circuit design", *IEEE Transactions on Electron Devices*, Vol. 57, No. 10, pp. 2515–2525, 2010.

[30] N. Singh, K. D. Buddharaju, S. K. Manhas, A. Agarwal, S. C. Rustagi, G. Q. Lo, N. Balasubramanian, and D.-L. Kwong, "Si, SiGe nanowire devices by top–down technology and their applications", *IEEE Transactions on Electron Devices*, Vol. 55, no. 11, pp. 3107–3118, 2008.

[31] R. Gandhi, Z. Chen, N. Singh, K. Banerjee, and S. Lee, "Vertical Si-nanowire n-type tunneling FETs with low subthreshold swing ($\leq$ 50 mV/decade) at room temperature", *IEEE Transactions on Electron Devices*, Vol. 32, No. 4, pp. 437–439, 2011.

[32] R. Gandhi, Z. Chen, N. Singh, K. Banerjee, and S. Lee, "CMOS compatible vertical silicon nanowire gate-all-around p-type tunneling FETs with $\leq$50 mV/decade subthreshold swing", *IEEE Transactions on Electron Devices*, Vol. 32, No. 11, pp. 1504–1506, 2011.

[33] K. Banerjee and A. Mehrotra, "A power-optimal repeater insertion methodology for global interconnects in nanometer designs", *IEEE Transactions on Electron Devices*, Vol. 49, No. 11, pp. 2001–2007, 2002.

[34] N. Magen, A. Kolodny, U. Weiser, and N. Shamir, "Interconnect-power dissipation in a microprocessor", *in Proceedings of the International Workshop on System Level Interconnect Prediction* (SLIP '04), pp. 7–13, February 2004.

[35] K. Banerjee, *NSF Workshop on Emerging Technologies for Interconnects (WETI)*, Available at http://weti.cs.ohiou.edu/ (accessed October 16, 2012), February 2–3, 2012.

[36] K. Banerjee and A. Mehrotra, "Analysis of on-chip inductance effects for distributed RLC interconnects", *IEEE Trans. Computer-Aided Design*, Vol. 21, No. 8, pp. 904–915, 2002.

[37] D. Edelstein, J. Heidenreich, R. Goldblatt, *et al.*, "Full copper wiring in a sub-0.25 μm CMOS ULSI technology", *in IEDM Tech. Dig.*, pp. 773–776, 1997.

[38] S. Venkatesan, A. V. Gelatos, S. Hisra, *et al.*, "A high performance 1.8 V, 0.20 μm CMOS technology with copper metallization", *Techn. Dig. IEEE International Electron Devices Meeting (IEDM)*, pp. 769–772, 1997.

[39] K. Banerjee, A. Amerasekera, G. Dixit, and C. Hu, "The effect of interconnect scaling and low-k dielectric on the thermal characteristics of the IC metal", in *Techn. Dig. IEEE International Electron Devices Meeting (IEDM)*, pp. 65–68, 1996.

[40] M. Tada, N. Inoue, and Y. Hayashi, "Performance modeling of low-k/Cu interconnects for 32-nm-node and beyond", *IEEE Transactions on Electron Devices*, Vol. 56, No. 9, pp. 1852–1861, 2009.

[41] K. Banerjee and A. Mehrotra, "Global (interconnect) warming", *IEEE Circuits and Devices Magazine*, Vol. 17, No. 5, pp. 16–32, September, 2001.

[42] N. Srivastava and K. Banerjee, "Interconnect challenges for nanoscale electronic circuits", *TMS Journal of Materials*, Vol. 56, No. 10, pp. 30–31, 2004.

[43] H. Li, C. Xu, N. Srivastava, and K. Banerjee, "Carbon nanomaterials for next generation interconnects and passives: Physics, status, and prospects", *IEEE Trans. Electron Devices*, Vol. 56, no. 9, pp. 1799–1821, 2009.

[44] *ISSCC Trends Report, ISSCC*, 2012.

[45] S. Lai, "Non-volatile memory technologies: The quest for ever lower cost", *Techn. Dig. IEEE International Electron Devices Meeting (IEDM)*, p. 1, 2008.

[46] H. Kanaya, K. Tomioka, T. Matsushita, *et al.*, "A 0.602 μm^2 nestled 'Chain' cell structure formed by one mask etching process for 64 Mbit FeRAM", *Dig. Techn. P., Symposium on VLSI Technology*, p 150, 2004.

[47] B. Dieny, R. Sousa, S. Bandiera, *et al.*, "Extended scalability and functionalities of MRAM based on thermally assisted writing", *Techn. Dig. IEEE International Electron Devices Meeting (IEDM)*, p 1.3.1, 2011.

[48] A. Driskill-Smith, S. Watts, D. Apalkov, *et al.*, *Non-volatile Spin-Transfer Torque RAM (STT-RAM): An analysis of chip data, thermal stability and scalability*, Non-Volatile Memories Workshop, p 1. Available at http://nvmw.ucsd.edu/2010/documents/Driskill-Smith_Alexander.pdf (accessed October 16, 2012), April 2010.

[49] K. Kinoshita, T. Tamura, H. Aso, *et al.*, "New model proposed for switching mechanisms of ReRAM", *Proceedings Non-Volatile Semiconductor Memory Workshop*, p 84, 2006.

[50] S. Lai, "Current status of the phase change memory and its future", in *Techn. Dig. IEEE International Electron Devices Meeting (IEDM)*, p. 255, 2003.

[51] K. Banerjee, S. J. Souri, P. Kapur, and K. C. Saraswat, "3-D ICs: A novel chip design for improving deep-submicrometer interconnect performance and systems-on-chip integration", *Proceedings of IEEE*, Vol. 89, pp. 602–633, May 2001.

[52] N. Sillon, A. Astier, H. Boutry, L. Di Cioccio, D. Henry, and P. Leduc, "Enabling technologies for 3D integration: From packaging miniaturization to advanced stacked ICs", in *Techn. Dig. IEEE International Electron Devices Meeting (IEDM)*, pp. 595–598, 2008.

[53] J. Van Olmen, A. Mercha, G. Katti, *et al.*, "3D stacked IC demonstration using a through silicon via first approach", in *Techn. Dig. IEEE International Electron Devices Meeting (IEDM)*, pp. 603–606, 2008.

[54] C. Xu, H. Li, R. Suaya, and K. Banerjee, "Compact AC modeling and performance analysis of through-silicon vias in 3-D ICs", *IEEE Trans. Electron Devices*, Vol. 57, No. 12, pp. 3405–3417, December 2010.

Chapter 12

Mixed-Signal Technologies and Integrated Circuits

Bin Zhao and James A. Hutchby

12.1 Introduction

Since the invention of integrated circuits (ICs) in the late 1950s, semiconductor technologies and various related ICs have enabled many electronic products which are essential in our daily life. These include not only microprocessors and memory chips but also wireless/mobile communication devices, portable multimedia devices, high-definition displays, entertainment systems, home appliances, control systems in automobiles, and medical instruments. A key enabler for these broad applications are the mixed-signal technologies and ICs which enable large scale integration of non-digital functions, such as analog/radio-frequency (RF) signal processing, data conversion between analog and digital functions, power management, sensors and actuators, to maximize value in diversified functionality, performance, cost, power efficiency, and convenience of use. As traditional digital CMOS scaling is approaching fundamental limits, mixed-signal technologies and ICs have more profound influence on the semiconductor industry and create new opportunities for growth (see also Chapter 14).

Figure 12.1 shows the functional blocks of a typical wireless/mobile communication system which incorporates a series of analog and digital signal processing functions. In the signal transmission path, signal

Guide to State-of-the-Art Electron Devices, First Edition. Edited by Joachim N. Burghartz.
© 2013 John Wiley & Sons, Ltd. Published 2013 by John Wiley & Sons, Ltd.

| International Rectifier Co. introduces a 400V, 25A power MOSFET | First demonstration of 0.1-μm MOS | ELECTRON DEVICES SOCIETY® | Alpha particles found to cause soft error rate (SER) in DRAMs | Vertical diffused MOSFET (DMOSFET) is invented |

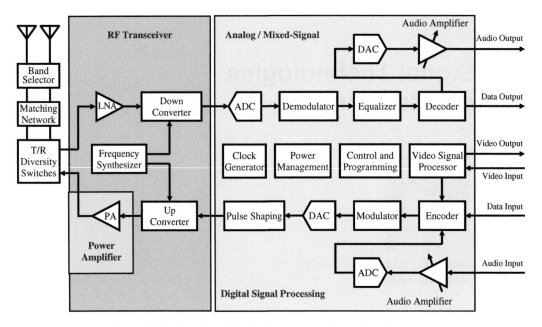

Figure 12.1   Functional blocks in a typical wireless/mobile system

encoding and modulation are performed in the digital domain after the original analog signal is converted into digital format by an analog-to-digital converter (ADC). The signal is then translated to the analog/RF domain and launched into a communication channel. In the signal receiving path, the analog/RF signal received from the communication channel is converted into digital format by an ADC followed by signal demodulation and decoding in the digital domain. The signal is then converted to the analog domain by a digital-to-analog converter (DAC) to recover the original information. These sophisticated signal processing functions are employed to realize efficient and effective communications, such as channel and user management, data compression and decompression, as well as to overcome various deficiencies in the communication channels, such as signal attenuation, noise, interference, and signal distortion. For any communication system, the classical *Shannon Information Capacity Theorem* (see sidebar) [1] shows that there exists a theoretical upper bound in the communication rate. Design and development of a communication system are basically the engineering required to maximize the information transfer rate and make it approach the Shannon limit by various means of analog and digital signal processing at the transmitters and the receivers.

Shannon Information Capacity Theorem

For a given communication channel, the error-free maximum information transfer rate C is:

$$C = B Log_2 \left(1 + S/N\right)$$

where B is the allocated channel bandwidth and S/N is the signal-to-noise ratio at the receiver.

Shannon's theorem shows two fundamental limitations in any communication system: channel bandwidth and noise.

Today, communication technologies are leading a revolution in the world both economically and socially. There are two factors which have made this revolution a reality. The first is the advance in digital signal processing (DSP), which, together with the data conversion technology, has made it possible to perform various required signal processing functions in the digital domain rather than perform them in the analog domain. DSP makes signal processing much more reliable and repeatable. By taking advantage of digital logic IC technology scaling and advancement, DSP has made these signal processing operations quicker, cheaper, and integrated on a single chip. The other factor, making the modern communications revolution possible, is the development of various advanced modulation techniques for communications. These advanced modulation techniques are the foundation of many communication standards and protocols, which have made communication more efficient, affordable, and manageable. Similarly, in consumer electronics, data compression, image/video signal processing, displays, and intelligent control are all dependent on advanced DSP techniques to provide the required features and performance. All these require mixed-signal technologies and circuits which have enabled the critical interfaces between the real world of analog signals and the processors/controllers of digital signals.

Analog/Mixed-Signal IC Technologies

Different from the mainstream Si digital logic and memory ICs (μP, DRAM, Flash, ASIC), the correlation between device feature size and circuits' gures-of-merit is weak for analog/mixed-signal ICs. Speed and density are usually not the only major considerations while precision, linearity, power ef ciency are more critical and demanding. In many applications, trade-offs must be made in technology development and circuit design in analog/mixed-signal ICs.

12.2 Analog/Mixed-Signal Technologies in Scaled CMOS

Advancements of analog/mixed-signal IC and related VLSI technologies are key factors for continued semiconductor market growth driven by consumer electronics, wireless/mobile communications, and other applications. In comparison to digital logic and memory ICs, device feature size scaling is not the main driving force in analog/mixed-signal ICs and related technologies. Instead, they are driven by many application aspects, such as signal accuracy and fidelity, feature sets, communication carrier frequency, channel bandwidth, modulation schemes, coding algorithm, appropriate (not necessarily the best) speed performance, power consumption, sizes of the boards and end-products, manufacturability, and cost. These key considerations lead to some unique requirements in analog/mixed-signal ICs and VLSI technologies. Furthermore, analog/mixed-signal ICs depend on many different materials and devices to provide optimal solutions. The devices include not only silicon (Si)-based CMOS and bipolar transistors but also compound semiconductor devices in some applications.

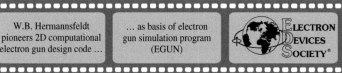

| W.B. Hermannsfeldt pioneers 2D computational electron gun design code … | … as basis of electron gun simulation program (EGUN) | ELECTRON DEVICES SOCIETY® | Invention of compact disc (CD) at Philips and Sony |

An industry-driven technology roadmap for analog/mixed-signal and RF ICs has become quite useful to forecast future technology requirements based on industry consensus. The roadmap helps to accelerate technology development and deployment because resources and investments are better guided and focused. Furthermore, the roadmap helps identify technical challenges and presents them to the research community for innovative solutions. All these aspects significantly assist individual semiconductor companies as well as the industry to prioritize investments and resources, to set up development plan, and to predict and execute for growth. However, it is very difficult to develop a roadmap for analog/mixed-signal and RF IC technologies because of the diversity in related products, systems, standards, and protocols. Consideration of cost versus performance determines what materials (Si, SiGe, GaAs, InP, GaN, etc.) and what device structures (CMOS, bipolar, LDMOS, HBT, HEMT, etc.) can be used for optimum solutions (see Chapters 1, 2, 14, 15). Other technologies such as Micro-Electro-Mechanical Systems (MEMS), bulk acoustic wave (BAW) devices, and integrated passive components need to be considered as well (see Chapters 4 and 18).

By expanding the initial work in the analog/mixed-signal design roadmap [2], focusing on wireless communications, and using a "divide and conquer" method [3], a dedicated roadmap on analog/mixed-signal and RF IC technologies for wireless communications has been developed and incorporated in the 2003 International Technology Roadmap for Semiconductors (ITRS) [4, 5]. Key objectives and achievements for development of this roadmap include identifying technical challenges and requirements for analog/mixed-signal and RF IC technologies in wireless communications with carrier frequencies between 0.8–100 GHz. They also include a study of the intersection and tradeoffs between Si-based devices, compound semiconductor devices, and other non-conventional technologies, such as MEMS and BAW. As an example, Figure 12.2 shows a portion of the developed roadmap for analog/mixed-signal IC technologies which shows the critical device parameters and highlights key technology characteristics. Since its first release in 2003, this roadmap has been updated on a yearly basis by a team of experts from both industry and academia; the latest edition can be found in [6].

In comparison to RF and power amplifier (PA) circuits [7–13], analog/mixed-signal circuits typically process signals at a relatively low frequency. As the speed of digital signal processing increases with the advance of digital technology, the operational frequency of analog circuits increases in certain applications. Signal swing, headroom, distortion, settling time, noise, offset errors, etc. are the performance concerns for analog/mixed-signal ICs, which determine the technology requirements of analog/mixed-signal devices [14–16]. Analog/mixed-signal technologies are often based on mainstream digital CMOS with some process tweaks to provide relatively higher-voltage analog transistors of high precision, good matching, and adequate speed [17–20]. High-quality, high-density, and highly linear integrated passive elements, signal isolation technology, and other active devices, such as bipolar and input/output (I/O) devices, are added [21–26]. Because of their intrinsic advantages in gain, noise, and matching, bipolar devices continue to remain a viable choice for some analog/mixed-signal IC applications [27–33]. The I/O transistors operating at a relatively high voltage need to be integrated seamlessly to provide appropriate interfaces at the required speed. Transconductance, 1/f noise, leakage current, and matching are the key device performance parameters for many analog/mixed-signal IC applications. High-quality and low-parasitic on-chip integrated passive components, such as capacitors, varactors, resistors, and inductors, are critical elements for various analog/mixed-signal circuit building blocks used in different circuit configurations (see Chapter 4). In addition, device breakdown voltage and substrate isolation are important requirements.

| First publication of an InP-based VCSEL operating at Room temperature ... | ... by H. Soda, K. Iga, C. Kitahara and Y. Suematsu | ELECTRON DEVICES SOCIETY® | Takashi Mimura invents the high-electron mobility transistor (HEMT) | Margret Thatcher becomes first woman prime minister of Britain |

		90	80	70	65	57	50	45
DRAM ½ Pitch (nm)		90	80	70	65	57	50	45
MPU / ASIC ½ Pitch (nm)		107	90	80	70	65	57	50
MPU Printed Gate Length(nm)		65	53	45	40	35	32	28
MPU Physical Gate Length(nm)		45	37	32	28	25	22	20
Physical Lgate (Low Operating Power)		65	53	45	37	32	28	25
Minimum Supply Voltage	Digital Design (V)	1	1	0.95	0.9	0.85	0.8	0.75
	Analog Design (V)	3.3 - 1.8	2.5 - 1.8	2.5 - 1.8	2.5 - 1.8	2.5 - 1.8	2.5 - 1.8	2.5 - 1.8
nMOS Analog Speed Device	T_{ox} (nm)	1.2 - 1.8	1.1 - 1.6	1.1 - 1.6	1.1 - 1.6	0.7 - 1.2	0.7 - 1.2	0.7 - 1.2
	g_m/g_{ds} @ $5.L_{min\text{-}digital}$	100	100	100	100	100	100	100
	1/f-noise ($\mu V^2.\mu m^2 / Hz$)	300	200	200	200	150	150	150
	σV_{th} matching (m V.μm)	5	4	4	4	3	3	3
nMOS Analog Precision Device	T_{ox} (nm)							5 - 3
	Analog V_{th} (V)							0.3 - 0.2
	g_m/g_{ds} @ $10.L_{min\text{-}digital}$							300
	1/f Noise ($\mu V^2.\mu m^2 / Hz$)							200
	σV_{th} matching (m V.μm)							9
Analog Capacitor	Density ($fF/\mu m^2$)							4
	Voltage linearity (ppm / V^2)							< 100
	Leakage ($fA / [pF.V]$)							7
	σ Matching (%.μm)							0.5
Analog Resistor	Parasitic capacitance ($fF/\mu m^2$)							0.1 - 0.02
	Temp. linearity ($ppm / °C$)							30 - 60
	1/f-current-noise per current^2 ($1/[\mu m^2.Hz]$)							$> 6 \times 10^{-19}$
	σ Matching (%.μm)							2
Bipolar Analog Device	g_m/g_{ce} @ $W_{e\text{-}min}^*$							1100
	1/f-noise ($\mu V^2.\mu m^2/Hz$)	5	3	3	3	2	2	2
	σ current Matching (%.μm^2)	20	20	20	20	20	20	20

Overlay box (obscuring center of table):

✓ Focus on analog/mixed-signal applications
✓ MOS device scaling with relatively high voltage
✓ nMOS – speed
✓ nMOS – precision
✓ On – chip capacitors and resistors
✓ Bipolar for analog functions
✓ Precision and matching performance
✓ 1/f noise
✓ capacitor nonlinearity
✓ leakage

Figure 12.2 Mixed-signal technology evolution along with CMOS scaling

12.3 Data Converter ICs

Data converters have played a critical and enabling role in mixed-signal communications and consumer electronics. Although the data conversion concept and practice can be traced back to the sixteenth-century, the emergence of electronic data converters has been primarily driven by the development of telephone systems. About 70 years after Bell's invention of the telephone in 1875, it was eventually realized in the 1940s that pulse code modulation (PCM) along with time division multiplexing (TDM) was much more effective to increase communication capacity in telephone systems than the analog technique using frequency division multiplexing (FDM). PCM is a digital communication technique which provided telephone systems with much improved immunity to noise and distortion. It involved signal sampling, quantization using an ADC, data transmission over the channel, and signal reconstruction from the quantized data using a DAC. Along with the PCM telephone system's development in the 1950s and deployment in the 1960s, data converters also started to find their usage in digital computing, voltage measurement, industrial process control, and military applications. The early standalone commercial ADCs were vacuum-tube based and, thus, very bulky and expensive. In addition, they consumed a lot of energy.

Following invention of the transistors and, later, the integrated circuit, electronic circuit design began to switch from vacuum tubes to transistors in the late 1950s and 1960s. Data converters based on solid-state discrete devices and building block ICs started to emerge with much reduced cost, size and power

consumption in the same time period. Although data converters in the 1970s initially appeared in multi-chip modules and/or packages, advancement in device technology and process integration along with innovations in circuit design eventually led to single-chip monolithic data converters in the second half of the 1970s. These monolithic data converters included a 10-bit SAR ADC, a high-speed flash ADC offering 8 bits running at 30-MSPS (mega samples per second), and a 10-bit current-mode R-2R DAC [34–36]. In the 1980s, data converters had high growth driven by applications such as instrumentation, data acquisition, medical imaging, audio and video, and computer graphics. For example, ADCs with 8–10 bit resolution running at 20–100 MSPS emerged to support digital video. Voice and audio signal processing required 16–18 bit ADCs and DACs to support applications like compact disk (CD) players. A monolithic Σ-Δ ADC came to the market in 1988, which provided 16-bit resolution at 20 kSPS (kilo samples per second) for voice signal digitization [37]. Meanwhile, along with rapid development of CMOS technology, integrated monolithic data converters started to shift their process technology from pure bipolar to BiCMOS and CMOS for more integrated functions, low power, and low cost.

From the mid 1990s, high speed internet and wireless communication have become the major forces to drive development of low-cost, low-power, and high-performance data converters in applications like modems, cell phones handsets, and base-stations. Dynamic performance of the data converters became important due to the increase in frequency-domain signal processing backed by modern communication techniques for high speed and high capacity. Lower power pipelined subranging ADCs became prevalent and the higher power flash ADCs became a building block in the pipelined ADCs. CMOS became the preferred technology for general-purpose data converters while BiCMOS was for high-end products. The Σ-Δ architectures were popular in audio and high precision applications due to their much higher oversampling ratio, relaxed output filter requirement, high dynamic range, and lower distortion. CMOS is especially suitable for Σ-Δ architectures in data conversion. In recent years, power dissipation continues to drop as well as the power supply voltages in data converters. Data converters with power supply at 5 V, 3.3 V, 2.5 V, and 1.8 V, 1.2 V, and sub-1V have been developed as the CMOS device geometry shrunk to 0.35 µm, 0.25 µm, 0.18 µm, 0.13 µm, 90 nm, 65 nm, and 45 nm.

Data converters have been highly diversified – their development and evolution history can be found in [38–40] and references therein. For example, ADCs were developed with a wide variety of resolutions, sampling rates, power and cost levels – each has its own merit for

A Figure-of-Merit for ADC

This FoM, energy per conversion step, for ADC is de ned as:

$$FoM = P / \left(f_{sample} \times 2^{ENOB} \right)$$

where P is the ADC power consumption, f_{sample} is the ADC sampling rate (number of samples taken by ADC per unit time), and ENOB is the effective number of bits and it is related to ADC's signal-to-noise-and-distortion ratio (SNDR) by:

$$ENOB = (SNDR - 1.76)/6.02$$

ENOB and SNDR indicate the dynamic performance of an ADC in real applications.

| 1980s | | Focus on manufacturing | | 1980s |

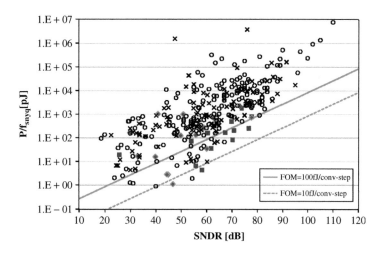

Figure 12.3 ADC conversion energy versus achieved dynamic performance (SNDR) [40]

certain applications. Although comparing different ADCs is a difficult task, a figure-of-merit (FoM) for ADC, energy per conversion step, has been used to compare various ADCs and track their progress and evolution [39]. Although, as shown in Figure 12.3 [40], a FoM of sub-10 pJ/conversion-step has been achieved lately, most state-of-the-art ADCs have this FoM in the range of 100–1000 pJ/conversion-step.

As frequency of the input signal increases, the overall non-idealities in an ADC increase; for example, signal distortion becomes worse at high input frequency. This degrades ADC's dynamic performance, such as SNDR (signal-to-noise-and-distortion ratio) or ENOB (effective number of bits), at high input frequency. Figure 12.4 [40] illustrates this tradeoff from published ADC performance data and shows the theoretical

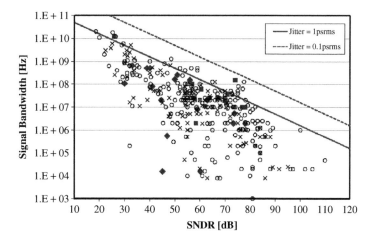

Figure 12.4 ADC input signal bandwidth versus achieved dynamic performance (SNDR) [40]

| Direct step-and-repeat lithography comes up | Plasma processing emerges | ELECTRON DEVICES SOCIETY[a] | Invention of metal organic vapor phase epitaxy (MOVPE) | Silicided contacts become part of CMOS process technology |

limit of the tradeoff at sampling aperture jitter of 1 ps (rms) and 0.1 ps (rms), respectively. As the technology is improved in device precision and matching, innovation in circuit topology and new design techniques continue to push data converters to approach their theoretical performance limits.

12.4 Mixed-Signal Circuits for Low Power Displays

Along with the continued advances in IC technologies, much increased performance and more diversified feature sets have been realized in related electronic products. Power consumption has emerged as a critical challenge in many applications under the constraints of energy conservation, chip/system thermal limitation, and product usage time before recharging.

Flat Panel Displays are now a vital human-machine visual interface in many electronic products, such as televisions (TVs), desktop computers, laptops, smartphones, and public displays, as a result of the advancement in liquid crystal display (LCD) and plasma display technologies (see Chapter 17). Because of its advanced features and application benefits, LCD is the preferred technology in both portable applications and large-screen displays. Figure 12.5 shows a typical LCD display system. After decoding or processing of the input video data, display data is further processed in digital domain before they are sent to drive the display apparatus. These processing functions include: image scaling, de-interlacing (even/odd fields in SD/HDTV combined for one frame), temporal noise reduction, motion compensation, and cross color reduction, OSD (on screen display), PIP (picture in picture), various image enhancement (adaptive contrast control and adaptive color management), Gamma correction (to covert the input video data to the corresponding output data suitable for human eyes), response time compensation (to compensate motion blur caused by slow response time of LCD panel), and so on. Then, the display and synchronization data are

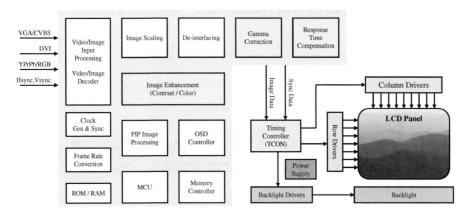

Figure 12.5   LCD display system

sent to the timing controller (TCON) to generate the control signals for LCD driver and backlight apparatus to display the video.

The backlight apparatus usually takes a majority of the power dissipation in a LCD display because LCD is a passive display technology. Tremendous backlight (energy) is wasted in the liquid crystal configuration due to light blocking (polarizers) and color filtering for image formation. If uniform and time-constant backlighting is used, only the liquid crystal provides the image contrast. In order to save energy and provide improved image contrast, non-uniform and time-varying backlighting (local dimming) can be employed as shown in Figure 12.6.

Light emitting diodes (LEDs) often are used for backlighting sources in LCD and other displays, where LEDs are arranged as parallel strings driven by a common voltage source and each LED string has an equal number of LEDs connected in series (see Chapter 20 on LEDs). To provide consistent light output between the LED strings, each string typically is driven at a regulated current of equal magnitude. Because of process variation in fabrication and manufacturing of the LEDs, there often is considerable variation in the bias voltages needed to drive each LED string at the regulated current due to variation in forward-voltage drop of every individual LED. Pulse width modulation (PWM) is usually used to control backlight intensity by controlling the PWM duty cycle. When the local dimming technique is used, not all LED strings are on at the same time. If the supply voltage is fixed to meet the requirement for the LED string of maximum voltage drop, there is a large LED tail voltage for the LED string of smaller voltage drop, which results in unnecessary power dissipation and thermal issue to the LED driver. Figure 12.7 shows a smart LED driver IC where it monitors the tail voltages of the active LED strings to identify the minimum tail voltage and adjusts the output voltage of the voltage source based on the lowest tail voltage [41].

The smart LED driver adjusts the output voltage so as to maintain the lowest tail voltage at a predetermined non-zero threshold voltage and to ensure that the output voltage is sufficient to properly drive each active LED string with a regulated current in view of pulse width modulation (PWM) timing requirements without excessive power consumption. To ensure fast response time and control loop stability of the backlighting, a mixed-signal control technique is used by employing ADC, DAC, and control intelligence enabled by digital signal processing.

Uniform Backlighting   Displayed Image   Non-Uniform Backlighting

Figure 12.6 Local dimming in LCD display system

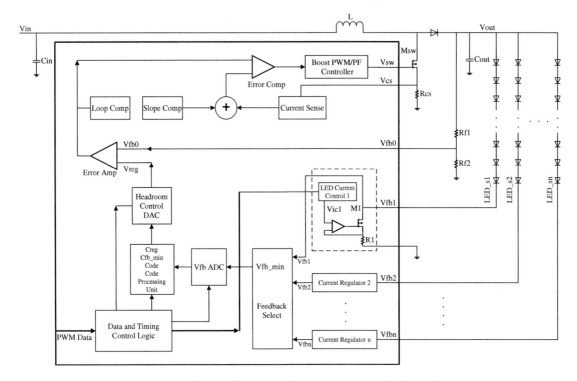

Figure 12.7 Smart LED driver IC for low-power LCD display system

12.5 Image Sensor Technologies and Circuits

Since the first consumer digital camera with 250,000 pixels came to the market in 1995, solid-state image sensors have experienced explosive growth and supported the continuously increased demands in digital still camera, digital video, mobile imaging, video conferencing, surveillance, and security camcorders. As the sensor pixel count continues to grow, digital still cameras of 18 megapixels have been available in the market. Although charge-coupled devices (CCDs) have been the dominant image-sensor technology initially (see Chapter 20 on CCDs), CMOS technologies have become a viable alternative due to continued CMOS scaling and improvement in sensor design. Today, CMOS image sensors not only can be found in high-volume products, such as mobile phones and PC cameras where cost is critical, they also appear in high-end products, such as high-quality professional digital cameras.

As a critical element in a digital imaging system, an image sensor comprises a two-dimensional array of pixels which converts the light incident at its surface into an array of electrical signals. A color filter array (CFA) is usually formed in a red/green/blue/green pattern (Bayer CFA) on top of the image sensor pixel array. Photodiodes or other kinds of photodetectors are used to convert the light signal received by each pixel into electrical signals corresponding to each of the three colors (red, blue, and green).

These analog pixel signals are then read out from the image sensor and digitized by an ADC for further signal processing, such as white balancing, color correction and others to reduce the adverse effects of faulty pixels and imperfect optics.

In passive pixel sensors (PPS), a simple switch was utilized in the pixel to read out the photodiode-generated charge. The PPS suffered from lower performance and larger pixel size than CCDs. Conceived originally in the late 1960s, the active pixel sensors (APS), where the pixel contains an amplifier, started to be developed for modern CMOS image sensors in the 1990s [42, 43]. APS can overcome many short-comings in PPS, such as low speed and poor signal-to-noise ratio. Along with the CMOS technology scaled down to deep submicron size, the use of pinned photodiode technology, and the advance in micro lens technologies, APS has made CMOS image sensors achieve a level of performance that is comparable to, or exceeds, that of CCD in some aspects. Figure 12.8 shows a recent 33-megapixel 120-frames/s CMOS image sensor with high speed LVDS (low voltage differential signaling) data interface [44].

In a CCD image sensor, photodetector-generated charge is shifted out by using vertical and horizontal CCDs, converted into a voltage

CCD versus CMOS Image Sensors

A major difference between CCD and CMOS image sensors is in their image signal readout.

In a CCD image sensor, photodetector generated charge is shifted out by using vertical and horizontal CCDs, converted into voltage by a follower amplifier, and then read out serially.

In a CMOS image sensor, charge voltage signals are read out one row at a time using row and column select circuit, resulting in advantages in high speed and low power.

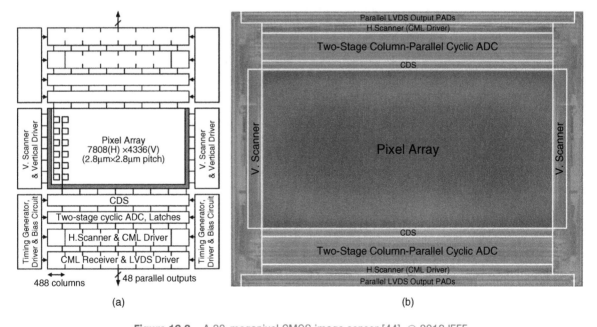

(a)   (b)

Figure 12.8 A 33-megapixel CMOS image sensor [44], © 2012 IEEE

First publication of monolithically integrated photoreceiver by R.F. Leheny and co-authors

Start of the IEEE Electron Devices Letters …

ELECTRON DEVICES SOCIETY®

… with George E. Smith of Bell Labs as Editor-in-Chief

IEEE ELECTRON DEVICE LETTERS

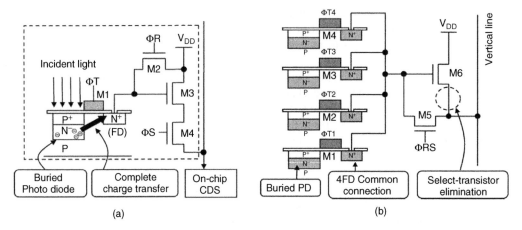

Figure 12.9   Pixel Configuration: (a) 4-transistor/pixel; (b) 1.5-transistor/pixel [46], © 2004 IEEE

signal by a follower amplifier, and then read out serially, resulting in low pixel overhead and small pixel size. Charge transfer is a passive process and, thus, introduces very little temporal noise and fixed-pattern noise (FPN). But, charge transfer readout is serial and, thus, limits the readout speed. It consumes high power because both high speed clock and high voltage are required for the charge transfer. In a CMOS image sensor, charge voltage signals are read out one row at a time using a row and column select circuitry, resulting in advantages in high speed and low power. Because the signal readout passes through active devices, it introduces temporal noise and FPN [45]. Techniques such as double/multiple sampling and delta reset sampling have been developed to reduce these limitations in CMOS imagers.

In applications with limited space for image sensors, such as high-end smart phones, small pixels are required to increase spatial resolution for high-quality images. CMOS image sensor pixel sizes have been reduced by taking advantage of CMOS technology scaling and using novel pixel structures that share some of the transistors in neighboring pixels. Figure 12.9(b) shows an example of 1.5-transistor (1.5T) per pixel on average [46]. In comparison, the conventional APS pixel uses 3T or 4T per pixel as shown in Figure 12.9(a). Along with CMOS scaling, technology issues need to be resolved specifically for image sensor applications. For example, shallow junctions and high doping led to low photoresponsivity. Shallow trench isolation and thin gate oxide led to high dark current. As the number of interconnet layers increases to support complex peripheral circuits on the same chip, the distance from the chip surface to the photodiode increases. This introduces issues in light collection, reduced efficiency, and signal distortion due to shadow effect. Several configurations of backside illumination have been developed to resolve these issues and to reduce pixel size [47–52].

References

[1] C. E. Shannon, "A Mathematical Theory of Communication", *The Bell System Technical Journal*, vol. 27, p. 379, July, p. 623, October, 1948.
[2] R. Brederlow, W. Werner, J. Sauerer, S. Donnay, P. Wambacq, and M. Vertregt, "A Mixed-Signal Design Roadmap", *IEEE Design & Test of Computers*, vol. 18, p. 34, 2001.

[3] B. Zhao, "Analog/Mixed-Signal and RF Integrated Circuit Technologies for Wireless Communications", *Proc. International Conference on Solid-State and Integrated-Circuit Technology*, vol. 2, p. 1220, 2004.

[4] "Radio Frequency and Analog/Mixed-Signal Technologies for Wireless Communications", *International Technology Roadmap for Semiconductors*, 2003 Edition, International SEMATECH, Austin, TX, 2003.

[5] H. S. Bennett, R. Brederlow, J. C. Costa, M. Huang, A. A. Immorlica, J.-E. Mueller, M. Racanelli, C. E. Weitzel, and B. Zhao, "Radio-Frequency and Analog/Mixed-Signal Circuits and Devices for Wireless Communications", *IEEE Circuits and Devices Magazine*, vol. 20, p. 38, 2004.

[6] "Radio Frequency and Analog/Mixed-Signal Technologies for Communications",*International Technology Roadmap for Semiconductors*, [Online]. Available at: http://www.itrs.net (accessed October 16, 2012).

[7] A. Matsuzawa, "RF-SoC – Expectations and Required Conditions", *IEEE Trans. Microwave Theory Tech.*, vol. 50, p. 245, 2002.

[8] T. H. Lee, H. Samavati, and H. R. Rategh, "5-GHz CMOS Wireless LANs", *IEEE Trans. Microwave Theory Tech.*, vol. 50, p. 268, 2002.

[9] B. Razavi, "RF CMOS Transceivers for Cellular Telephony", *IEEE Commun. Magazine*, vol. 41, p. 144, 2003.

[10] L. L. Larson, "Silicon Technology Tradeoffs for Radio-Frequency/Mixed-Signal Systems-on-a-Chip", *IEEE Trans. Electron Devices*, vol. 50, p. 683, 2003.

[11] A. A. Abidi, "RF CMOS Comes of Age", *IEEE Journal of Solid-State Circuits*, vol. 39, p. 549, 2004.

[12] C. E. Weitzel, "RF Power Amplifiers for Wireless Communications", *GaAs IC Symp. – Digest*, p. 127, 2002.

[13] A. M. Niknejad, D. Chowdbury, and J. Chen, "Design of CMOS Power Amplifiers", *IEEE Trans. Microwave Theory Tech.*, vol. 60, p. 1784, 2012.

[14] P. A. Stolk, H. P. Tuinhout, R. Duffy, *et al.*, "CMOS Device Optimization for Mixed-Signal Technologies", *IEDM Digest*, p. 215, 2001.

[15] A. Vandooren, *et al.*, "Mixed-Signal Performance of Sub-100nm Fully-Depleted SOI Devices with Metal Gate, High K (HfO$_2$) Dielectric and Evevated Source/Drain Extensions", *IEDM Digest*, p. 975, 2003.

[16] Y. Yasuda, T. K. Liu, and C. Hu, "Flicker-Noise Impact on Scaling of Mixed-Signal CMOS With HfSiON", *IEEE Trans. Electron Devices*, vol. 55, p. 417, 2008.

[17] K. Mikashita, *et al.*, "A High Performance 100 nm Generation SOC Technology [CMOS IV] for High Density Embedded Memory and Mixed Signal LSIs", *IEEE Symp. on VLSI Technology – Digest*, p. 11, 2001.

[18] T. Schafbauer, J. Brighten, Y. Chen, *et al.*, "Integration of High-performance, Low-leakage and Mixed Signal Features into a 100nm CMOS Technology", *IEEE Symp. on VLSI Technology – Digest*, p. 62, 2002.

[19] C. H. Diaz, D. D. Tang, J. Sun, "CMOS Technology for MS/RF SoC", *IEEE Trans. Electron Devices*, vol. 50, p. 557, 2003.

[20] M.-T. Yang, *et al.*, "RF and Mixed-Signal Performances of a Low Cost 28nm Low-Power CMOS Technology for Wireless System-on-Chip Applications", *IEEE Symp. on VLSI Technology – Digest*, p. 40, 2011.

[21] J. N. Burghartz and B. Rejaei, "On the Design of RF Spiral Inductors on Silicon", *IEEE Trans. Electron Devices*, vol. 50, p. 718, 2003.

[22] S. J. Kim, B. J. Cho, M.-F. Li, *et al.*, "Engineering of Voltage Nonlinearity in High-K MIM Capacitor for Analog/Mixed-Signal ICs", *IEEE Symp. on VLSI Technology – Digest*, p. 218, 2004.

[23] K. C. Chiang, C. H. Lai, A. Chin, *et al.*, "Very High-Density (23 fF/um^2) RF MIM Capacitors Using High-k TaTiO as the Dielectric", *IEEE Electron Device Lett.*, vol. 26, p. 728, 2005.

[24] C. H. Ng, C. S. Ho, S. S. Chu and S. C. Sun, "MIM Capacitor Integration for Mixed-Signal/RF Applications", *IEEE Trans. Electron Devices*, vol. 52, p. 1399, 2005.

[25] A. Afzali-Kusha, M. Nagata, N. K. Verghese, and D. J. Allstot, "Substrate Noise Coupling in SoC Design: Modeling, Avoidance, and Validation", *Proceedings of the IEEE*, vol. 94, p. 2109, 2006.

[26] S. Uemura, *et al.*, "Isolation Techniques Against Substrate Noise Coupling Utilizing Through Silicon Via (TSV) Process for RF/Mixed-Signal SoCs", *IEEE Journal of Solid-State Circuits*, vol. 47, p. 810, 2012.

IBM PC (Model 5150) is brought to market | Tak H. Ning invents and demonstrates ... | ELECTRON DEVICES SOCIETY® | ... the double-poly self-aligned bipolar transistor structure

[27] J. D. Cressler, "SiGe HBT Technology: A New Contender for Si-Based RF and Microwave Circuit Applications", *IEEE Trans. Microwave Theory Tech.*, vol. 46, p. 572, 1998.

[28] A. J. Joseph, D. L. Harame, B. Jagannathan, *et al.*, "Status and Direction of Communication Technologies – SiGe BiCMOS and RFCMOS", *Proceedings of the IEEE*, vol. 93, p. 1539, 2005.

[29] J. Long, "SiGe Radio Frequency ICs for Low-Power Portable Communication", *Proceedings of the IEEE*, vol. 93, p. 1598, 2005.

[30] P. Chevalier, F. Pourchon, T. Lacave, *et al.*, "A Conventional Double-Polysilicon FSA-SEG Si/SiGe:C HBT Reaching 400 GHz fmax", *Proc. IEEE Bipolar/BiCMOS Circuits and Technology Meeting*, p.1, 2009.

[31] S. Van Huylenbroeck, A. Sibaja-Hernandez, R. Venegas, *et al.*, "A 400 GHz fmax Fully Self-Aligned SiGe:C HBT Architecture", *Proc. IEEE Bipolar/BiCMOS Circuits and Technology Meeting*, p.5, 2009.

[32] M. J. W. Rodwell, M. Le, and B. Brar, "InP Bipolar ICs: Scaling Roadmaps, Frequency Limits, Manufacturable Technologies", *Proceedings of the IEEE*, vol. 96, p. 271, 2008.

[33] M. Zaknoune, E. Mairiaux, Y. Roelens, *et al.*, "480-GHz fmax in InP/GaAsSb/InP DHBT With New Base Isolation μ-Airbridge Design", *IEEE Electron Device Lett.*, vol. 33, p. 1381, 2012.

[34] P. Holloway and M. Norton, "A High-Yield Second-Generation 10-Bit Monolithic DAC", *ISSCC Digest of Technical Papers*, p. 106, 1976.

[35] A. P. Brokaw, "A Monolithic 10-Bit A/D Using I^2L and LWT Thin-Film Resistors", *IEEE Journal of Solid-State Circuits*, vol. 13, p. 736, 1978.

[36] J. G. Peterson, "A Monolithic Video A/D Converter", *IEEE Journal of Solid-State Circuits*, vol. 14, p. 932, 1979.

[37] "16-Bit, 20 kHz Oversampling A/D Converter", *Datasheet CSZ5316*, Crystal Semiconductor, Austin, TX, 1987.

[38] *The Data Conversion Handbook*, edited by Walt Kester, Analog Devices, 2005.

[39] R. H. Walden, "Analog-to-Digital Converter Survey and Analysis", *IEEE J. Select. Areas Commun.*, vol. 17, p. 539, 1999.

[40] B. Murmann, "ADC Performance Survey 1997–2012", [Online]. Available at: http://www.stanford.edu/~murmann/adcsurvey.html (accessed October 16, 2012).

[41] B. Zhao, J. W. Cornish, B. B. Horng, V. K. Lee, A. M. Kameya, "LED Driver with Dynamic Power Management", *US Patent 7,825,610*, 2010.

[42] P. Noble, "Self-Scanned Silicon Image Detector Arrays", *IEEE Trans. Electron Devices*, vol. 15, p. 202, 1968.

[43] E. R. Fossum, "CMOS Image Sensors: Electronic Camera-On-A-Chip", *IEEE Trans. Electron Devices*, vol. 44, p. 1689, 1997.

[44] K. Kitamura, *et al.*, "A 33-Megapixel 120-Frames-Per-Second 2.5-Watt CMOS Image Sensor with Column-Parallel Two-Stage Cyclic Analog-to-Digital Converters", *IEEE Trans. Electron Devices*, vol. 59, p. 3426, 2012.

[45] A. El Gamal, and H. Eltoukhy, "CMOS Image Sensors", *IEEE Circuits and Devices Magazine*, vol. 21, p. 6, 2005.

[46] H. Takahashi, M. Kinoshita, K. Morita, *et al.*, "A 3.9-um Pixel Pitch VGA Format 10-b Digital Output CMOS Image Sensor with 1.5 Transistor/Pixel", *IEEE Journal of Solid-State Circuits*, vol. 39, p. 2417, 2004.

[47] T. Joy, S. Pyo, S. Park, *et al.*, "Development of a Production-Ready, Back-Illuminated CMOS Image Sensor with Small Pixel", *IEDM Digest*, p. 1007, 2007.

[48] J. Ahn, C.-R. Moon, B. Kim, *et al.*, "Advanced Image Sensor Technology for Pixel Scaling Down Toward 1.0 μm", *IEDM Digest*, p. 1, 2008.

[49] V. Suntharalingam, R. Berger, S. Clark, *et al.*, "A 4-Side Tileable Back Illuminated 3D-Integrated Mpixel CMOS Image Sensor", *ISSCC Digest of Technical Papers*, p. 38, 2009.

[50] H. Wakabayashi, K. Yamaguchi, M. Okano1, *et al.*, "A 1/2.3-inch 10.3Mpixel 50frame/s Back-Illuminated CMOS Image Sensor", *ISSCC Digest of Technical Papers*, p. 410, 2009.

[51] S. G. Wuu, C. C. Wang, B. C. Hseih, *et al.*, "A Leading-Edge 0.9 μm Pixel CMOS Image Sensor Technology with Backside Illumination: Future Challenges for Pixel Scaling", *IEDM Digest*, p. 332, 2010.

[52] S. Lee, K. Lee, J. Park, *et al.*, "A 1/2.33-inch 14.6M 1.4 μm-Pixel Backside-Illuminated CMOS Image Sensor with Floating Diffusion Boosting", *ISSCC Digest of Technical Papers*, p. 416, 2011.

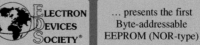

| 1982 | G. Yaron of National Semiconductors … | ELECTRON DEVICES SOCIETY® | … presents the first Byte-addressable EEPROM (NOR-type) | Yannis P. Tsividis makes the first benchmark tests for detecting problems with MOSFET models |

Chapter 13

Memory Technologies

Stephen Parke, Kristy A. Campbell and Chandra Mouli

13.1 Semiconductor Memory History

Early random-access memory (RAM) used in mainframe computers was magnetic core memory, incorporating tiny donut-shaped ferrite cores at the intersections of an X-Y array of copper wires (Figure 13.1).

Large semiconductor memories were foreseen by engineers and scientists at the very start of the integrated circuit era (see also chapter 3). But they could not have imagined that the semiconductor memory market would grow to $250 billion today! In 1960, Fairchild Semiconductor introduced the Micrologic family of ICs, and stated, "Until (size) reductions are possible in non-logic portions of a computer – especially the memory system – progress will be slow. There's not much point in a "pea-sized" logic chip with a "barrel-sized" memory" [1]. But, it would be 10 more years before magnetic core memories would be replaced by semiconductor memory chips. In 1961, Bob Norman from the Fairchild group proposed creating solid-state memory by putting multiple flip-flops on an IC, but it was still considered impractical at that time. In 1963, he patented a bipolar static random access memory (SRAM) cell that was later used by IBM [2, 3]. In October 1961, Texas Instruments (TI) delivered a Series 51 computer with "a few hundred bits" of semiconductor memory to the US Air Force [4]. In Gordon Moore's famous 1965 paper establishing Moore's Law [5], he predicted that "memories built of integrated electronics may be distributed throughout the (computer) instead of being concentrated in a central unit." Also in 1965, IBM Fellow Bob Heale wrote an internal report entitled, *Potential for Monolithic Megabit Memories*.

Later that year, Scientific Data Systems and Signetics partnered to produce a fully-decoded 8-bit bipolar SRAM [6], and IBM engineers Agusta, Bardell, and Castrucci developed the SP95 16-bit bipolar SRAM

Guide to State-of-the-Art Electron Devices, First Edition. Edited by Joachim N. Burghartz.
© 2013 John Wiley & Sons, Ltd. Published 2013 by John Wiley & Sons, Ltd.

The LIGA process for high aspect ratio metal MEMS structures ... | ... is invented at Karlsruhe Research Center in Germany | ELECTRON DEVICES SOCIETY® |  | Introduction of 1.5 μm silicon technology node in manufacturing

Figure 13.1   Magnetic core memory [9]. Reproduced with permission from http://commons.wikimedia.org/wiki/File: Magnetic_core.jpg. H.J. Sommer III, Professor of Mechanical Engineering, Penn State University

used in the IBM System/360 Model 95 for NASA [7]. In 1966, a team at Transitron, led by Tom Longo, built the TMC3162 16-bit TTL SRAM for the Honeywell Model 4200 minicomputer. This chip was also manufactured by Fairchild, Sylvania, and TI and was soon followed by 64-bit bipolar SRAMs. In 1969 the IBM East Fishkill, NY facility produced a 128-bit bipolar SRAM (Figure 13.2) for the 1971 shipment of System/370 Model 145, the company's first computer to employ semiconductor main memory instead of ferrite cores [8].

Figure 13.2   IBM 128-bit bipolar TTL SRAM for System/370 [2]

| The RESURF power MOSFET is introduced | VLSI Co. defines the "standard cell" in IC layout | **E**LECTRON **D**EVICES **S**OCIETY® | Kurt E. Peterson published his seminal paper "Silicon as Structural Material" in the Proceedings of IEEE | First IEDM held in San Francisco, first outside of Washington, D.C. |

In 1965, Sylvania produced a 256-bit bipolar TTL Read Only Memory (ROM) for Honeywell that was programmed one bit at a time by a skilled technician at the factory who physically scratched metal link connections to selected diodes [8]. Volume production orders were satisfied with custom-mask programming. By the early 1970s, Fairchild, Intel, Motorola, Signetics, and TI offered 1 kb TTL ROMs, while AMD, AMI, General Instrument, National, Rockwell and others produced 4 kb MOS ROMs.

In 1967, Andrew Bobeck of Bell Labs developed non-volatile magnetic bubble memory, which led several companies to develop this technology up to megabit densities by the mid-1970s before it succumbed to faster technologies [8].

The first commercial 256-bit TTL SRAM, the Fairchild 4100, (Figure 13.3) was used in the Burroughs Illiac IV computer in 1970, which was the first semiconductor main memory in a supercomputer [8]. The Cray 1 supercomputer introduced in 1976 used 65 000 Fairchild 1 kb ECL SRAM chips and a lot of power!

Dr. Robert H. Dennard at the IBM Thomas J. Watson Research Center, invented the one-transistor MOS dynamic RAM (DRAM) cell in 1967 (see sidebar "The One-Transistor DRAM Cell"). Dennard and his team, led by Dr. Dale Critchlow, were working on early field-effect transistors and integrated circuits. One day in fall 1966, his attention was captured after hearing another team's research presentation on memory. At that time, early DRAM cells contained three or more transistors. Dennard went to work, and within a few months had invented a simpler memory cell that used only a single transistor

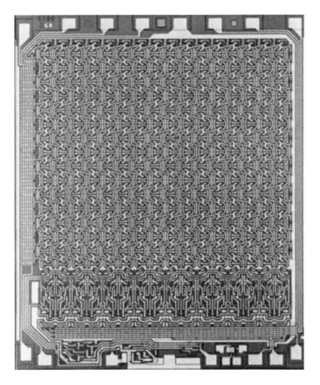

Figure 13.3 Fairchild 4100 256-bit commercial TTL SRAM [2]

and a single capacitor (1T1C) [9]. On June 4, 1968, US Patent #3,387,286 was issued to IBM for the MOS 1T1C DRAM cell [10]. MOS memories would soon beat bipolar memories in density and power.

Boyle and Smith of Bell Labs invented the Charge-Coupled Device (CCD) in 1969, for which they would receive the Nobel Prize in 2009 (see Chapters 12 and 20). The CCD, shown in Figure 13.4, was originally applied to memory technology, and by 1975 a 16-kb CCD memory had been achieved, but the introduction of fast NMOS 64-kb DRAMs in 1978 eclipsed these slower CCD serial access memories. The CCD was applied to imaging by Gil Amelio and Jim Early, who had moved from Bell Labs to Fairchild.

Radiation, Inc. (later Harris Semiconductor) introduced the first user-fuse-programmable PROM in 1969, which became popular because designers could easily debug and change their firmware.

In 1969, a new company named Intel introduced its first successful product, the 3101 Schottky TTL 64-bit bipolar SRAM shown in Figure 13.5, which was twice as fast as previous TTL SRAMs [8].

Intel's charter was to design, manufacture and market semiconductor memory components. The company wanted to replace core memories by producing low-cost, standardized chips in high volume. Later in 1969, Intel introduced the 1101, a 256-bit MOS SRAM. This was the world's first high-volume MOS semiconductor memory using silicon gate technology. The same year, Honeywell worked with Intel to develop the 1102, a 1 kb three-transistor cell dynamic RAM (DRAM). However, the 1102 had many problems, prompting Intel to begin secret work on their own improved design to avoid conflict with Honeywell. In 1970, this became the first commercially-available DRAM memory,

The One-Transistor DRAM Cell

The dynamic random access memory (DRAM) is based on this compact memory cell structure, invented by Robert Dennard in 1967, and consisting only of a MOS transistor switch and a capacitor for charge storage. Now, in one year, almost eight billion of these DRAM bits (=1GByte) are manufactured for every man, woman, and child in the world, with ten times more NAND flash bits manufactured!

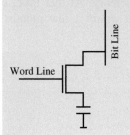

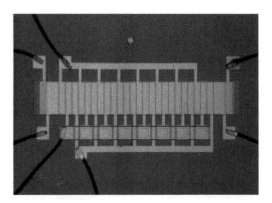

Figure 13.4   First 8-bit CCD memory [8]

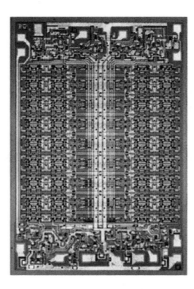

Figure 13.5   Intel 3101 64-bit Schottky TTL SRAM [2]

the Intel 1103 1 kb MOS DRAM shown in Figure 13.6 [8]. It proved that semiconductor memories were not only viable, but were a vast improvement over core memories. The introduction of the 1103 was a turning point in the history of integrated circuits. For the first time, a significant amount of information could be stored on a single chip. It began to replace core memories and was nicknamed "The Core Killer." In 1972, it was the largest selling semiconductor in the world. MOS SRAM also rapidly replaced much

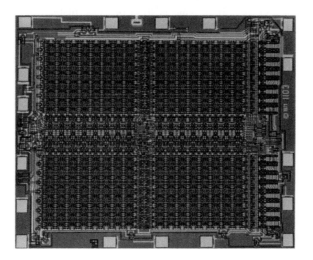

Figure 13.6   Intel 1103 1 kb DRAM "The Core Killer" [2]

| Initiation of the Paul Rappaport Award … | for the best annual paper in the Transactions on Electron Devices | ELECTRON DEVICES SOCIETY® | Macintosh computer is launched | A. Nakagawa demonstrates the non-latch-up IGBT |

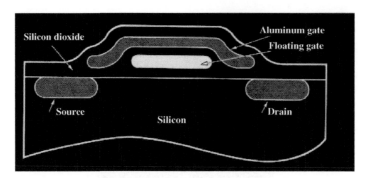

Figure 13.7 Floating gate MOSFET [9]

of the bipolar SRAM market and became one of the foundation product families for the development of MOS technology and the introduction and growth of the microprocessor market.

In 1971, Dov Frohman designed the Intel 1702 2 kb erasable and reprogrammable EPROM. This was the first application of Bell Labs' Kahng and Sze's floating gate MOS structure shown in Figure 13.7 [11]. These first EPROMs were erased by exposure to UV light through a quartz window on the top of the package.

Also in 1971, Fairchild introduced a 256-bit Oxide-Isolated bipolar TTL SRAM [8]. It provided a 50% size reduction and 30% speed improvement over junction isolation. This technology allowed Fairchild to dominate the high-performance memory market for many years.

The first memory with multiplexed row and column address pins was the Mostek MK4096 4 kb DRAM designed by Robert Proebsting and introduced in 1973 [8]. This addressing scheme, which eventually became the industry standard for DRAM, enabled it to fit into a 16-pin DIP package. The MK4096 proved to be a very robust design for customer applications. The 4 kb DRAM generation was the first to utilize Dennard's 1T1C memory cell and the silicon gate NMOS process, instead of a three-transistor cell.

The MK4116 16kb DRAM shown in Figure 13.8, introduced in 1976, achieved greater than 75% worldwide DRAM market share. Intel had introduced the first 16-kb DRAM (the 2116) in 1975, but it was the MK4116 that succeeded in the marketplace, because it was faster and more reliable [8].

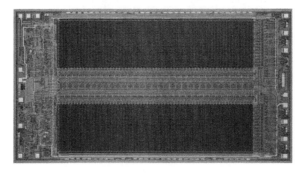

Figure 13.8 Mostek 4116 16-kb DRAM [8]

| Introduction of box or shallow trench isolation by N. Mikoshiba *et al.* of NEC Co. | D.A.B. Miller *et al.* publish the first electro-optical modulator | ELECTRON DEVICES SOCIETY® | First Japanese AdCom Member of EDS (Tatsuo Sugano) | |

IBM has made many important innovations in semiconductor technology. In 1975, when IC companies were struggling to deliver the first 16 kb DRAMs, IBM surprised the industry with the creation of a 64 kb DRAM, mass-produced in IBM's Essex Junction. VT plant two to three years ahead of other companies. This was soon to be followed by a 288 kb DRAM [8], shown in Figure 13.9.

In October of 1978, engineers from Inmos (Joe and Ward Parkinson, Dennis Wilson, and Doug Pitman) who had worked earlier in Mostek, started Micron Technology, in Boise, Idaho. Entering an intensely competitive DRAM market, Micron introduced a low-cost 64 kb DRAM. It has since grown to become one of the top three worldwide memory companies and the only remaining US memory maker. "In business as in art, the individual vision prevails over the corporate leviathan; the small company confounds the industrial policy; the entrepreneur dominates the hierarch… That is the message of Micron, and it is a central lesson of human life and history" [12].

In 1983, Intel developed the first CMOS memory, the 1 Mb DRAM, but ironically soon exited the DRAM business, due to collapsing prices. 1 Mb DRAMs in 1985 contained the last planar capacitors. All 4 Mb DRAMs, in 1988, used either a stacked or trench 3D capacitor to achieve adequate charge storage in the small cell area. 16 Mb DRAMs were introduced in 1991 and 64 Mb in 1994. High-k dielectrics were introduced in 256 Mb DRAMs in 1998 (Table 13.1).

The memory market grew from \$15m in 1970 to \$2.4bn in 1980 to \$15bn in 1990 to \$80bn in 2000 to \$250bn in 2010 [13]. It experienced plunges in 1981 and 1985 due to Japanese expansion, after which most US companies exited the market. After the US-Japan 1986 trade agreement, there was huge growth in 1988–1989, but another big plunge in 1990. HVM infrastructure growth was driven by Japanese companies during the 1980s which in turn fueled the growth of the market. Substantial innovations in manufacturing technology for yield and reduced cost were made. The 1990s saw explosive growth in Korean memory companies, while the 2000s have seen global consolidation and joint development partnerships.

In 1984, Fujio Masuoka (Figure 13.10) of Toshiba reported his invention of an electrically programmable and erasable, non-volatile EEP-ROM memory that could be rapidly erased in blocks (named *Flash* for flash erase) [14].

The Flash architecture required only a single transistor per memory cell rather than two transistors per cell as in EEPROMs. It used quantum tunneling effects induced by relatively high voltages for both writing and erasing. In 1985, Toshiba introduced the first 256 kb flash memory. In 1986, the ETOX-style 256 kb NOR Flash was introduced by

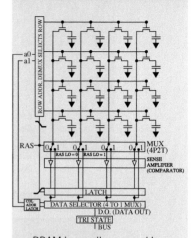

How DRAM works

DRAM is usually arranged in a rectangular array of word lines and bit lines with each bit cell containing one transistor and one capacitor. A sense amplifier (a pair of cross-connected inverters forming a latch) is connected to the end of each bit line. A row address decoder/demux is used to activate only one of the word lines at a time. This turns on all access transistors along this word line, causing the charge stored in each cell capacitor along this row to be dumped onto its respective bit line. Although this charge is only 20–30fC, it causes a slight change in the bitline voltage which can be subsequently sensed or detected by the sense amplifier and amplified into a full rail logic signal. This output logic signal is then fed out through a column address decoded data selector to the output data bus.

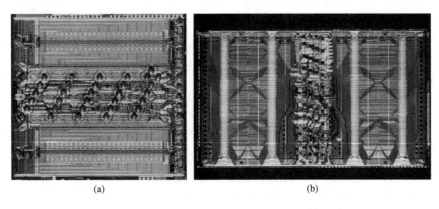

Figure 13.9   IBM 64 kb DRAM (a), IBM 288 kb DRAM (b) [8]

Figure 13.10   Fujio Masuoka, inventor of Flash memory and a 64 Gb MLC NAND Flash chip [15]

Stefan Lai of Intel. NAND Flash architecture was introduced by Toshiba at the 1987 IEDM. NAND flash architecture offers higher densities and larger capacities at lower cost with faster speeds, but sacrifices the random-access advantage of the NOR architecture, since NAND allows only page access. NOR and NAND flash get their names from the structure of the interconnections between memory cells. In NOR flash, cells are connected in parallel to the bit lines, allowing cells to be read and programmed individually. The parallel connection of cells resembles the parallel connection of transistors in a NOR gate. In NAND flash, cells are connected in series, resembling the transistors in a NAND gate. This series connection scheme consumes less space than NOR, thus reducing the cost of NAND flash.

13.2 State of Mainstream Semiconductor Memory Today

Significant advancements in process and design technology have enabled aggressive Moore's Law scaling of memories (see also Chapter 3). This has driven DRAM prices down to about $3 per Gb and NAND flash prices down to about $0.30 per Gb, thus enabling today's affordable large solid-state drives (SSD).

| Scaling impact on device reliability becomes a major concern | Introduction of 1.0-μm silicon technology node in manufacturing | **E**LECTRON **D**EVICES **S**OCIETY® | M. Green of the University of New South Whales demonstrates a 20% efficient silicon solar cell | Y. Takemae presents a 1Mb DRAM using poly-insulator-poly (PIP) cell capacitors |

Table 13.1 Mainstream memory progress

Year	DRAM density	NAND density	Feature size
1970	1 kb	–	11 µm
1973	4 kb	–	8 µm
1976	16 kb	–	5 µm
1979	64 kb	–	3 µm
1982	256 kb	–	2 µm
1985	1 Mb	–	1.2 µm
1988	4 Mb	–	0.8 µm
1991	16 Mb	–	0.5 µm
1994	64 Mb	–	0.35 µm
1998	256 Mb	–	0.25 µm
2000	512 Mb	–	0.18 µm
2002	1 Gb	2 Gb	0.13 µm
2004	–	4 Gb	90 nm
2006	2 Gb	8 Gb	65 nm
2008	–	16 Gb	45 nm
2010	–	32 Gb	32 nm
2012	Memory Cube	64 Gb	25 nm

| **1986** | John C. Bean of AT&T Bell Labs publishes comprehensive paper … | ELECTRON DEVICES SOCIETY® | … on silicon-germanium alloys | Silicon wafer bonding is developed |

About 10 times more NAND bits than DRAM bits are produced each year, resulting in nearly equal $100bn per-year markets for each memory type.

As DRAM and NAND flash cells are scaled deep into the nanoscale, many challenges arise. In DRAM, it is difficult to maintain adequate cell capacitance, minimum pass transistor leakage, and high-performance periphery logic transistors. In NAND flash, it is difficult to meet scaling and reliability requirements for multi-level cells using the conventional planar, floating gate technology.

13.2.1 DRAM

Even though the basic design of the 1T1C DRAM cell has not changed over the years, evolutionary changes have produced complex 3D capacitor and non-planar transistor structures (see Figure 13.11)[16]. In order to scale DRAM to the nanometer regime, $6F^2$ open-bitline cells have recently replaced the traditional $8F^2$ folded-bitline design. The pass transistor has also undergone changes from a planar device to more complex, non-planar, complex 3D geometries [17]. Trap-assisted and band-to-band tunneling induced leakages dominate retention time characteristics, particularly in the tail distribution. Newer materials with higher k dielectric constant have been progressively introduced in the DRAM cell capacitor but the fundamental inverse relationship between k and the magnitude of bandgap energy has placed restrictions on the choice of materials to meet the capacitance and maximum allowable leakage requirements. Variable retention time (VRT) occurs when a cell randomly toggles from a high retention to a low retention state and vice-versa. This effect gets worse as temperature increases. Careful optimization is required to minimize VRT through optimal passivation to avoid metastable states, and reduced electric fields.

MIM DRAM capacitors can achieve about 25 fF using atomic layer deposition (ALD) of hafnium oxide, aluminum oxide, and zirconium oxide high-k "sandwich" dielectrics, such as zirconium /aluminum/zirconium (ZAZ). The tall aspect ratio of capacitors can cause them to become mechanically unstable and short to neighbor cells. Significant process technology complexity places a limit on cost-effective scaling of stacked capacitors (see Chapter 4).

Figure 13.11 A cross section of DRAM cells showing cell capacitors and access transistors. Notice the tall, skinny, very high aspect ratio capacitors. Source: Micron Technology

Future DRAMs may need to use vertical surround-gate pillar access transistors for SCE control. Such structures enable $4F^2$ cells but major challenges in both process and design complexity remain.

Double data rate SDRAM (DDR) was used in PC memory beginning in 2000. DDR SDRAM uses a double data rate interface to transfer one output word on each clock edge. DDR2 and DDR3 increased this factor to four times and eight times, respectively, delivering four-word and eight-word bursts over two and four clock cycles, respectively. Pseudostatic RAM (PSRAM) is DRAM with built-in refresh and address-control circuitry to make it behave similar to SRAM. It combines the high density of DRAM with the ease of use of SRAM. PSRAM is used in the Apple iPhone and the Nintendo Wii. To overcome process complexity and cost, several 1T DRAM cells have been proposed without the need for a storage capacitor. Such "capacitor-less" designs store data in the floating body capacitor that is present in SOI devices or in vertical FET (VFET) devices. While they have not entered main stream commodity applications, they have the potential to serve low density embedded applications.

Conventional 1T1C DRAM technology is expected to continue scaling well past the 20 nm node. Meanwhile, other technologies such as Hybrid Memory Cubes (HMC) [18] that use through-silicon-vias (TSV) to enable 3D memory chip stacking, are being developed to improve data-rate performance.

13.2.2 NAND Flash

Typical NAND flash architecture has a series of several floating gate cells stacked in between the bitline select gate and the source line. Due to simple orthogonal patterns, NAND has a lithography-friendly design, allowing aggressive scaling as seen in the NAND scaling history of Figure 13.12.

The advent of media management for NAND Flash, introduced by SanDisk, was an important development that assured the success of NAND technology and scaling. The advent of multi-level cell (MLC) NAND flash, where two or even four bits per cell may be stored, has enabled continued density

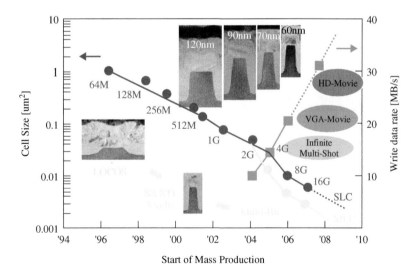

Figure 13.12   NAND Flash – Scaling History [19], © 2006 IEEE

advancements. However, as the space between floating gates (FG) is reduced, cell-to-cell interference or disturbance becomes a dominant issue, since the threshold voltage of the device is influenced by capacitive coupling from the adjacent cells. The number of electrons stored on a floating gate is shrinking to the point that the threshold distributions are overlapping. Trap-assisted tunneling and stress-induced leakage currents have placed serious limits on the scaling of the tunnel oxide thickness in NAND flash. Variations in cell current due to random telegraph noise (RTS) is also a big concern. These issues place major limitations on the sensing margins for MLC NAND flash cells. There have been several proposals to find alternatives to the FG cell, for example, charge-trap flash (CTF), metal nano crystals (MNC), but due to numerous process and device challenges the effort continues to shrink FG cells below the 20 nm node.

NAND flash technology will encounter serious scaling challenges below 20 nm, due to both electrical and physical limits. To overcome these challenges, there are various vertical 3D cell structures being researched. In essence, they involve flipping the NAND string to the vertical axis with multiple, vertically stacked transistors. Various schemes are being pursued (Figure 13.13) such as P-BICS (Toshiba), T-CAT (Samsung), VSAT/VG (UCLA). Most of these structures involve using polysilicon as the channel material instead of crystalline silicon. This poses major challenges such as poor mobility, lateral charge spilling, and so on. Key requirements for NAND technology, such as boosting, program/erase efficiency etc. will need to be met for successful adoption of these 3D vertical structures. However, such structures could potentially lower cost and continue increase in bit density.

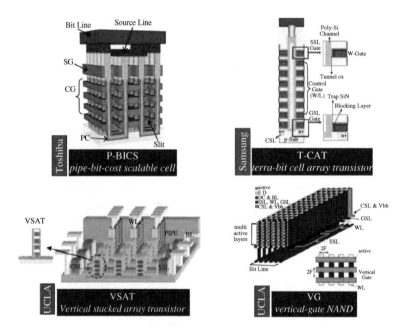

Figure 13.13 Vertical 3D NAND Flash structures being developed [16]. Reproduced with permission from Mouli *et al.*, "Trends in memory technology – reliability perspectives, challenges and opportunities", *Proc. of IPFA*, 2007

First presentation of a Si/SiGe HBT (grown by MBE) at IBM

M.A. Maher and C.A. Mead introduce the inversion charge based MOSFET model

ELECTRON DEVICES SOCIETY®

H. Koezuga *et al.* report on the first organic field-effect transistor

13.3 Emerging Memory Technologies

Emerging memory technologies are moving away from the charge storage mechanisms of DRAM and NAND flash, and instead are looking at new methods of non-volatile data storage, such as storing the data state via a change in device resistance. Resistance can be altered in several ways, such as through a change in material magnetization (spintronics devices such as MRAM and STT-MRAM), a change in material phase (PCRAM), a change in material structure or creation of oxygen vacancies (RRAM), as well as the creation of a conductive bridge between two electrodes (CBRAM).

13.3.1 Magnetization-Based Devices

The giant magneto-resistance effect (GMR), which is the change in electrical resistance of a stack of alternating ferromagnetic and non-ferromagnetic layers when the magnetization of the ferromagnetic layers is switched between a parallel and anti-parallel orientation, was discovered in 1988 by Peter Grünberg at Jülich Research Center, and simultaneously at the University of Paris-Sud by Albert Fert. For this discovery, both Grünberg and Fert were awarded the Nobel Prize in Physics in 2007. With this technology, and the breakthrough discovery of room temperature tunnel magneto-resistance (TMR) in 1995 [20, 21], the theoretical prediction of spin-transfer in 1996 [22, 23] and the subsequent experimental observations of spin-transfer in 1998, 1999, and 2000 [24, 26], the emerging spintronic memories, MRAM and spin-transfer torque RAM (STT-MRAM), have become realizable. In 2004, Freescale Semiconductor began offering MRAM on its standard product line. However, due to scalability issues associated with MRAM [27], different approaches for writing the MRAM cell were investigated. One of the most promising uses STT to switch the storage layer, allowing the device to be scaled down with a corresponding decrease in the total current required to write to the cell. Because of the advantages of STT-MRAM, it is currently widely under investigation as one of the leading emerging memory technologies.

13.3.2 PCRAM Devices

Phase-change memory devices change their resistance through a change in material phase, typically from an amorphous (high resistance) phase to a crystalline or polycrystalline (lower resistance) phase. In 1968, Stanford Ovshinsky reported a "rapid and reversible transition between a highly resistive and a conductive state effected by an electric field" [28]. From this discovery, phase-change memory gradually emerged. While optical phase-change memory, which comprises modern DVD and CD storage devices, has been successfully commercialized for over 30 years, it has taken much longer to commercialize electrical phase-change memory devices.

Compared to Flash, phase-change memories have:

- Better Scalability
- Faster programming speeds
- Direct overwrite capability without erasing first
- Better endurance
- Easier embedding with CMOS logic
- A more flexible access device: can be MOS, BJT, or a diode.

A challenge for PCRAM is to minimize the value of the reset/programming/melting current, and to minimize the value of the low set-state resistance. In 2006, BAE Systems announced the first commercially available PCRAM memory chip, a radiation hardened 4 Mb chip comprised of eight 512 Kb arrays. Only a few months after the BAE Systems chip was released, Samsung announced a 90 nm, 512 Mb, $5.8F^2$, 0.047 μm^2 cell PCRAM with one million endurance cycles and 10 years retention at 85°C, using a vertical BJT access transistor to avoid cell-cell disturbance. This has been followed by the rapid product development of PCRAM technology, with products being offered by several companies.

13.3.3 CBRAM Devices

The CBRAM device (also referred to as an ion-conducting device or a programmable metallization cell device), changes resistance by the reversible formation of conductive metallic channels between the two device electrodes (Figure 13.14).

The key material layers in this device structure are the insulating material layer, typically a chalcogenide glass, and an easily oxidized metal layer, typically silver or copper. The easily oxidized metal layer can also serve as the electrode. The conductive channels are generated when a positive potential is applied to the oxidizable electrode thus generating metal ions that migrate through the insulating material of the device towards the more negative electrode. Upon reaching the more negative electrode, the charged metal ions

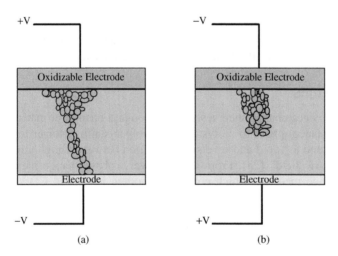

Figure 13.14 CBRAM device figure. The insulating layer (shown in red) is sandwiched between the two electrodes. (a) Conductive filament formed when a more positive potential is applied to the oxidizable electrode, creating a low resistance state. (b) Dissolution of the conductive filament when the oxidizable electrode is at the lower potential (high resistance state)

are reduced to a neutral state and begin to form a dendrite-like structure through the insulating material back towards the more positive electrode. Eventually, the dendrite growth terminates when it becomes long enough to make contact with the more positive electrode. It is at this point that the resistance instantly falls and the device is considered to have changed its resistance state from a high to a low state. When a potential with the same polarity as used to cause this dendrite growth is applied to the device, the resistance will appear low ($k\Omega$ or less). However, if the potential polarity is reversed and a negative potential is applied to the oxidizable electrode, the opposite effect will be observed: the dendrite will appear to dissolve as the electrode now at the more positive potential causes the generation of metal ions from the dendrite that migrate to the more negative potential electrode (the oxidizable electrode in Figure 13.14b), thus removing the conductive bridge between electrodes and causing the resistance to instantly increase (typically $M\Omega$).

This effect was first reported by H. Hirose and Y. Hirose in 1976 [29] in a device with an As_2S_3 insulating layer which had been photodoped with Ag and which also contained an Ag electrode as the source of metal ions. Memory devices based on this technology were largely ignored until the late 1990s when there was a resurgence in this technology leading to research and development by companies throughout the 2000s and into 2010s. While still not commercialized, the CBRAM devices have been demonstrated to exhibit lower power operation than either the MRAM or PCRAM technologies.

13.3.4 Resistive Devices

While the previously discussed devices store data in the resistance state, "resistive devices" are defined as those devices that do not use electron spin, a change in phase, or develop a conductively bridged channel in order to change resistance. This class of devices includes devices comprising transition metal oxide materials, including binary and mixed-valence oxides and/or perovskites. Electrical switching between resistance states in amorphous materials, such as oxides, has been studied since the early 1960s [30]. Therefore, it is difficult to give credit to any one person or research group for the discovery of this type of memory. Hewlett-Packard announced work on a TiOx-based device in 2008, and published a paper highlighting evidence of "memristive" behavior [31].

An advantage of all these devices is the comparative ease of incorporation into the CMOS fabrication process. Typically, the processing temperatures can be much higher than the PCRAM, CBRAM, and MRAM technologies without degrading the oxide material. However, many of these device technologies require bipolar operation, like in the case of the CBRAM device. This precludes the use of a one diode one resistor (1D1R) cell technology to prevent sneak path currents in a memory array, thus causing the need for more sophisticated array access designs.

13.4 Closing Remarks

SRAM, DRAM, NOR FLASH, NAND FLASH, PCRAM, RRAM, CBRAM, STTRAM, MRAM, QUAN-TUM, MOLECULAR, CNT...? While it is difficult to predict which of these semiconductor memory technologies will thrive or even survive in the years ahead, the search for a "universal" memory technol-

ogy will no doubt continue for many years. Researchers will continue searching for memory technologies with the lowest Energy-Volume-Delay product, as well as the best tradeoff among these desired traits:

- High Density
- Scalability
- High Endurance
- High Reliability
- Long Retention Time/Non-Volatility
- High Bandwidth/Data-rate
- Low Latency
- High Programming/Erasing Speed
- Low Programming Current
- Low Cost

References

[1] *Solid State Journal*, Sep/Oct 1960.
[2] Laws, D.A., McClure, J. and Riordan, M. *"The Silicon Engine: A Timeline of Semiconductors in Computers"*, Available at http://www.computerhistory.org/semiconductor/timeline (accessed October 16, 2012) 2007.
[3] Norman, R. "Solid State Switching and Memory Apparatus", U.S. Patent (#3562721) Filed March 5, 1963.
[4] Kilby, J. "Invention of the Integrated Circuit", *IEEE Trans. on Electron Devices*, Vol. ED-23, No.7, p. 653, July 1976.
[5] Moore, G. "Cramming more components onto integrated circuits", *Electronics*, Vol. 38, No. 8, April 1965.
[6] Perkins, H.A. and Schmidt, J.D. "An integrated semiconductor memory system", *Fall Joint Computer Conference. AFIPS Proc.*, Vol. 27, pp. 1053–1064, Nov. 1965.
[7] Agusta, B., Bardell, P., and Castrucci, P. *"Sixteen bit monolithic memory array chip"*, *IEEE Electron Devices Meeting, 1965 International*, Vol. 11, p. 39, 1965.
[8] Augarten, S. *"The Chip Collection: State of the Art"*, Available at http://smithsonianchips.si.edu (accessed October 16, 2012), Ticknor & Fields, 1983.
[9] Adee, S. "Thanks for the Memories", *IEEE Spectrum*, May 2009.
[10] Dennard, R.H., *"Field-Effect Transistor Memory DRAM"*, U.S. Patent (#3387286), issued 1968.
[11] Kahng, D. and Sze, S.M. "A floating-gate and its application to memory devices", *The Bell System Technical Journal*, Vol. 46, No. 4, pp. 1288–1295, 1967.
[12] Gilder, G., *"The Spirit of Enterprise"*, pp. 243–244, Simon and Schuster, 1984.
[13] Prince, B., *"Semiconductor Memories: A Handbook of Design"*, Manufacture and Application, John Wiley & Sons, Inc., 1992.
[14] Masuoka, F., Asano, M., Iwahashi, H., Komuro, T. and Tanaka, S., "A new Flash EEPROM cell using triple polysilicon technology", *Techn. Dig. IEEE International Electron Devices Meeting (IEDM)*, pp. 464–467, 1984.
[15] Santo, Brian. "25 Microchips That Shook the World", *IEEE Spectrum*, May 2009. Available at http://spectrum.ieee.org/semiconductors/processors/25-microchips-that-shook-the-world.
[16] Mouli, C., Prall, K. and Roberts, C., "Trends in memory technology – reliability perspectives, challenges and opportunities", *Proc. International Symposium on Physical Failure Analysis of Integrated Circuits (IPFA)*, pp. 130–134, 2007.
[17] Chung, S.-W., Yoo, M.-S., Kim, K.-O., *et al.*, "Highly Scalable Saddle-Fin (S-Fin) Transistor for Sub-50 nm DRAM Technology", *Dig. Techn. P. Symposium on VLSI Technology*, 2006.

[18] Pawlowski, T. *"Hybrid memory cube (HMC)"*, *Hot Chips*, 23, available at http://www.hotchips.org/wp-content /uploads/hc_archives/hc23/HC23.18.3-memory-FPGA/HC23.18.320-HybridCube-Pawlowski-Micron.pdf (accessed October 16, 2012), 2011.

[19] Kim, K. and Choi, J., "Future Outlook of NAND Flash Technology for 40 nm Node and Beyond", Non-Volatile Semiconductor Memory Workshop, *IEEE*, pp. 9–11, 2006.

[20] Miyazaki, T. and Tezuka, N., "Giant magnetic tunneling effect in $Fe/Al_2O_3/Fe$ junction", *J. Magn. Magn. Mater.*, 139, L231–L234, 1995.

[21] Moodera, J. S., Kinder, L. R., Wong, T. M., and Meservey, R., "Large magnetoresistance at room temperature in ferromagnetic thin film tunnel junctions", *Phys. Rev. Lett.*, 74, 3273–3276, 1995.

[22] Berger, L., "Emission of spin waves by a magnetic multilayer traversed by a current", *Phys. Rev. B*, 54, 9353–9358, 1996.

[23] Slonczewski, J., "Current-driven excitation of magnetic multilayers", *J. Magn. Magn. Mater.*, 159, L1–L17, 1996.

[24] Tsoi, M., Jansen, A. G. M., Bass, J., Chiang, W.-C., Seck, M., Tsoi, V., and Wyder, P., "Excitation of a magnetic multilayer by an electric current", *Phys. Rev. Lett.*, 80, 4281–4284, 1998.

[25] Wegrowe, J.-E., Kelly, D., Jaccard, Y., Guittienne, Ph., Ansermet, and J.-P.H., "Current-induced magnetization reversal in magnetic nanowires", *Europhys. Lett.*, 45, 626–632, 1999.

[26] Katine, J. A., Albert, F. J., Buhrman, R. A., Myers, E. B., and Ralph, D. C., "Current-driven magnetization reversal and spin-wave excitations in Co/Cu/Co pillars", *Phys. Rev. Lett.*, 84, 3149–3152, 2000.

[27] Dieny, B., Sousa, R. C., Herault, J., Papusoi, C., Prenat, G., Ebels, U., Houssameddine, D., Rodmacq, B., Auffret, S., Prejbeanu-Buda, L., Cyrille, M. C., Delaet, B., Redon, O., Ducruet, C., Nozieres, J. P., and Prejbeanu, L. "Spintronic devices for memory and logic applications", *Handbook of Magnetic Materials*, 19, 107–127, 2011.

[28] Ovshinsky, S. R. "Reversible electrical switching phenomena in disordered structures", *Phys. Rev. Lett.*, 21, 1450–1455, 1968.

[29] Hirose, Y. and Hirose, H. "Polarity-dependent memory switching and behavior of silver dendrite in silver-photodoped amorphous arsenic(3+) sulfide films", *J. Appl. Phys.*, 47, 2767–2772, 1976.

[30] Dearnaley, G., Stoneham, A. M., and Morgan, D. V. "Electrical phenomena in amorphous oxide films", *Rep. Prog. Phys.*, 33, 1129–1191, 1970.

[31] Strukov, D. B., Snider, G. S., Steward, D. R., and Williams, R. S. "The missing memristor found", *Nature*, 453, 80–83, 2008.

| Researchers at Motorola and IBM describe threshold voltage shift ... | ... due to boron penetration in PMOS transistors | ELECTRON DEVICES SOCIETY® | B. Ankele, W. Holzl and P. O'Leary introduce the first multi-region, single-piece DC MOSFET model | Fall of the Berlin Wall |

Chapter 14

RF and Microwave Semiconductor Technologies

Giovanni Ghione, Fabrizio Bonani, Ruediger Quay and Erich Kasper

14.1 III-V-Based: GaAs and InP

In the mid-1960s the first microwave bipolar transistors appeared, so that, by 1968, silicon (Si) and germanium (Ge) bipolar junction transistors (BJTs) were capable of low-noise and power performances up to X band (see also Chapter 1). However, by the beginning of the 1970s a real breakthrough in microwave transistors was due to the introduction of a Schottky-barrier FET, the MESFET (see sidebar), on the innovative, high-mobility semiconductor gallium arsenide (GaAs). A proof of principle of the GaAs MESFET had been given by Mead in 1966 [1], and already in 1967 Hooper and Leherer demonstrated microwave operation of a device fabricated using a 2 μm thick epitaxial GaAs film with 2×10^{16} cm^{-3} doping grown on a semi-insulating substrate [2]; the gate contact was evaporated Al and the cutoff frequency (ft; see sidebar) was around 3 GHz (see Figure 14.1). In 1971–1972 GaAs MESFETs with gate length of 1 μm showed a maximum oscillation frequency (f_{max}; see sidebar) around 50 GHz and useful gain up to 20 GHz, and at the same time 3.5 dB noise figure with 6.6 dB associated gain at 10 GHz (see [3] and references therein).

Success of GaAs MESFETs was due to both the high electron mobility of electrons in GaAs (leading to high trans-conductance, low transit time in the channel, and low parasitic resistances) and to the availability of semi-insulating (S.I.) substrates minimizing capacitive parasitics and substrate losses (see also Chapter 4).

Guide to State-of-the-Art Electron Devices, First Edition. Edited by Joachim N. Burghartz.
© 2013 John Wiley & Sons, Ltd. Published 2013 by John Wiley & Sons, Ltd.

Incoherent scattering processes included in the non-equilibrium Green's function formalism as fictitious contacts (Supriyo Datta)

Tang, Nguyen and Howe of UC Berkeley present the lateral comb drive

ELECTRON DEVICES SOCIETY®

Charles Stark Draper Price awarded to Robert Noyce and Jack Kilby for their independent development of the integrated circuit

The S.I. substrate technology was strongly correlated with the progress of ingot growth techniques like the Liquid Encapsulated Czochralski (LEC) [4]. While Plessey already had a 4 μm gate length MESFET with 10 dB gain at 1 GHz commercially available in 1967 [4], in 1974 NEC released the first commercial low-noise GaAs MESFET while in 1974 Fujitsu released 1 W X-band multi-finger power MESFETs meant as a replacement for Traveling Wave Tubes (TWTs) in satellite applications.

The S.I. nature of GaAs wafers also made it possible to exploit it as a dielectric substrate for low-loss microwave passive lumped and distributed components, thus achieving GaAs-based Microwave Monolithic Integrated Circuits (MMICs) with microstrip or coplanar waveguide technology. The first example of realized MMIC is the X-band wideband amplifier developed at Plessey by Pengelly and Turner in 1976 [5]. The amplifier gain was around 5 dB between 7–11 GHz with a reverse isolation around 20 dB. During the 1980s GaAs MMICs became commercially available also through foundry services.

Metal Semiconductor FETs – MESFETs

A MESFET is a field-effect transistor whose insulating gate junction is a rectifying Schottky barrier in reverse bias (rather than a *pn* junction as in junction FETs). III-V semiconductors are particularly well-suited to implementing MESFETs since semi-insulating (non-intentionally doped or compensated) GaAs and InP substrates are available, on which a *n*-doped active layer can be grown by epitaxy.

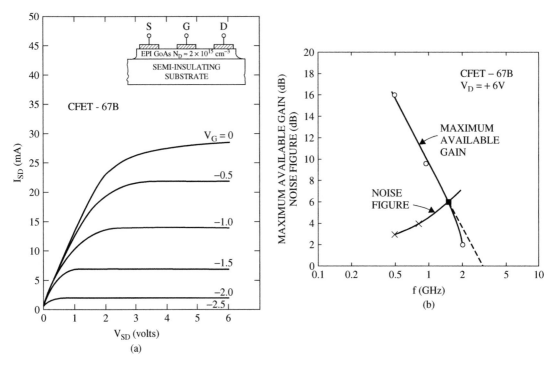

Figure 14.1 (a) Source-drain characteristics of a 1967 Schottky barrier gate GaAs FET and (b) maximum available gain and noise figure versus frequency with $V_{DS} = +6$ V, gate bias for maximum transconductance. © 1966 IEEE. Reprinted with permission from W.W. Hooper and W.I. Lehrer, "An Epitaxial GaAs Field-Effect Transistor," *Proceedings of the IEEE*, 55:7, 1237–1238, 1967

GaAs MESFET downscaling posed a key challenge and related consequences: to increase the cutoff and maximum oscillation frequencies the gate length had to be reduced, decreasing at the same time the active layer thickness and increasing the doping. Unfortunately, high doping also meant increased impurity scattering in the channel, and, thus, lower mobility. A very promising way out was offered after 1975 by the development of AlGaAs/GaAs heterojunctions grown by Molecular Beam Epitaxy (MBE) or Metal Organic Chemical Vapor Deposition (MOCVD). While single and double (AlGaAs/GaAs/AlGaAs) heterostructures where known (from the development of semiconductor lasers) to provide carrier and photon confinement through the potential barrier and refractive index jump introduced by the junction between a small- and a large-bandgap semiconductor, a new important property was discovered in heterojunctions between a doped AlGaAs epitaxial layer and a S.I. GaAs substrate (called Modulation Doping structure). In 1978 Dingle *et al.* [6] of Bell Labs presented evidence on the fact that, in such structures, electrons transferred from the doped AlGaAs epilayer to a conduction band potential well, located at the interface with the S.I. GaAs substrate. This creates a thin layer of electrons (2DEG) vertically confined by the well, able to operate as a FET channel. Since the 2DEG is located in undoped GaAs, it has high carrier density but also high mobility. The Modulation Doped FET (MODFET) also called High Electron Mobility Transistor (HEMT) (see Figure 14.2 and sidebar)

> **Speed, noise, power figures of merit: f_T, f_{max}, NF_{min}**
>
> Figures of merit for the microwave transistors are the cutoff frequency f_T (unit short-circuit current gain) and the maximum oscillation frequency f_{max} (unit maximum available power gain). RF power is proportional to the maximum current – breakdown voltage product while noise performances are denoted by the minimum noise figure NF_{min} as minimized versus bias point and generator impedance.

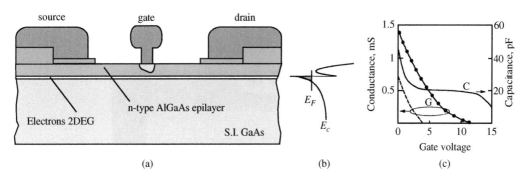

Figure 14.2 (a) Cross-section of the AlGaAs/GaAs HEMT with (b) corresponding profile of the conduction band energy versus depth. The interface potential well hosts an electron two-dimensional gas (2DEG) with high density and mobility. (c) Conductance and capacitance of the AlGaAs/GaAs HEMT (continuous line: conductance experimental data; dashed line, conductance evaluated as a conventional MESFET model; dots: conductance evaluated taking into account the 2DEG). © The IET. Reproduced by permission of the Institution of Engineering and Technology from D. Delagebeaudeuf, P. Delescluse, P. Etienne, M. Laviron, J. Chaplart, and N.-T. Linh, "Two-dimensional electron gas m.e.s.f.e.t. structure," *Electronics Letters*, 16:17, 667–668, 1980

| Nitridation of gate oxide is introduced … | … as a means to suppress boron penetration in PMOS transistors | ELECTRON DEVICES SOCIETY® | High brightness green and blue LED's | Terms 'nanoscale' and "nanotechnology" are being introduced |

was proposed almost at the same time in Japan by T. Mimura *et al.* (Fujitsu) [7], and in Europe by D. Delagebeaudeuf (Thomson-CSF, France) in 1979–1980 [8] (see Figure 14.2c).

During the 1980s, HEMT-based ICs were developed first for digital applications, but the real breakthrough came with low-noise and then power amplifiers, exploiting the better HEMT NF_{min} and associated gain versus MESFETs, first proposed in 1983 by Niori *et al.* with a 20 GHz four-stage X-band HEMT LNA [9]. During the following years new concepts emerged, such as the PHEMT (pseudomorphic HEMT) exploiting a strained epitaxial Quantum Well (QW) as the channel (e.g., S.I. GaAs substrate, undoped InGaAs strained channel, *n*-type AlGaAs supply layer), with advantages in terms of 2DEG confinement. At the same time InP-based HEMTs and PHEMTs (e.g., S.I. InP substrate, undoped InGaAs strained or lattice matched channel, *n*-type InAlAs supply layer lattice matched to InP [10]), provided millimeter wave operation due to the InP superior electron maximum speed versus GaAs. Due to its higher cost the InP technology was typically developed for mm-wave applications, while GaAs PHEMTs found their way also in consumer applications (e.g., power amplifiers for GSM or UMTS cellphones) gradually replacing everywhere GaAs MESFETs. At the same time, the mm-wave performances of InP-based PHEMTs reached in 2010 f_{max} above 1.2 THz and f_T above 600 GHz, with LNA implementations at almost 0.5 THz [11] (see Figure 14.3).

Bipolar transistors traditionally have provided front run performance in terms of device speed, mainly because the dimensions controlling transit time are defined by doping and growth processes having an excellent dimensional control, whereas in FETs the impact of lithography is much stronger. Furthermore, the vertical current conduction

High-Electron Mobility Transistors (HEMTs)

The HEMT is a heterostructure FET where the conducting channel is an electron quantized two-dimensional gas (2DEG) trapped in a potential well arising when a doped *n*-type widegap (e.g., AlGaAs) layer is grown on a narrower gap (e.g., GaAs) S.I. substrate. Since the 2DEG is in the S.I. GaAs no impurity scattering limits the mobility.

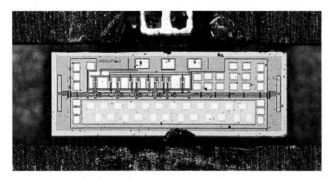

Figure 14.3 Microphotograph of coplanar 480 GHz LNA in split block housing. The signal is coupled to the chip by monolithically integrated dipole probes. © 2010 IEEE. Reprinted with permission from W.R. Deal *et al.*, "Demonstration of a 0.48 THz Amplifier Module Using InP HEMT Transistors," *IEEE Microwave and Wireless Components Letters*, 20:5, 289–291, 2010

| Rapid thermal processing becomes part of integrated device technology | Stress engineering is introduced for increasing carrier mobility |  **E**LECTRON **D**EVICES **S**OCIETY® | Halo implants are used to reduce short channel effects in CMOS | Cu interconnects emerge |

provides a favorable output current capability per unit area, and the exponential relationship between the output current and the input voltage guarantees excellent transconductance, proportional to the collector current. Finally, the device turn-on voltage is defined by the base emitter *pn* junction, and therefore is not plagued by variations across the wafer, and the input capacitance is proportional to the input current, as far as it is dominated by the diffusion component. On the other hand, bipolar devices suffer from some disadvantages with respect to FETs, chiefly a lower input impedance (even in DC), the lack of depletion mode transistors (the turn on voltage cannot be tailored), and the complexity of the fabrication process, in particular if heterostructures are involved.

From the device standpoint, one of the major limitations of homojunction bipolar transistors for high frequency applications, is the combined requirement to increase the base doping in order to control the base resistance, and to keep the emitter doping even larger to guarantee a high injection efficiency and ultimately, current gain. For deeply scaled devices, the emitter doping level may suffer from bandgap narrowing, thus impairing the gain advantages because of the increased effective intrinsic concentration. This issue was recognized in the early days of semiconductor devices and the proposed solution was the use of a wide bandgap material for the emitter [12, 13], leading to the Single Heterostructure Bipolar Transistor (SHBT) structure. The advantage of the SHBT is the spatial confinement of the holes in the base, made possible by a properly tailored emitter base junction, thus enabling adequate injection efficiency to be attained with very high base and low emitter doping. The reduction in the base resistance significantly improves f_{max}, and the Early voltage increases considerably. The reduced storage of holes in the emitter, finally, improves f_T.

The practical realization of the SHBT structure had to wait for the development of a reliable technology: III-V materials provided an excellent choice since their superior performance in terms of carrier mobility and saturation velocity are combined with the availability of lattice matched combinations of materials with large bandgap discontinuities. The availability of GaAs substrates provided the basis for the development of the AlGaAs/GaAs HBT, first demonstrated in 1972 [14]. Currently, InP wafers are also used as substrates for the lattice matched InGaAs base (and InP or AlInAs emitter): in this case, an abrupt heterojunction is also present at the base–collector contact (Double HBT – DHBT). InP devices operate at substantially lower base emitter bias, thus leading to significant advantages for some power sensitive applications. The DHBT device provides a symmetric structure with respect to the SHBT case, allowing for improved switching

Heterojunction Bipolar Transistors (HBTs)

In *npn* HBTs the emitter–base (EB) junction is a heterojunction where the emitter is widegap and the base narrowgap. A potential barrier arising at the EB heterojunction stops holes from being injected back into the emitter, thus enabling the emitter efficiency to be almost one even if the base is highly doped – a way to reduce the input base resistance that is a limiting factor for both device noise and f_{max}.

| Cu damascene process for interconnects and vias is developed | | ELECTRON DEVICES SOCIETY® | Low-k dielectrics for interconnects are being used | Deep UV lithography emerges |

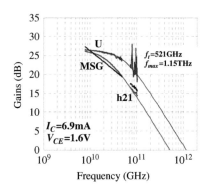

Figure 14.4   RF gains of 120 nm base width InP DHBT with 1.15 THz maximum frequency of oscillation. © 2011 IEEE. Reprinted with permission from M. Urteaga, R. Pierson, P. Rowell, V. Jain, E. Lobisser, M. J. W. Rodwell, "130 nm InP DHBTs with ft >0.52 THz and fmax >1.1 THz," *69th Annual Device Research Conference (DRC)*, 2011, pp. 281–282, 20–22 June 2011

speed from saturation to cutoff, although the design of the base layer in this case is quite critical. The HBT applications field partly overlaps with the one of III-V FETs in low-noise, high-gain and power applications, where HBTs allow for a more compact layout at the expense of a more complex technology and some additional criticality in power dissipation issues. On the other hand the HBT remains the device of choice in low-phase noise oscillators, due to the favorable low-frequency noise typical of vertical conduction devices.

Currently, the efforts in the development of advanced III-V HBT structures are leading to record values of high frequency performance figures of merit such as the half-THz f_T and above-THz f_{max} reported in Figure 14.4 for a InP DHBT.

14.2 Si and SiGe

The development of Si-based RF and microwave devices took during the last 20 years some unexpected paths. Downscaling to nanometer gate length has made the Si MOSFET not only a fast device for logical applications but also as a byproduct, a good RF and microwave device, perhaps today the most common RF device for applications well below 10 GHz. On the other hand, the Si heterojunction bipolar transistor concept, already present in the BJT original patent (see Ref. [12], Figure 12), took a very long time to implement from a technological standpoint by giving rise to the SiGe HBT, initially developed by IBM as a digital device, but then to become one of the key microwave transistors (see Chapter 1). Silicon monolithic millimeter wave integrated circuits (SIMMWIC) were first suggested in 1981 by A. Rosen [16] using high resistivity substrates. Now this term is used for monolithic integration of waveguides, passive devices, diodes and transistor circuits, and even antenna. Annual IEEE Symposia [17] report about progress in this field.

The development of RF MOSFETs was in a way a quiet revolution, fostered by the huge technological effort of the Si world to obtain faster and low-power, low voltage digital circuits. Nanometer scale MOSFETs finally demonstrated not only cutoff frequencies above 100 GHz, but also fairly good RF noise performances

1990   Supriyo Datta and Biswajit Das propose the spin-based field effect transistor   **E**LECTRON **D**EVICES **S**OCIETY®   First self-aligned Si/SiGe HBT transistors and circuits at IBM

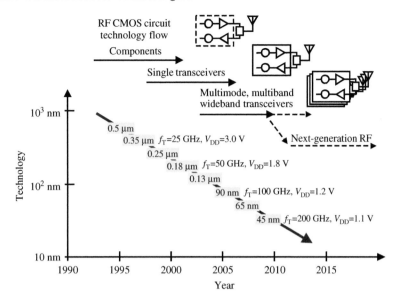

Figure 14.5 Development of MOSFET RF technology: performances versus technological node versus year. Adapted from [19], Figure 1

(see Figure 14.5). Though being widespread in all RF applications, RF MOSFETs still have some basic problems, that seem to be inevitably connected to the device structure: a comparatively high input resistance that limits the device f_{max} (typically lower than f_T), and low maximum specific output power. For this reason low-power applications (like Bluetooth or WLAN transceivers) are dominated by the MOS IC technology, while applications requiring output powers already of the order of 1 W (like GSM or UMTS) must rely on III-V power stages. The typical high-power MOSFET technology, the LDMOS, is unfortunately limited to frequencies below 5 GHz, even if it is today dominating medium power (e.g., wireless base station) applications. Despite this, recent developments show that clever circuit design can bring MOSFETs up to the millimeter wave range, as in the 60 GHz receiver proposed by Razavi in 2006 [18].

The Si-SiGe heterostructure has a potential for application in both heterojunction bipolar and field-effect transistors. In particular, SiGe HBTs have a Si emitter, a SiGe pseudomorphic base and a Si collector; they therefore belong to the DHBT class. The first SiGe transistors were presented only in 1985–1987, when the MBE growth technique became good enough, by IBM (HBT [20]), Bell Labs (*p*-channel MODFET) and AEG (strained *n*-channel MODFET [21]). The HBT presented in [20] had a base thickness down to 100 nm (see Figure 14.6) and showed a substantial increase in current gain when compared to the traditional BJT.

Two ideal design philosophies can been applied to SiGe HBTs (see also Chapter 1): (a) the base Ge fraction is constant, and (b) the Ge fraction is linearly graded from the emitter to the collector [22]. Case (b) is not a true HBT but rather a graded base BJT. Design (a) improves the electron injection into the base, while the current gain increases due to the better emitter efficiency resulting from suppression of

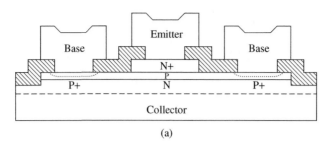

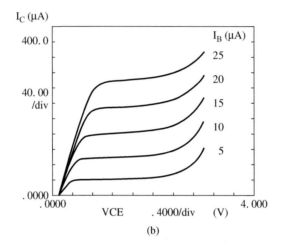

(b)

Figure 14.6 (a) Mesa structure and (b) output characteristics of the first SiGe HBT. The pseudomorphic base was 100 nm thick with composition $Si_{0.88}Ge_{0.12}$. © 1987 IEEE. Reprinted with permission from S.S. Iyer, G.L. Patton, S. L. Delage, S. Tiwari and J. M. C. Stork, "Silicon-germanium Base Heterojunction Bipolar Transistors by Molecular Beam Epitaxy", *International Electron Devices Meeting*, 1987, 33, pp. 874−876, 1987

hole BE injection through a valence band potential barrier; this enables to increase base doping, thus allowing us to decrease the base thickness and distributed resistance, with an improvement of the transit time (thus of f_T) and of the input resistance (thus of f_{max}). The graded design (b) introduces a base quasi-field that reduces the base transit time, and therefore increases (but less than in case a ") the current gain. Furthermore, decreasing the collector width while increasing the doping is another way to increase the speed at the expense of the breakdown voltage and therefore of the output power. For designs (a) and (b) the gain advantage over a Si BJT is given by the factors $\exp(r)$ and $r/\left[1 - \exp(-r)\right]$, $r = \Delta E_g/k_B T$, respectively, with ΔE_g the bandgap reduction, k_B the Boltzmann constant, and T the ambient temperature. Real devices are a compromise between the high gain design (a) and the high-speed design (b) and also a compromise between high speed and high breakdown voltage (see Figure 14.7). SiGe HBTs are now on the market since 1999 in a number of RF, microwave and mm wave applications.

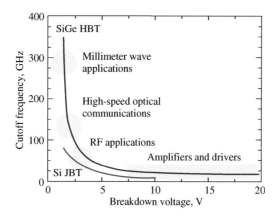

Figure 14.7 Cutoff frequency – breakdown voltage compromise and application field for SiGe HBT

14.3 Wide Bandgap Devices (Group-III Nitrides, SiC and Diamond)

Wide bandgap devices based on group III-nitride (III-N) heterostruc-tures, silicon Carbide (SiC) and diamond, especially FETs with channel bandgap energies of more than 2.5 eV, paved the way to comparably higher operating voltages (for a given device area), thus leading to an increase of RF power densities by orders of magnitude. SiC-based FETs were the first widegap devices to be investigated [23], leading to the development of MESFETs, JFETs and MOSFETs for frequencies in the MHz and lower GHz ranges. SiC MESFETs showed operat-ing bias above 50 V, with impressive progress in terms of power. Bipolar junction SiC devices gained also great interest for opera-tion to at least 2 GHz thanks to the increased power density; for an overview see for example [24]. In the study of group III-N widegap semiconductors, early examples of material growth were reported in the beginning of the 1970s; however, the real interest began in the early 1990s with FETs, that is, GaN MESFETs [25] and AlGaN/GaN HEMTs [26] coinciding with the material development taking place in "blue" optoelectronics. By exploiting the progress in the material and heterostructure growth for blue LEDs and lasers, it was observed by 1995 that heterostructure-based widegap transistors might be very attractive contenders for any form of power electronics, and would also be suitable for (at least) X-band RF applications [27]. The lack

Polarization Doping

Group III-N semiconductors are effectively doped by using the electrical polarization effects at heterointerfaces, e.g., AlGaN/GaN. The difference in both spontaneous and piezoelectric polarization at the interface leads to considerable sheet carrier densities which can be exploited by additional bandgap engineering. The source of the carriers which form the actual conducting layers are impurities that do not interfere with the transport properties.

1992   Grating light modulator was introduced by Olav Solgaard

Chris Pister shows the first micromachined hinge enabling pseudo 3D structures and assembly

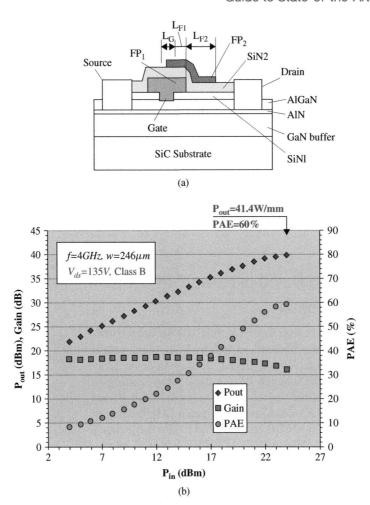

Figure 14.8   (a) Cross section of the double field-plated GaN HEMT with 41.4 W/mm output power at 4 GHz; (b) power sweep, gain and efficiency. © 2006 IEEE. Reprinted with permission from Y.-F. Wu, M. Moore, A. Saxler, T. Wisleder, P. Parikh, "40-W/mm Double Field-plated GaN HEMTs," *64th Device Research Conference, 2006*, June 2006, pp. 151–152

of a native substrate led to the extensive use of semi-insulating SiC substrates for good heat spreading. The polarization doping based on both spontaneous and piezoelectric effects. [28] (see sidebar) was another critical aspect to be analyzed and optimized to obtain reliable devices without having to use impurity doping. The steady increase of output power in continuous-wave operation culminated in double field-plated HEMTs with power densities of more than 40 W/mm at 4 GHz at an operation bias as high as 135 V (Wu *et al.* from University of California and Cree [29], see Figure 14.8). Later developments (2006) demonstrated the RF power potential of GaN HEMTs up to at least 100 GHz (e.g., by Micovic *et al.* from HRL [30]), with impressive MMIC performance up to W-band frequencies (75–110 GHz) associated to

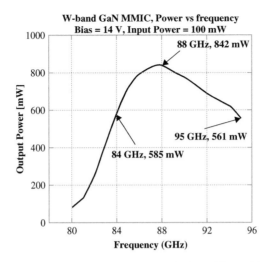

Figure 14.9 Output power versus frequency of a power module including a W-band GaN MMIC measured in continuous wave mode (drain bias 14 V, input RF power 100 mW). © 2010 IEEE. Reprinted, with permission, from M. Micovic, A. Kurdoghlian, K. Shinohara, S. Burnham, I. Milosavljevic, M. Hu, A. Corrion, A. Fung, R. Lin, L. Samoska, P. Kangaslahti, B. Lambrigtsen, P. Goldsmith, W. S. Wong, A. Schmitz, P. Hashimoto, P. J. Willadsen, D. H. Chow, "W-Band GaN MMIC with 842 mW output power at 88 GHz," *IEEE MTT-S International Microwave Symposium Digest (MTT)*, 2010, pp. 237–239, 23–28 May 2010

output power levels of more than 800 mW (2 W/mm) at 88 GHz unprecedented by any other material system [30] (see Figure 14.9). Diamond-based RF power electronics [31], exploiting the exceptionally high heat conductivity of the substrate, poses the next technological challenge with microwave diamond FETs based on *p*-type accumulation layers, to be used as a surface channel, with hole mobilities in the range of 90 cm^2/Vs, see for example, [32].

References

[1] C.A. Mead, "Schottky barrier gate field effect transistor", *Proceedings of the IEEE*, vol. 54, no. 2, pp. 307–308, Feb. 1966.

[2] W.W. Hooper and W.I. Lehrer, "An Epitaxial GaAs Field-Effect Transistor", *Proceedings of the IEEE*, vol. 55, no. 7, pp. 1237–1238, Jul. 1967.

[3] C.A. Liechti, "Microwave Field-Effect Transistors – 1976", *IEEE Transactions on Microwave Theory and Techniques*, vol. 24, no. 6, pp. 279–300, Jun. 1976.

[4] I.D. Robertson, S. Lucyszyn (eds), *RFIC and MMIC Design and Technology*, IEE Circuits, Devices and Systems Series, IEE, 2001.

[5] R.S. Pengelly and J.A. Turner, "Monolithic broadband GaAs f.e.t. amplifiers", *Electronics Letters*, vol. 12, no. 10, pp. 251–252, May 13, 1976.

[6] R. Dingle, H.L. Störmer, A.C. Gossard, and W. Wiegmann, "Electron mobilities in modulation doped semiconductor heterojunction superlattices", *Appl. Phys. Lett.*, vol. 33, pp. 665, 1978.

[7] T. Mimura, S. Hiyamizu, T. Fujii and K. Nanbu, "A New Field-Effect Transistor with Selectively Doped GaAs/n-Al_xGa_{1-x} As Heterojunctions", *Jpn. J. Appl. Phys.*, 19, pp. L225–L227, 1980.

[8] D. Delagebeaudeuf, P. Delescluse, P. Etienne, M. Laviron, J. Chaplart, and N.-T. Linh, "Two-dimensional electron gas m.e.s.f.e.t. structure", *Electronics Letters*, vol. 16, no. 17, pp. 667–668, Aug. 14, 1980.

[9] M. Niori, T. Saito, K. Joshin, T. Mimura, "A 20 GHz high electron mobility transistor amplifier for satellite communications", *Dig. of Tech. P. IEEE International Solid-State Circuits Conference (ISSCC)*, vol. XXVI, pp. 198–199, Feb, 1983.

[10] C.Y. Chen, A.Y. Cho, K. Alavi, and P.A. Garbinski, "Short channel $Ga_{0.47}In_{0.53}As/Al_{0.48}In_{0.52}$ As selectively doped field effect transistors", *IEEE Electron Device Letters*, vol. 3, no. 8, pp. 205–208, Aug. 1982.

[11] W.R. Deal, X.B. Mei, V. Radisic, *et al.*, "Demonstration of a 0.48 THz Amplifier Module Using InP HEMT Transistors", *IEEE Microwave and Wireless Components Letters*, vol. 20, no. 5, pp. 289–291, May 2010.

[12] W. Schockley, "Circuit Element Utilizing Semiconductive Material", *United States Patent* 2,569,347, 1951.

[13] H. Kroemer, "Theory of a wide-gap emitter for transistors", *Proc. of the IRE*, vol. 45, pp. 1535–1537, 1957.

[14] W. Dumke, J. Woodall, and V. Rideout, "GaAs-GaAlAs heterojunction transistor for high frequency operation", *Solid-State El.*, vol. 15, pp. 1339–1343, 1972.

[15] M. Urteaga, R. Pierson, P. Rowell, V. Jain, E. Lobisser, M.J.W. Rodwell, "130 nm InP DHBTs with f_t >0.52 THz and f_{max} >1.1 THz", *69th Annual Device Research Conference (DRC)*, 2011, pp. 281–282, 20–22 Jun. 2011.

[16] A. Rosen, M. Caulton, P. Stabile, *et al.*, "Silicon as a millimeter-wave monolithically integrated substrate", *RCA Review*, vol. 42, pp. 633–660, 1981.

[17] SiRF, *Topical Meeting on Silicon Integrated Circuits in RF Systems* (SiRF), 2013 (available at http://www.silicon-rf.org/sirf2013/, accessed October 9, 2012).

[18] B. Razavi, "A 60-GHz direct-conversion CMOS receiver", *Dig. Techn. P. IEEE International Solid-State Circuits Conference (ISSCC)*, 2005, vol. 1, pp. 400–402, Feb. 2005.

[19] N. Ishihara, S. Amakawa, and K. Masu, "RF CMOS integrated circuits: history, current status and future prospects", *IEICE Trans.*, vol. E94-A no. 2 pp. 556–567, Feb. 2011.

[20] S.S. Iyer, G.L. Patton, S.L. Delage, S. Tiwari and J.M.C. Stork, "Silicon-germanium Base Heterojunction Bipolar Transistors by Molecular Beam Epitaxy", *Technical Digest IEEE International Electron Devices Meeting (IEDM)*, 1987, vol. 33, pp. 874–876, 1987.

[21] H. Dämbkes, H.-J. Herzog, H. Kibbel, H. Jorke, and E. Kasper, "Fabrication and Properties of n-Channel SiGe/Si Modulation Doped Field-Effect Transistors Grown by MBE", *Techn. Dig. IEEE International Electron Devices Meeting (IEDM)*, 1985, vol. 31, pp. 768–770, 1985.

[22] E. Kasper, and D.J. Paul, *Silicon Quantum Integrated Circuits*, Springer, 2005.

[23] J.W. Palmour, S.T. Sheppard, R.P. Smith, S.T. Allen, W.L. Pribble, T.J. Smith, Z. Ring, J.J. Sumakeris, A.W. Saxler, and J.W. Milligan, "Wide Bandgap Semiconductor Devices and MMICs for RF Power Applications", *International Electron Devices Meeting*, 2001, vol. 33, pp. 17.4.1–17.4.4, 2001.

[24] F. Zhao, "Current Status and Future Prospects of 4H-SiC Power RF Bipolar Junction Transistors", *Proc. IEEE Conference on Industrial Electronics and Applications (ICIEA)*, pp. 2060–2064, 2009.

[25] M. Asif Khan, J.N. Kuznia, A.R. Bhattarai, and D.T. Olson, "Metal semiconductor field effect transistor based on single crystal GaN", *Appl. Phys. Lett.*, 62, pp. 1786–1787, 1993.

[26] M.A. Khan, A. Bhattarai, J.N. Kuznia, and D.T. Olson, "High electron mobility transistor based on a GaN-$Al_xGa_{1-x}N$ heterojunction", *Appl. Phys. Lett.*, vol. 63, no. 9, pp. 1214–1215, 1993.

[27] U.K. Mishra, L. Shen, T.E. Kazior, and Y.-F. Wu, "GaN-Based RF Power Devices and Amplifiers", *Proceedings of the IEEE*, vol. 96, no. 2, pp. 288–305, Feb. 2008.

[28] O. Ambacher, J. Smart, J.R. Shealy, N.G. Weimann, K. Chu, M. Murphy, W.J. Schaff, L.F. Eastman, R. Dimitrov, L. Wittmer, M. Stutzmann, W. Rieger and J. Hilsenbeck, "Two-dimensional electron gases induced by spontaneous and piezoelectric polarization charges in N- and Ga-face AlGaN/GaN heterostructures", *J. Applied Physics*, vol. 85, no. 6, pp. 3222–3233, 1999.

1993

Larry Hornbeck of Texas Instruments Co. pioneers the digital MEMS micro mirror

ELECTRON DEVICES SOCIETY

IBM and Motorola introduce the Reduced Instruction Set Computer (RISC) processor

[29] Y.-F. Wu, M. Moore, A. Saxler, T. Wisleder, and P. Parikh, "40-W/mm Double Field-plated GaN HEMTs", *64th Device Research Conference*, 2006, pp. 151–152, Jun. 2006.

[30] M. Micovic, A. Kurdoghlian, K. Shinohara, S. Burnham, I. Milosavljevic, M. Hu, A. Corrion, A. Fung, R. Lin, L. Samoska, P. Kangaslahti, B. Lambrigtsen, P. Goldsmith, W.S. Wong, A. Schmitz, P. Hashimoto, P.J. Willadsen, and D.H. Chow, "W-Band GaN MMIC with 842 mW output power at 88 GHz", *IEEE MTT-S International Microwave Symposium Digest (MTT)*, 2010, pp. 237–239, 23–28 May 2010.

[31] M. Kasu, "Diamond Field-effect Transistors as Microwave Power Amplifiers", *NTT Technical Review*, vol. 8, no. 8, pp. 1–5, Aug. 2010.

[32] P. Calvani, A. Corsaro, F. Sinisi, M.C. Rossi, G. Conte, E. Giovine, W. Ciccognani, and E. Limiti, "Diamond MESFET technology development for microwave integrated circuits", *Proceedings of the 4th European Microwave Integrated Circuits Conference*, pp. 148–151, 2009.

| First 40-nm MOSFET (Toshiba) | First GaN MOSFET | ELECTRON DEVICES SOCIETY® | First production of an accelerometer in MEMS technology at Analog Devices Co. | Publication of the first optical switch by A.L. Lentine and D.A.B. Miller |

Chapter 15

Power Devices and ICs

Richard K. Williams, Mohamed N. Darwish, Theodore J. Letavic and Mikael Östling

15.1 Overview of Power Devices and ICs

What do trains, cars, cellphones, computers, HDTVs, games and consumer electronics share in common? They all rely on power devices – an eclectic mix of discrete devices and ICs addressing diverse applications, voltages and currents (see Figure 15.1), operating from DC to multi-megahertz frequencies to precisely control the energy flow between a power source and one or more electrical loads (see sidebar).

Originating in the 1950s, the first semiconductor devices able to operate at high current densities and scale to large areas – the PNPN thyristor (or SCR) [1] and subsequently, the TRIAC and the GTO [2], revolutionized electric motor control and power distribution globally. The "latching" nature of the thyristor (i.e., losing on-state "gate" control) however, limited its operation and application to low frequencies. Yet, despite predating thyristors, bipolar junction transistors sufficiently robust to facilitate power switching did not appear until the early 1960s coinciding with silicon epitaxy and the "diffused" base [3], (see Chapters 1 and 10).

> **Power Devices and Packages**
>
> While the generally accepted term for *power device* describes any semiconductor component, discrete or integrated, *delivering* more than one watt of power to an electrical load, informal usage often includes all high-voltage devices, i.e., > 30 V, regardless of a load's power consumption. Power packages (such as the DPAK, TO220 or IPMs) refer to packages capable of *dissipating* high power without overheating.

Guide to State-of-the-Art Electron Devices, First Edition. Edited by Joachim N. Burghartz.
© 2013 John Wiley & Sons, Ltd. Published 2013 by John Wiley & Sons, Ltd.

| EDS Distinguished Service Award established | **1994** | ELECTRON DEVICES SOCIETY® | Supriyo Bandyopadhyay, Biswajit Das and Albert Miller propose ... | ... the single spin logic scheme, the first spin based logic gates |

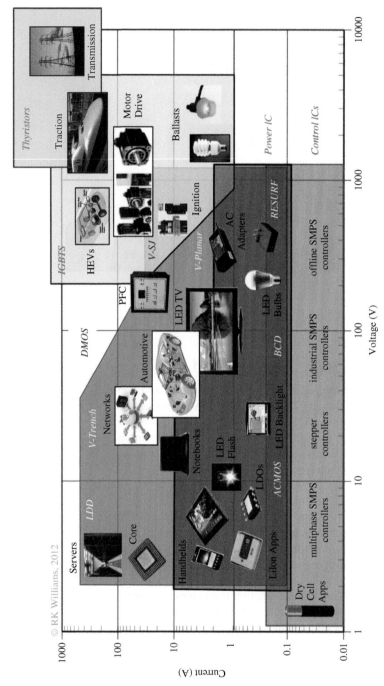

Figure 15.1 The current and voltage range of various power devices and applications

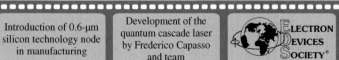

| Introduction of 0.6-μm silicon technology node in manufacturing | Development of the quantum cascade laser by Frederico Capasso and team | **E**LECTRON **D**EVICES **S**OCIETY® | Franz Laermer and Andrea Urban of Robert Bosch GmbH in Germany … | … pioneer high-aspect ratio etching technique in silicon, the "Bosch process" |

The advent of the insulated gate doubled diffused MOSFET (VMOS, DMOS) in the late 1970s heralded the dawn of high-frequency gate controlled power semiconductors [2, 4], (see Chapter 2). Virtually free from second breakdown and hotspots, DMOS transistors could easily be paralleled for high-current low on-resistance operation with minimal drive losses. By the 1980s, fueled by the emergence of personal computers, hard disk drives and automotive electronics, the planar vertical DMOS switching at hundreds of kilohertz enabled a dramatic size reduction in power supplies and small motor drives. Soon after, the insulated gate bipolar transistor (IGBT) [5–6], capable of switching high currents at high frequencies (>20 kHz) in offline and HVDC circuits, quickly emerged as the preeminent gate-controlled high voltage device with applications in inverters for motor drive, and later in UPS, electronic ignition and traction.

Contemporaneously, lateral DMOS [7] emerged in audio amps [8], as integrated high-voltage lightly doped drain LDD and "RESURF" devices [9] in displays and offline power supply controller ICs, and as low-voltage power devices in motor drive and multi-channel voltage regulator BCD power ICs.

By the early 1990s, the commercial premiere of the vertical trench gate DMOS [10] represented a significant milestone in the power handling capability of low voltage devices, soon thereafter tripling vertical planar DMOS cell densities and halving best-in-class device on-resistance [11]. Fortuitously, the concurrent emergence of lithium ion battery protection in cellphones, switching voltage regulators in notebooks, and ABS and airbags in automobiles, accelerated industry adoption of the trench DMOS into ubiquity.

In the 2000s, power semiconductor fabrication began its inexorable move into 200-mm submicron facilities [12] offering a tenfold increase in DMOS cell densities, and moving submicron BCD ICs [13] into integrated systems. The introduction of high voltage superjunction DMOS [14] expanded unipolar device ratings beyond the "silicon limit", so improving PFC and offline supplies.

In the 2010s, power device focus turned toward energy efficiency including photovoltaic (PV) inverters, LED lighting, next-gen superjunction DMOS, and commercial development of wide bandgap semiconductor devices [15] comprising silicon carbide, gallium nitride and other heterogeneous materials.

15.2 Two-Carrier and High-Power Devices

Carrying current by the simultaneous conduction of both majority and minority carriers, two-carrier ("bipolar") power devices comprise an expansive range of discrete and integrated components. Conveniently categorized by their number of constituent P and N (conductivity type) layers, two-carrier power devices include two-layer *diodes*, three-layer *bipolar transistors*, and four- (or-more) layer *IGBTs* and *thyristors* (see Figure 15.2). During conduction, at least one P-N junction is forward biased, injecting minority carriers throughout portions of the device. These carriers "diffuse" through the crystal lattice exhibiting minimal power losses until they are "collected" (or recombine), even in thick lightly doped materials common to high voltage devices. Because "bulk" conduction is relatively immune to the presence of surface defects, large area (even whole-wafer) power devices can be volume manufactured with high yields, making two-carrier devices well suited for high power operation up to hundreds of amps.

Unfortunately, forward-biased P-N junctions need time for injected carriers to recombine (i.e., the reverse recovery time t_{rr}) limiting a diode's upper switching frequency. Even so, to rectify AC into DC in switching regulators or to clamp inductive spikes, junction and PIN diodes remain indispensable. Power bipolars require high base currents (10% of a load's current) to operate as high current switches, and have been largely displaced by power MOSFETs and IGBTs.

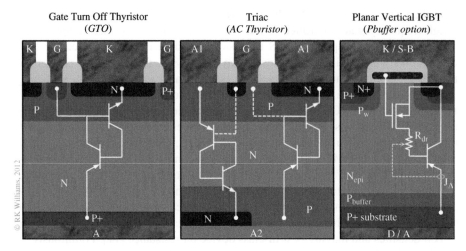

Figure 15.2   Two-carrier power devices

A dramatic improvement over bipolar transistors, the N-channel insulated-gate bipolar transistor (IGBT) [16–18] derives its base drive through an integrated N-channel vertical DMOS *in-series* with the base of a high voltage PNP bipolar, combining high-impedance DMOS gate drive, low conduction losses, and high speed (20–100 kHz) [19] non-latching switching into a single vertical discrete device. Holes injected by IGBT's anode, conductivity-modulate the lightly doped N_{epi} layer, lowering the DMOS drain resistance R_D (see inset). In contrast, for-layer PNPN thyristors contain both NPN and PNP bipolars [20] and, once activated by its gate or light, latch into a low voltage state flooding the entire device with minority carriers. In the GTO or gate turn-off thyristor, a negative gate pulse diverts holes away from the cathode, shutting off the device. The TRIAC is a symmetric AC thyristor limited to line frequencies with a four-quadrant gate (dotted lines) offering a variety of options for controlling turn on.

> ## Conductivity Modulation
>
> Based on the physical principle of charge neutrality, regions of a semiconductor device ooded with minority carriers exhibit *conductivity modulation* – a reduced resistance and voltage drop resulting from a localized increase in majority carriers, an effect bene cial in improving the on-state current density in IGBTs, but adversely invoking snapback breakdown in some lateral high voltage devices

15.3 Power MOSFET Devices

Silicon power MOSFETs, insulated gate field effect transistors with large channel widths (where a single device may integrate up to several *meters* of gate width), represent one of the most versatile elements in modern power electronics, offering simple high-impedance (capacitive) gate drive, low on-state conduction losses, high frequency (>1 MHz) switching ability, and high off-state breakdowns. As a

switch, a MOSFET's negative temperature coefficient-of-current and its intrinsic current-saturation provides easy paralleling of integrated or multiple discrete devices while maintaining immunity from hotspots and thermal runaway. Relying on majority carrier conduction, switching speed is not limited by minority carrier recombination. In linear circuits such as linear regulators, the device can maintain a constant current over a wide voltage range.

Collectively, power MOSFETs represent a diverse range of discrete and integrated components as specified by their maximum rated gate and drain voltages, channel polarity (N, P, CMOS), construction (planar or trench gate, uniform or DMOS channel, drain engineering), current flow (lateral, vertical, quasi-vertical), and by threshold voltage including enhancement or depletion (normally-on) mode types. Drain voltage ratings vary from 3 V for submicron planar devices to over 1200 V for vertical DMOS. Power MOSFETs may briefly survive operating in avalanche (see sidebar), but even short gate spikes can damage a device. As unipolar devices in large signal switching circuits, gate-charging current ideally limits a MOSFET's maximum frequency. In DC/DC converters, however, gate drive losses $Q_G V_G f$ must be managed [21] either by minimizing gate charge Q_G (i.e., reducing a device's input capacitance) or slowing the clock. Most vertical power MOSFETs comprise double diffused, that is "DMOS" channel construction, having a P-type well or "body" diffused or alternatively implanted at high energies [12] into a lighter doped epitaxial drain, with an (N+) source formed inside said well (see Figure 15.3). When biased,

Operation in Avalanche

A power MOSFET may reliably operate in avalanche breakdown provided it is designed to prevent turn-on of its intrinsic parasitic (NPN) bipolar transistor and that its avalanche energy rating (as measured in mJ) is not exceeded. A "rugged" device is one where no snapback (second breakdown) or parasitic conduction occurs, even at the point of melting. Power MOSFETs in a car's ABS brakes routinely operate in avalanche.

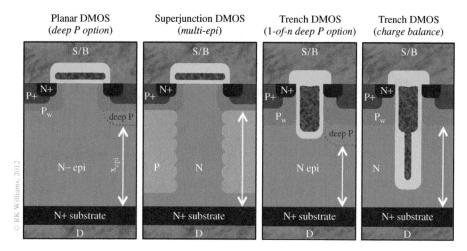

Figure 15.3 Various types of vertical DMOS devices

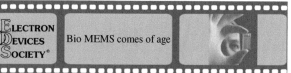

the DMOS channel's negative concentration gradient pushes depletion spreading into the drain facilitating a *short channel device* free from channel punch-through. Although planar gates predominate high voltage vertical DMOS (>100 V) designs, a tighter cell pitch and trench gate are uniquely advantageous at lower voltages where channel resistance is a significant factor.

In planar vertical DMOS, deep P suppression of parasitic bipolar turn-on [22] was the subject of extensive 1980s patent litigation. Deep P was later used to shield fragile trench gate oxides from damage, first in every cell [23] and then as an avalanche clamp in one of every "n" cells [11]. Today, charge balancing in vertical trench and superjunction silicon DMOS facilitates safely increasing epitaxial drain concentrations, while pushing R_{DS}·A specific on-resistance (mΩ-cm^2) beyond the "silicon limit". Below 20 V, lateral LDD NMOS with sidewall spacers [24] offer best-in-class milliohm devices with low Q_G.

Long before its vertical device successors, charge balancing was first realized using the "RESURF" principle (an acronym for *re*duced *surf*ace *fields*). In RESURF, charge in the lateral drift region is supported by depletion from the vertical PN junction, dramatically reducing the peak electric field [9] to create essentially uniform lateral electric field, with a higher breakdown and a lower DMOS specific on-resistance (see Figure 15.4). A constant electric field in the "drift" region also beneficially allows the breakdown voltage of a lateral DMOS to be scaled by mask design, increasing breakdown linearly with drift length L_D without the need for modifying fabrication processes. Adapting RESURF to silicon-on-insulator (SOI) lateral DMOS, a nearly ideal uniform electric field profile [25] is as shown achieved by monotonically increasing the charge profile from source-to-drain while absorbing depletion spreading in the handle and polysilicon electrodes (Figure 15.4). Reliable operation also requires minimal charge trapping in the drift passivant.

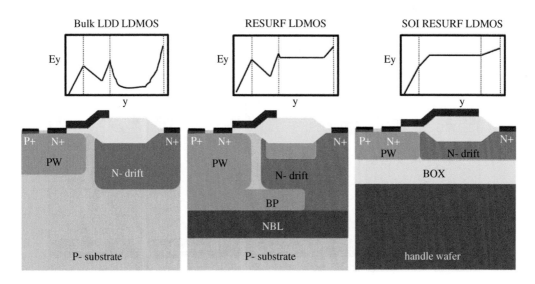

Figure 15.4   Comparison of surface fields in bulk LDD versus epitaxial and SOI RESURF LDMOS

Supriyo Datta *et al.* start to develop Huckel theory based model for conduction in aromatic thiols

1996

ELECTRON DEVICES SOCIETY®

Luc Berger and John Slonczewski propose the spin transfer torque idea …

… that later forms the basis of magnetic random access memory (MRAM)

15.4 High-Voltage and Power ICs

The remarkable development of high voltage (HVICs) and power ICs (PICs) is sometimes called the "quiet" revolution in electronics. Without its benefit on cost, size, efficiency, battery life, and reliability, many of the products we take for granted today (e.g., smartphones, tablets, and notebook computers) simply could not exist. Increased levels of integration made possible by reuse of legacy submicron digital [26] and ex-DRAM wafer fabs integrating disparate analog and digital control with power devices, programmable logic, memory and even sensors, now enable power system-on-chip (PSoC) solutions thought impossible even one decade ago. Examples include PMICs for smartphones integrating up to 20 different linear and switching regulated voltages with a switching charger and LiIon fuel gauge, and HVICs including a three-phase motor inverter for consumer air conditioners and miniature universal input AC adapters for USB phone charging.

One unique and distinct technology widely used for power integration is "BCD" technology. BCD processes, combining *b*ipolar, *C*MOS, and *D*MOS device arsenals (see Figure 15.5), benefit directly both from advances in CMOS scaling [27] and feature density and from tremendous progress made in lateral, quasi-vertical and vertical power devices through high-density planar and trench gates, drain engineering using LDD, multi-RESURF, and superjunction concepts, shallow and deep trench isolation, eliminating high temperature diffusions using MeV high-energy ion implantation [28], and integrating two-carrier high-voltage devices such as lateral IGBTs. Significant advances have been accomplished to overcome some major challenges to HVICs and PICs, particularly issues involving isolation and interconnection. The large scale of BCD integration invariably results in unwanted parasitic MOS devices due to polysilicon/metal interconnects, NPN and PNP bipolar transistors, and thyristors.

To suppress parasitic conduction, process methods including "*j*unction *i*solation" (JI), *d*ielectric *i*solation (DI), *s*ilicon *o*n *i*nsulator (SOI), and *d*eep *t*rench *i*solation (DTI) are used to isolate power devices and sensitive circuit blocks into silicon "islands". Layout methods to suppress parasitics include locally shorting

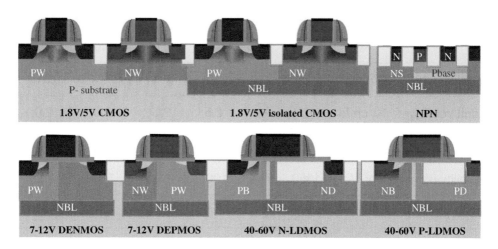

Figure 15.5   Cross-section of submicron BCD technology

Y. Cheng, M. Chan, K. Hui, M.-C. Jeng, Z. Liu, J. Huang, K. Chen, J. Chen, R. Tu, P.-K. Ko, and C. Hu …

… release of the MOSFET compact model BSIM3v3.0

ELECTRON DEVICES SOCIETY®

J.J. Yao *et al.* present the first micromachined switch

D.J. Young and B.E. Boser show RF MEMS tunable capacitors

base–emitter junctions, and utilizing guard rings, field plates, and DTI in sensitive areas. As Nobel laureate Jack Kilby once commented [29], "the challenge of integration is preventing conduction in devices you never intended to make." Interconnections pose another major challenge to HVICs and PICs as the need for *fine line metal* conflicts with the *thick metal* (>3 μm) required by power devices for high current capability. Multi-layer metal using a dry-etched thick copper (or Al-Cu) top layer and chip-scale packaging (CSP) are now common in power ICs.

15.5 Wide Bandgap Power Devices

Despite its low cost and mature manufacturing, the electrical and thermal properties of silicon diminish its future prospects in power. For one, silicon's relatively small bandgap (1.1 eV) and high leakage currents at elevated temperatures prohibits its reliable operation above 200°C – conditions demanded in engine compartment applications and by new generation high-power devices. The optimum specific on-resistance of a power device [30] as given by the relation $R_{on}A = 4V_B^2/\varepsilon\mu E_c^3$ depends most strongly on its critical electric field E_c, [31] and to a lesser degree on its dielectric constant and carrier mobility. Brief inspection of Table 15.1 comparing silicon to compound and wide bandgap (WBG) materials reveals among the semiconductors shown, the critical electric field E_c of GaN and SiC exceed silicon by roughly an order of magnitude, meaning a given length drift region in WBG material can support at least 10 times the voltage of silicon. Conversely, a WBG device having drift length one tenth as long can support the same voltage as silicon but with 10% the resistance. In some WBG materials, for example, in GaN, high-electron mobility transistor (HEMT) designs may furthermore, substantially improve mobility and reduce channel resistance.

Table 15.1 Comparison of silicon to compound and WBG semiconductor materials

Property (at 300 K)	Units	Si	GaAs	GaN	3C-SiC	6H-SiC	4H-SiC	Diamd
Bandgap E_g	eV	1.12	1.43	3.4	2.4	3.0	3.2	5.5
Thermal conductivity λ	W/cm K	1.5	0.5	1.3	3–4	3–4	3–4	19.5
Sat. elec. drift vel. v_{sat}	10^7 cm/s	1	1	2.5	2.5	2.0	2.0	2.5
Dielectric constant ε_r	–	11.9	13.0	9.5	9.7	10	10	5.6
Electron mobility μ_n	cm^2/V·s	1350	8500	1000	1000	500	950	1700
Critical electric field E_c	10^5 V/cm	2.5	3.0	30	20	25	22	73
Critical field ratio to Si	–	1	1.2	12	8	10	9	29

| 1997 | 3D Intracortical Electrode Array (E.M. Maynard, C.T. Nordhausen, R.A. Normann) | ELECTRON DEVICES SOCIETY® | Mark Reed measures currents through BDT molecules … | … in a break junction geometry (assumed single molecules, but not yet demonstrated) |

Despite the material, device and fabrication challenges, clearly a compelling motivation exists to develop power devices based on WBG materials. To date, the first WBG devices comprising silicon carbide (SiC) are now commercially [32] available from 100 mm wafers including high voltage Schottky diodes, bipolar junction transistors (BJTs), JFETs and power MOSFETs. Development of GaN materials and devices, less mature than SiC, focus on lower power ratings using lateral FET designs up to 1 kV and on integrating GaN devices on top of silicon integrated circuits using silicon wafer fabrication facilities.

References

[1] Nick Holonyak, "The Silicon p-n-p-n Switch and Controlled Rectifier (Thyristor)", *IEEE Trans. Power Elec.*, vol. 16, no. 1, pp. 8–16, Jan. 2001.

[2] Michael S. Adler, King W. Owyang, B. Jayant Baliga, and Richard A. Kokosa, "Evolution of Power Device Technology", *Trans. Elec. Dev.*, vol. 31, no. 11, pp. 1570–1591, Nov. 1984.

[3] John N. Ellis and Vince S. Osadchy, "The 2N3055: A Case History", *IEEE Trans. Elec. Dev.*, vol. 48, no. 11, pp. 1477–2484, Nov. 2001.

[4] Siliconix, "MOSFET Power Soars to 60W with Currents Up to 2A", *Elec. Des.*, vol. 21, no. 11, p. 103, Nov. 1975.

[5] B.J. Baliga, M.S. Adler, P.V. Gray, R.P. Love, and N. Zommer, "The Insulated Gate Rectifier (IGR) – A New Power Switching Device", *Tech. Dig. IEEE International Electron Devices Meeting (IEDM)*, pp. 264–267, 1982.

[6] J.P. Russell, A.M. Goodman, L.A. Goodman, and J.M. Neilson, "The COMFET – A New High Conductance MOS-gate Device", *IEEE Elec. Dev. Lett.*, vol. 4, pp. 63–65, 1983.

[7] Michael J. Declercq and James D. Plummer, "Avalanche Breakdown in High-Voltage D-MOS Devices", *IEEE Trans. Elec. Dev.*, vol. 23, no. 1, pp. 1–4, 1976.

[8] Ben Duncan, *High Performance Audio Amplifiers*, Newnes, Oxford, Chapter 5.2 – Power Devices, pp. 177–179, 1996.

[9] J.A. Appels and H.M.J. Vaes, "High Voltage Thin Layer Devices (RESURF Devices)", *Techn. Dig. IEEE International Electron Devices Meeting (IEDM)*, pp. 238–241, 1979.

[10] Richard K. Williams, King Owyang, Hamza Yilmaz, Michael Chang and Wayne Grabowski, "Complementary Trench Power MOSFETs Define New Levels of Performance", *16th Internationale Fachmesse für Bauelemente und Baugruppen der Elektronik (Electronica)*, Munich, Germany, pp. 1–7, Nov. 1994.

[11] Richard K. Williams, Wayne Grabowski, Mohamed Darwish, Michael Chang, Hamza Yilmaz, and King Owyang, "1 Million-cell 2.0-mΩ 30-V TrenchFET Utilizing 32 Mcell/in^2 Density with Distributed Voltage Clamping", *Techn. Dig. IEEE International Electron Devices Meeting (IEDM)*, pp. 363–366, 1997.

[12] Richard K. Williams, "Beyond Y2K: Technology Convergence as a Driver of Future Low-Voltage Power Management Semiconductors", *plenary session (invited), Proc. 12th Int. Symp. Power Semi. Dev. & ICs (ISPSD'00)*, pp. 19–22, Toulouse France, 2000.

[13] A. Andreini, C. Conterio, and P. Galbiati, "BCD Technologies for Smart Power ICs", in *Smart Power ICs, Technologies and Applications*, (eds B. Murari, F. Bertotti, G.A. Vignola), Springer, 1997.

[14] G. Deboy, N. Marz, J.P. Stengl, H Strack, J. Tihanyi, and H. Weber, "A New Generation of High Voltage MOSFETs Breaks the Limit Line of Silicon", *Techn. Dig. IEEE International Electron devices Meeting (IEDM)*, pp. 683–685, 1998.

[15] T.P. Chow and R. Tyagi, "Wide Bandgap Compound Semiconductors for Superior High-Voltage Power Devices", *Proc. 5th Int. Symp. Power Semi. Dev. & ICs (ISPSD'93)*, pp. 84–88, Monterey CA, 1993.

[16] Yamagami, *Japan patent no. 47-021739-B*, issued Jun. 1972 (filed 1968).

[17] B.J. Baliga, "Enhancement and Depletion Mode Vertical Channel MOS Gated Thyristors", *Elect. Letters*, vol. 15, pp. 645–647, Sep. 1979.

| The first plasma screen television is marketed by Pioneer Co., Japan | Kyoto protocol convention on climate change |  ELECTRON DEVICES SOCIETY® | Nitta *et al.* demonstrate gate control of … | … Rashba spin-orbit interaction in the channel of an FET |

[18] Hans W. Becke and Carl F. Wheatley, "Power MOSFET with an Anode Region", *U.S. patent no. 4,364,073*, issued Dec. 1982 (filed Mar. 1980).

[19] Carl Blake and Chris Bull, "IGBT or MOSFET: Choose Wisely", *Int. Rect.* (available at www.irf.com/technical-info/whitepaper/choosewisely.pdf, accessed October 9, 2012).

[20] J.J. Ebers, "Four terminal P-N-P-N transistors", *Proc. IRE*, vol. 40, pp. 1361–1364, Nov. 1952.

[21] R.K. Williams, R. Blattner, and B.E. Mohandes, "Optimization of Complementary Power DMOSFETs for Low-Voltage High-Frequency DC-DC Conversion", *10th Appl. Pwr. Elect. Conf (APEC'95)*, vol. 2, pp. 765–772, Mar. 1995.

[22] Alex Lidow and Tom Herman, "High Power MOSFET with Low On-Resistance and High Breakdown Voltage", *U.S. Patent no. 4,376,286*, issued Mar. 1983 (filed Feb. 1981).

[23] Constantin Bulucea and Rebecca Rossen, "Trench DMOS Power Transistor with Field-Shaping Body Profile and Three-Dimensional Geometry", *U.S. Patent no. 5,072,266*, issued Dec. 1991 (filed Dec. 1988).

[24] Z.J. Shen, D. Okada, F. Lin, A. Tintikakis, and S. Anderson, "Breaking the Scaling Barrier of Large Area Lateral Power Devices: a 1 mΩ Flip-Chip Power MOSFET with Ultra Low Gate Charge", *Proc. 16th Int. Symp. Power Semi. Dev. & ICs (ISPSD'04)*, pp. 387–390, May 2004.

[25] T. Letavic, E, Arnold, M. Simpson, R. Aquino, *et al.*, "High Performance 600V Smart Power Technology Based on Thin-Layer SOI", *Proc. 9th Int. Symp. Power Semi. Dev. & ICs (ISPSD'97)*, pp. 49–52, May 1997.

[26] T. Efland, S. Malhi, W. Bailey, O.-K. Kwon, W.-T. Ng, M. Torreno, and S. Keller, "An Optimized RESURF LDMOS Power Device Module Compatible with Advanced Logic Processes", *Techn. Dig. IEEE International Electron Devices Meeting (IEDM)*, pp. 237–240, 1992.

[27] S. Bach, L. Atzeni, A. Molfese, A. Dundulachi, E. Castellana, G. Croce, and C. Contiero, "HVCMOS8A: 22V-42V Rated MOS Integration in a 0.18 µm Technology Platform for High Voltage Applications", *Proc. 20th Int. Symp. Power Semi. Dev. & ICs (ISPSD'08)*, pp. 56–59, May 2008.

[28] R.K. Williams and M.E. Cornell, "The Emergence & Impact of DRAM-Fab Reuse in Analog and Power-Management Integrated Circuits", (invited) *Proc. Bipolar/BiCMOS Crkts & Tech. Mtg. (BCTM2002)*, Monterey CA, pp. 45–52, Sep. 2002.

[29] Jack Kilby, Nobel Laureate, *in private conversation with Richard K. Williams at University of Illinois at Urbana-Champaign (UIUC) ECE Distinguished Alumni Awards banquet*, Oct. 2000 (two months prior to winning the Nobel Prize for the invention of the integrated circuit).

[30] Chenming Hu, "Optimum Doping Profile for Minimum Ohmic Resistance and High-Breakdown Voltage", *IEEE Trans. Elec. Dev.*, vol. 26, no. 3, pp. 243–244, 1979.

[31] B.J. Baliga, "Semiconductors for High-voltage, Vertical Channel FET's", *J. Appl. Phys.*, vol. 53, no. 3, pp. 1759–1764, Mar. 1982.

[32] Mikael Östling, Martin Domeij, Carina Zaring, *et al.*, "SiC Bipolar Power Transistors - Design and Technology Issues for Ultimate Performance", in *Silicon Carbide 2010 – Materials, Processing, and Devices*, (eds S.E. Saddow, E. Sanchez, F. Zhao, and M. Dudley) *Mater. Res. Soc. Symp. Proc.*, vol. 1246, Warrendale, 2010.

| Introduction of 0.3-µm silicon technology node in manufacturing | First publication of uni-carrier travelling photodiodes ... | ELECTRON DEVICES SOCIETY® | ... with high performance up to 1 THz by T. Ishibshi and co-workers | EDS Chapter of the Year Award is introduced |

Chapter 16

Photovoltaic Devices

Steven A. Ringel, Timothy J. Anderson, Martin A. Green, Rajendra Singh and Robert J. Walters

16.1 Introduction

Photovoltaic (PV) technologies in their most common form are based on semiconducting devices that directly convert optical energy from sunlight to electrical energy. The key physical steps for a typical *p-n* junction solar cell are optical absorption, through which excess electron and hole charge carriers are generated within the device and separated by the junction built-in potential, followed by electron and hole transport to their respective contacts. When connected to a resistive load, the photovoltage and photocurrent generated by the solar cell dissipates power, the amount of which depends on the solar energy conversion efficiency of the solar cell and the incident solar intensity.

The apparent simplicity of a solar cell, coupled with the enormous abundance of clean, renewable and relatively non-geopolitical solar energy resources available to our planet (approximately 89,000 Terawatts at the Earth's surface), has driven interest in PV for many years. The impetus has been particularly strong since the early 2000s in the face of growing global industrialization and its effect on access to and use of traditional fossil fuels. However, the initial development of the modern day solar cell, first patented alongside the *p-n* junction in 1946 [1], occurred at Bell Laboratories in the 1950s and was actually driven by the need to power future (at the time) telecommunication satellites. This work led to the invention of the first modern solar cell in 1954 [2]. Since then, an extraordinary number of innovations have occurred at the materials, design, manufacturing, and system (concept) levels. Figure 16.1 shows a timeline of verified

Guide to State-of-the-Art Electron Devices, First Edition. Edited by Joachim N. Burghartz.
© 2013 John Wiley & Sons, Ltd. Published 2013 by John Wiley & Sons, Ltd.

| IEEE Journal of Technology Computer Aided Design (TCAD) starts publications | 1998 | ELECTRON DEVICES SOCIETY® | Superjunction MOSFET emerges | First demonstration of 20-nm MOSFET at Intel |

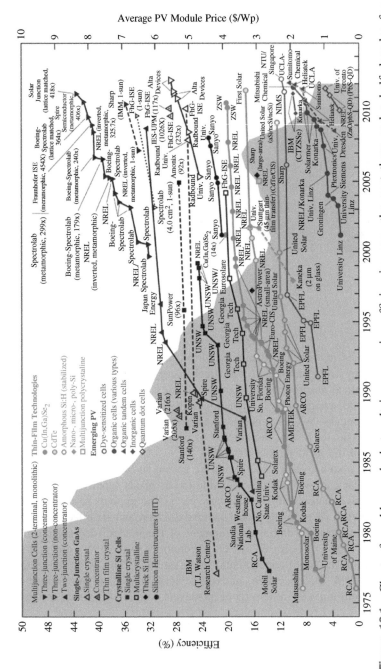

Figure 16.1 Chart of world-record photovoltaic conversion efficiencies over the past three and a half decades for a variety of solar cell technologies [3]. The background area plot (grey) tracks average PV module prices over the same time-span [4] (Solar cell efficiency data compiled by L. L. Kazmerski, NREL based on background cost chart http://nreldev.nrel.gov/ncpv/images/efficiency_chart.jpg.)

| Time dependent oxide breakdown (TDDB) observed in low-k interconnect dielectrics | Introduction of 0.25-μm silicon technology node in manufacturing | ELECTRON DEVICES SOCIETY® | 1999 | M. Green (UNSW) demonstrates silicon cell one-sun efficiency of 24.7% |

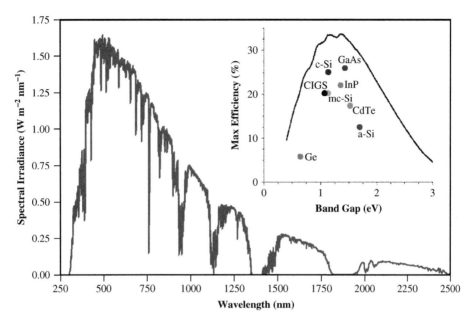

Figure 16.2   Plot of spectral irradiance versus wavelength for the standard AM1.5G solar spectrum. The inset shows the Shockley–Queisser [5] AM1.5G single-junction solar cell conversion efficiency limit versus bandgap, along with champion single-junction cell performances for a variety of PV materials

record efficiencies as a function of many PV technologies at varying stages of maturity; the grey background plot tracks the average cost of PV modules over the same time-frame, demonstrating the continued reduction in cost and improvement in performance.

The grand challenge for PV is becoming cost competitive with conventional energy sources. The economies of scale will continue to reduce cost directly, but performance must also be increased. The non-uniform (i.e., wavelength-dependent) intensity distribution of the solar spectrum creates considerable challenges in reaching theoretical energy conversion efficiency limits due to tradeoffs between sub-bandgap transmission losses and energy thermalization losses for absorbed photons with energy in excess of the solar cell bandgap. The inset of Figure 16.2 displays the theoretical limits and maximum experimental efficiencies for single bandgap solar cells. This chapter provides an historical perspective on the development of the primary PV technologies – crystalline Si, polycrystalline Si, thin films, and III-V multijunctions, as well as future concepts that have the potential to extend well beyond the single bandgap limit.

16.2 Silicon Photovoltaics

Silicon (Si), as the second most abundant material in the Earth's crust, is the dominant commercial PV material. Currently, about 90% of solar cells are based on Si, and many predict this prevalence to continue for years to come [6]. Polycrystalline solar cells can use material with impurity levels of up to 10 parts per million, while crystalline cells possess much more stringent requirements, with maximum allowable impurity concentrations of one part per trillion or less. Using polysilicon as the feedstock,

high-quality crystalline silicon ingots are grown by either the Czochralski (CZ) or float-zone (FZ) method, while a lower-cost directional solidification process can be used to produce polysilicon ingots or even an intermediate "quasi-monocrystalline" grade. Multi-wire saws are used to obtain Si wafers of about 150–180 μm thickness (see Chapter 10). The push to further reduce the Si material cost will result in even thinner solar wafers in the future.

For optimal device performance, n^+-p-p^+ or p^+-n-n^+ device structures are preferred. Generally diffusion is used for junction formation. For reducing reflection losses and creating multiple internal reflections of incident photons, the Si surface is generally textured, followed by the application of a silicon nitride antireflection coating. For current collection, a metallic grid is applied on the front side with back ohmic contacts applied to the cell rear. Figure 16.3(a) shows the image of a polycrystalline Si solar cell. A module made up of "hybrid" poly-/mono-crystalline silicon solar cells is shown in Figure 16.3(b).

Significant fundamental improvements of laboratory Si solar cells occurred from the 1977–2000 timeframe, with the highest demonstrated AM1.5G efficiency for crystalline silicon solar cells of 25% [9]. Figure 16.4(a) shows the schematic diagram of a passivated emitter with rear locally-diffused (PERL) silicon solar cell, which holds this record. Using standard, commercial-grade p-type silicon wafers recent work has led to the development of a commercial version of the PERL cell with a record production-level AM1.5G efficiency of 20.3% [10].

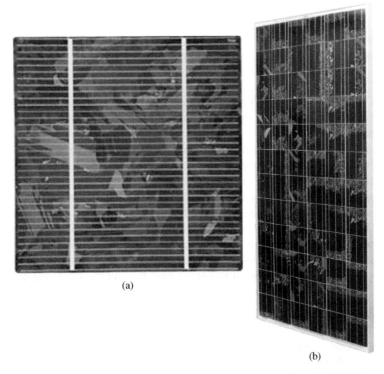

(a)

(b)

Figure 16.3   Photographs of (a) A typical polycrystalline Si solar cell [7]; (b) a Suntech 290 W "quasi-monocrystalline" PV module [8]

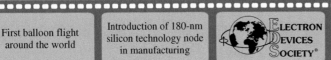

| First balloon flight around the world | Introduction of 180-nm silicon technology node in manufacturing | ELECTRON DEVICES SOCIETY® | MIS-structured high-k DRAM >1Gb (S.J. Won *et al.* of Samsung) | First MEMS optical network switch by Lucent Technologies Co. |

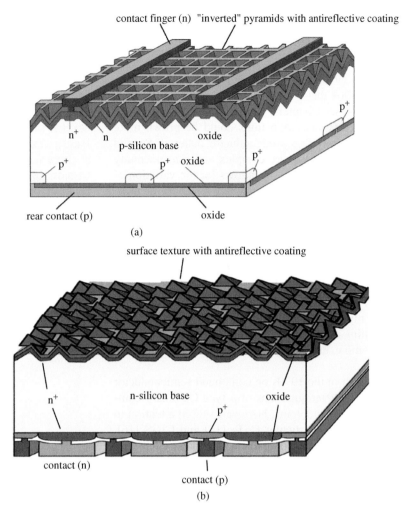

contact finger (n)   "inverted" pyramids with antireflective coating

p⁺

n⁺   n   oxide
p-silicon base
p⁺   oxide   p⁺
p⁺   oxide

rear contact (p)   oxide

(a)

surface texture with antireflective coating

n⁺   n-silicon base   oxide
p⁺

contact (n)

contact (p)

(b)

Figure 16.4 Schematic diagram of (a) passivated emitter with rear locally-diffused (PERL) and (b) rear-contact silicon solar cells (Courtesy of M. Green, UNSW)

Rear contact solar cells (Figure 16.4b) have all electrical contacts located on the back side of the cell. Advantages of this design include a grid-less front surface, as well as *n*-type starting material, which does not suffer the initial light-induced degradation of commonly-used *p*-type wafers. Using *n*-type CZ Si as the substrate, the highest demonstrated efficiency for a rear contact c-Si cell is 24.2% [11].

Use of larger wafer sizes reduction of materials defect densities, introduction of new process equipment with higher throughput, and reduced end-use footprint with lower cost of ownership are just some of the directions that the industry is expected to pursue for further cost reduction.

16.3 Polycrystalline Thin-Film Photovoltaics

Si-based devices are often termed "first generation" technologies as they were the first to be commercialized. A "second generation" of absorber materials emerged in the 1970s with higher absorption coefficients, allowing the use of thin films that could be deposited on inexpensive and even flexible substrates (see Chapter 17). These materials promised to reduce manufacturing costs since thin films could be grown more rapidly, while using significantly less semiconductor material. Second generation materials are chemically more complex and fundamentally less understood than Si, but this complexity allows for the variation of materials properties to optimize devices structures. A variety of these thin-film materials have been explored as photovoltaic absorbers, but amorphous Si (a-Si), CdTe, and $CuIn_xGa_{1-x}Se_2$ (CIGS) are the only ones that have progressed to significant commercialization [12]. It is noted, however, that solar cells based on both organic and so-called earth-abundant thin-film absorbers, such as Cu_2ZnSnS_4 (CZTS), have recently shown encouraging results (~10% efficiency).

a-Si: The bandgap of a-Si is ~1.8 eV and *p-i-n* type photovoltaic devices are typically only ~0.5 μm thick. Key challenges include increasing the rate of film deposition, reducing light-induced degradation, and integrating more efficient photon management schemes. [13, 14].

CdTe and CIGS: Research in the 1970s on compound semiconductor thin film devices focused on heterojunctions of *p*-type Cu binary compounds with *n*-CdS. Reliability problems, however, shifted attention to other *p*-type semiconductors, in particular CdTe [15] and CIGS [16]. Figure 16.5 shows a cross-sectional comparison of typical CIGS and CdTe device structures. Both devices typically use a glass substrate,

Higher Efficiency a-Si

Light-induced degradation in a-Si (Staebler–Wronski effect) limits device ef ciency, but the addition of H can reduce this effect. Higher ef ciency multijunction devices can be formed by alloying with Ge, while the growth of nano- or micro-crystalline Si domains embedded in a-Si, termed mc-Si, also increases performance [13].

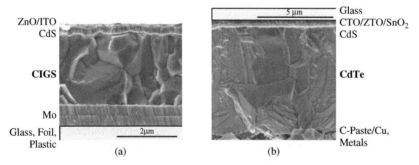

ZnO/ITO
CdS
CIGS
Mo
Glass, Foil, Plastic
2μm
(a)

5 μm
Glass
CTO/ZTO/SnO$_2$
CdS
CdTe
C-Paste/Cu, Metals
(b)

Figure 16.5   Cross-sectional scanning electron microscope images (false color) comparing typical (a) CIGS and (b) CdTe solar cell device structures [17] (Courtesy of R. Noufi and A. Hicks, NREL)

2000   Into the Third Millenium   2000

an *n*-type CdS heterojunction partner, an established transparent conducting oxide (TCO) top electrode, and a polycrystalline absorber. The most obvious difference besides the absorber thickness is that the CIGS device is grown with the back contact on the substrate, while the CdTe device is grown in the reverse order, with the top contact on the substrate (known as a superstrate, or inverted structure). The back contact to the *p*-type absorber is challenging in these materials, with strain-engineered Mo/MoSe$_2$ used for CIGS, while an interfacial layer is required to create a low resistance contact to CdTe.

Importantly, both CdTe and CIGS structures and processing requirements lend themselves to high-throughput, low-cost manufacturing, which reduces fabrication costs. Current research directions common to both technologies include identifying a Cd-free buffer layer material with a wider bandgap than CdS, decreasing absorption losses, and increasing conductivity in the TCO layer. Long-term module reliability is also of interest as CIGS, and to a lesser extent CdTe, modules do not have a long installation history.

16.4 III-V Compound Photovoltaics

Extremely high efficiency solar cells can be made using elements from the III and V columns of the periodic table (see Chapter 6). The first GaAs solar cell was reported in 1956 [19], and a GaAs cell now holds the single-junction world record at 28.8% [20]. Moreover, III-V multijunction (MJ) solar cells (Figure 16.6), in which each sub-junction is optimized for a specific portion of the solar spectrum, can far exceed the single-junction limit, with the first III-V dual-junction (AlGaAs/GaAs) reported in 1979 [21] and the first III-V triple-junctions (InGaP/GaAs/Ge) reported in 1998 [22]. III-V cells also show superior radiation hardness and have been developed for space applications. For terrestrial applications, III-V cells are being used primarily in concentrator systems, where the sunlight is focused onto a small solar cell using inexpensive optical components. This increases efficiency reduces the required cell area, thereby reducing the system cost.

The present industry MJ standard is the InGaP/GaAs/Ge cell, with a top demonstrated efficiency of 41.6% under 364 Suns [20]. The more optimal bandgaps of InGaP/GaAs/InGaNAs raised this to a world record efficiency of 44.0% under 942 Suns [20]. Another way to achieve optimum bandgap combinations is by using metamorphic buffers to allow mismatched growth, producing efficiencies as high as 42.3% under 406 Suns using InGaAs buffers [20].

CdTe Thin Film PV

Zincblende CdTe has an ideal direct bandgap (1.45 eV), high absorption coef cient, and low temperature coef cient. Heat treatment with CdCl$_2$ and O$_2$ is essential for high performance. Current research issues include decreasing the back contact resistance and reducing recombination defects.

CIGS Thin Film PV

CIGS's excellent adsorption coef cient and bandgap exibility have earned it the current thin lm solar cell record ef ciency (20.3%) [18]. Current research priorities are improving ef ciency at higher bandgap and increasing synthesis rates. The potential of CIGS will soon be known as production-scale plants are just coming on-line.

Nanostructures for Bandgap Engineering

Optimum bandgap combinations may also be attained through "bandgap engineering" achieved with nanostructures in the solar cell junction. For example, strain-balanced quantum wells (QW) or quantum dots (QD) can be used to effectively lower the bandgap, and single-junction cell ef ciencies of over 28% under concentration has been achieved via this approach [23].

| Immersion Lithography emerges | | ELECTRON DEVICES SOCIETY® | Marc Cahay, Phillippe Debray and others … | … take the first step towards all-electric spintronics |

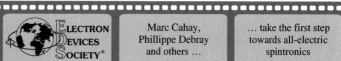

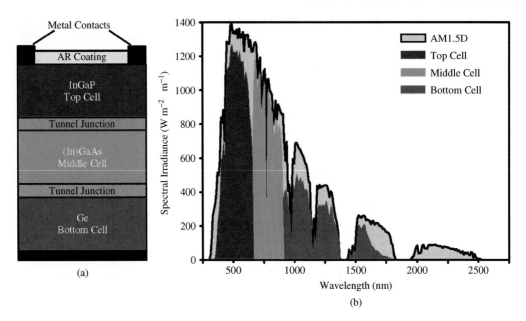

(a)

(b)

Figure 16.6 Generalized schematic diagram of (a) the current state-of-the-art lattice-matched triple-junction solar cell and (b) the partitioning of the solar spectrum amongst the individual sub-junctions

An alternate approach is the inverted metamorphic (IMM) architecture [24], in which the higher bandgap sub-cells are grown lattice-matched and the lower bandgap sub-cells are then grown lattice-mismatched. This effectively confines the deleterious effects of lattice-mismatched growth to the more forgiving low-bandgap junctions, and IMM MJ cells have achieved efficiencies over 40% at 326 Suns. Because of the inverted growth, the substrate must be removed either by etching or epitaxial lift-off, which has opened up entirely new application possibilities since the resultant IMM cell is extremely thin and flexible.

16.5 Future Concepts in Photovoltaics

Key losses for conventional cells (Figure 16.7a) restrict conversion efficiency to the Shockley–Queisser limit (31–34% for unfocused sunlight, depending on reference spectrum) [5]. Nonetheless, the photovoltaics industry will continue targeting ever increasing conversion efficiency. Thermodynamics suggest high efficiencies of over 70% are possible, increasing to 93% with concentration (although not all sunlight is easily concentrated). Figure 16.7(b) presents a number of proven photovoltaic approaches, as well as theoretical concepts and phenomena that hold potential to overcome the Shockley–Queisser single-junction limit.

Of the high-efficiency approaches shown in Figure 16.7(b), tandem stacks of different bandgap cells (see Section 16.4) have been most effective at increasing efficiency beyond the Shockley–Queisser limit [25]. However, apart from stacks of six or more cells, the potentially most efficient approach involves "hot carrier" cells, addressing loss mechanism (② in Figure 16.7a). Within picoseconds, electrons photoexcited

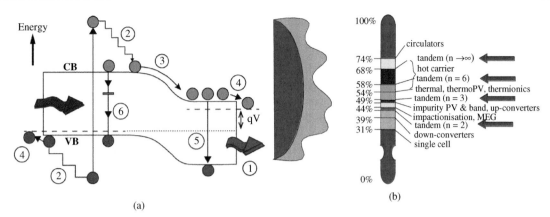

Figure 16.7 Diagrams describing (a) losses in standard solar cells [① non-absorption of sub-bandgap photons; ② thermalization losses; ③ and ④ junction and contact losses; ⑤ radiative and ⑥ non-radiative recombination losses] and (b) limiting efficiencies for various photovoltaic approaches and concepts (6000 K black body) (Courtesy of M. Green, UNSW)

from deep in the valence band (VB) to high in the conduction band (CB) lose any energy above the CB edge, with VB holes also quickly relaxing to the VB edge. Further energy loss occurs much more slowly (nanosecond to millisecond range).

Conventional cells target the latter timescale while hot carrier cells target the former, preserving photoexcited pair energy by slowing initial relaxation. Creating "hot phonon" pools via "phonon bottlenecks" can suppress net energy loss rates, one possible hot carrier approach.

Some approaches in Figure 16.7(b) with limits similar to four-cell tandem stacks involve heating the whole solar absorber to high temperature as in solar thermal electric conversion. For this conversion (e.g., "power towers"), fluids carry absorber energy to heat engines. In thermophotovoltaics, photons instead radiate absorber heat; they then are converted by a solar cell. In thermionics, energetic electrons transport the energy. All have identical limiting efficiency of 54%, increasing to 85% under focused sunlight, although the best experimental efficiency is 31% using Stirling engines and focused sunlight. Experimental thermionic and thermophotovoltaic devices perform even further from ideal, indicating higher losses than with ambient-temperature approaches.

Another group has limits between two- and three-cell tandem stacks. The impurity PV effect attempts turning loss (⑥ in Figure 16.7a) to advantage. Rather than photoexcited carriers recombining in a two-step process through mid-gap energy levels, photoexcitation occurs through such levels in two steps − excitation from the VB to the level by one photon and from there to the CB by a second. Design is relaxed using a band of states (impurity or intermediate band) rather than isolated levels [26], although limits are unchanged. Related photon "up-conversion" gives the same limit. Here the multi-step process occurs externally, with the excited electron relaxing by emitting a usable energy photon [25]. "Down-conversion", where an energetic photon converts to two lower energy photons addresses loss (② in Figure 16.7a), but has smaller theoretical impact.

| 2000 | Andrew Grove receives IEEE Medal of Honor and IEEE Presidents Award | ELECTRON DEVICES SOCIETY® | Russell Cowburn and Mark Welland propose nanomagnetic logic … | … with dipole coupled nanomagnets acting as logic switches |

The final process in Figure 16.7(b) involves creating multiple electron–hole pairs from a single energetic photon, such as by impact ionization. Here a strongly photoexcited electron uses surplus energy by exciting a second electron–hole pair, increasing photocurrent. In quantum dots, energetic photons can generate multiple bound electron–hole pairs (excitons). External quantum efficiency above unity has been reported using such multiple exciton generation, although power boosts are so far negligible [27].

Ongoing nanomaterial development will increase flexibility in implementing these concepts. By controlling light at the nanoscale, plasmonics may allow efficient conversion using very thin cells, importantly for concepts such as hot carrier cells.

For focused sunlight, with less critical cell costs, tandem III-V cell stacks have a clear path to continued improvement, with efficiencies above 50% likely. Mainstream photovoltaics, however, will continue using unfocused sunlight. With large recent cost reductions, silicon wafer-based cells are well-positioned to maintain dominance, and refinements in producing directionally-solidified monocrystalline ingots increase prospects for inexpensive monocrystalline cells prevailing by the end of this decade. Of these advanced concepts, thin tandem cells stacked on silicon again seem the most feasible in boosting performance, although silicon's awkward lattice constant presents challenges [28]. In the longer term, thin plasmonically-enhanced "hot carrier" cells or tandem stacks of thin-film cells using Earth-abundant materials might carry photovoltaics into the future.

References

[1] R. S. Ohl, *U.S. Patent No. 2,402,662* (June 25, 1946).

[2] D. M. Chapin, C. S. Fuller, and G. L. Pearson, "A New Silicon p-n Junction Photocell for Converting Solar Radiation into Electrical Power", *J. App. Phys.*, 25, pp. 676 (1954).

[3] NREL, *Best Research Cell Efficiencies*, Continuously updated at http://www.nrel.gov/ncpv/images/efficiency_chart .jpg (accessed October 9, 2012).

[4] P. Mints, Pre-2011 data from *Photovoltaic Manufacturer Shipments, Capacity, & Competitive Analysis 2009/2010.* Report # NPS-Supply4. Palo Alto, CA: Navigant Consulting Photovoltaic Service Program.; 2011 and 2012 numbers based on available market data. Prices given in contemporary U.S. dollars and are not adjusted for inflation (2011).

[5] W. Shockley and H. J. Queisser, "Detailed balance limit of efficiency of p-n junction solar cells", *J. App. Phys.*, 32, 510 (1961).

[6] R. Singh, "Why Silicon is and will Remain the Dominant Photovoltaic Material", *J. Nanophotonics*, 3, 032503 (2009).

[7] http://image.tradevv.com/2009/07/23/yybestsolar_432275_600/silicon-solar-cell-solar-cells.jpg.

[8] Suntech, *290 Watt Polycrystalline Solar Module*, available at http://am.suntech-power.com/images/stories/ pdf/datasheets/july2011/stp285_24vd_ul_superpoly285-290%20h4%20connector.pdf (accessed October 9, 2012) (2011).

[9] M. A. Green, "The Path to 25% Silicon Solar Cell Efficiency: History of Silicon Cell Evolution", *Prog. Photovolt.: Res. Appl.*, 17, 183 (2009).

[10] Z. Wang, P. Han, H. Lu, *et al.*, "Advanced PERC and PERL production cells with 20.3% record efficiency for standard commercial p-type silicon wafers", *Prog. Photovolt.: Res. Appl.*, 20, 260 (2012).

[11] P. J. Cousins, D. D. Smith, H. C. Luan, *et al.*, "Gen III: Improved Performance at Lower Cost", *35th IEEE Photovolt. Specialists Conf.*, 000275 (2010).

[12] C. A. Wolden, J. Kurtin, J. B. Baxter, *et al.*, "Photovoltaic manufacturing: Present status, future prospects, and research needs", *J. Vac. Sci. Technol. A*, 29, 030801 (2011).

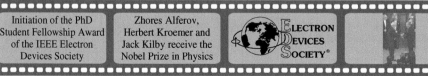

| Initiation of the PhD Student Fellowship Award of the IEEE Electron Devices Society | Zhores Alferov, Herbert Kroemer and Jack Kilby receive the Nobel Prize in Physics |  ELECTRON DEVICES SOCIETY® | | Replacement if the Jack A. Morton Award by the Andrew S. Grove Award |

[13] S. Guha and J. Yang, "Progress in amorphous and nanocrystalline silicon solar cells", *J. Non-Cryst. Solids*, 352, 1917 (2006).

[14] M. Konagai, "Present Status and Future Prospects of Silicon Thin-Film Solar Cells", *Jpn. J. Appl. Phys.*, 50, 030001 (2011).

[15] R. W. Birkmire and B. E. McCandless, "CdTe thin film technology: Leading thin film PV into the future", *Curr. Opin. Solid St. & Mat. Sci.*, 14, 139 (2010).

[16] B. J. Stanbery, "Copper Indium Selenides and Related Materials for Photovoltaic Devices", *Crit. Rev. Solid St. & Mat. Sci.*, 27, 73 (2002).

[17] Courtesy of R. Noufi and A. Hicks of NREL (n.d.).

[18] P. Jackson, D. Hariskos, E. Lotter, *et al.*, "New world record efficiency for Cu(In,Ga)Se$_2$ thin-film solar cells beyond 20%", *Prog. Photovolt.: Res. Appl.*, 19, 894 (2012).

[19] D. A. Jenny, J. J. Loferski, and P. Rappaport, "Photovoltaic effect in GaAs p-n junctions and solar energy conversion", *Phys. Rev.*, 101, 1208 (1956).

[20] M. A. Green, K. Emery, Y. Hishikawa, W. Warta, and E. D. Dunlop, "Solar cell efficiency tables (version 41)", *Prog. Photovolt.: Res. Appl.*, 21, 1 (2013).

[21] S. M. Bedair, M. F. Lamorte, and J. R. Hauser, "A two-junction cascade solar-cell structure", *Appl. Phys. Lett.*, 34, 38 (1979).

[22] N. H. Karam, L. Cai, D. E. Joslin, *et al.*, "High-efficiency GaInP$_2$/GaAs/Ge dual- and triple-junction solar cells for space applications", *Proc. 2nd World Conf. PVEC*, 3534 (1998).

[23] J. G. J. Adams, B. C. Browne, I. M. Ballard, *et al.*, "Recent results for single-junction and tandem quantum well solar cells", *Prog. Photovolt.: Res. Appl.*, 19, 865 (2011).

[24] M. W. Wanlass, S. P. Ahrenkiel, R. K. Ahrenkiel, *et al.*, "Lattice-mismatched approaches for high-performance photovoltaic energy converters", *Proc. 31st IEEE Photovolt. Specialists Conf.*, 530 (2005).

[25] M. A. Green, *Third Generation Photovoltaics: Advanced Solar Energy Conversion*, Springer-Verlag, Berlin (2003).

[26] A. Luque and A. Marti, "Increasing the efficiency of ideal solar cells by photon induced transitions at intermediate levels", *Phys. Rev. Lett.*, 78, 369 (1997).

[27] O. E. Semonin, J. M. Luther, S. Choi, *et al.*, "Peak External Photocurrent Quantum Efficiency Exceeding 100% via MEG in a Quantum Dot Solar Cell", *Science*, 334, 1530–1533 (2011).

[28] S. A. Ringel and T. J. Grassman, "III-V Solar Cells on Silicon", in *III-V Compound Semiconductors: Integration with Silicon-Based Microelectronics*, (ed. T. Li), CRC Press, Boca Raton, FL (2011).

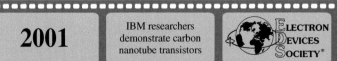

2001 — IBM researchers demonstrate carbon nanotube transistors

ELECTRON DEVICES SOCIETY®

Toshiba presents the first organic light emitting diode (OLED) display prototype

Chapter 17

Large Area Electronics

Arokia Nathan, Arman Ahnood, Jackson Lai and Xiaojun Guo

17.1 Thin-Film Solar Cells

17.1.1 Introduction

Solar cells fabricated using large area electronic technologies boasts a number of advantages (see Chapter 16). Through the use of large area thin-film materials and process technologies the cost of solar cells can be substantially reduced, when compared to bulk materials, making them an economically viable source of energy. Furthermore, thin-film technology lends itself to simplified system integration, opening the possibility of new energy-related application areas particularly with mobile devices (for examples, see [1, 2]). Additionally, as it will be discussed in this chapter, the possibility of depositing thin-films on flexible, lightweight, and low cost substrates, makes thin-film solar cells an attractive proposition from the design prospective. Cost per watt, typically presented as US$/watt, is an important figure of merit in the energy sector, often used to compare commercial viability of various energy sources. Figure 17.1. compares the efficiency and cost of three generations of solar cells, namely first-generation bulk solar cells, second-generation thin-film solar cells (TFSC) and third-generation emerging technologies (for examples see [3–7]. The dotted lines are the cost per watt benchmarks and the data points are based on market data for various solar cell technologies. It can be seen the TFSC provide a better-cost benefit compared with

Guide to State-of-the-Art Electron Devices, First Edition. Edited by Joachim N. Burghartz.
© 2013 John Wiley & Sons, Ltd. Published 2013 by John Wiley & Sons, Ltd.

First issue of the IEEE Transactions on Device Materials Reliability (T-DMR)

DEVICE AND MATERIALS RELIABILITY

ELECTRON DEVICES SOCIETY®

S.N. Keeny of Intel Co. presents 130-nm flash memory (32 Mb)

Symmetric linearization of bulk charge for MOSFET modeling by T.-L. Chen and G. Gildenblatt

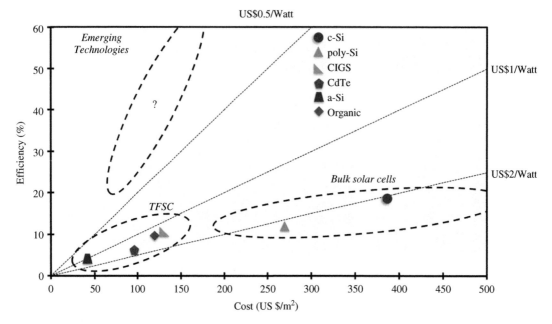

Figure 17.1 Efficiency and cost comparison for the three generations of solar cells; first-generation bulk solar cells, second-generation TFSCs, and the third-generation including tandem cells. Inserted data points are based on market research for the manufacturing costs and efficiencies of category of solar cells. Adapted from [8], additional data from [9, 10]

bulk solar cells with the emerging technologies projected to exceed both. It should be noted that many of the emerging technology designs are evolution of the TFSC [8].

The earliest observation of photovoltaic effect was reported by Becquerel in 1839 in a liquid based photoelectrochemical system [11], which forms the basis of what is now known as dye-sensitized solar cell (DSSC) [12, 13], an important category of thin-film solar cells (TFSC). Less than three years later the earliest solid-state photovoltaic effect was demonstrated using a platinum electrode contact pushed on selenium by Adams and Day in 1876 [14] followed by the fabrication of the first thin-film selenium solar cells by Fritts in 1883 [14, 15]. Indeed selenium also is one the earliest examples of commercialization of photosensitive thin-films, when it was used as analog light meters for photographic cameras in the 1890s, perhaps providing the precondition for its application in photovoltaic devices.

The solar cells made by Fritts and the subsequent early works used an electrostatic barrier formed at the junction of metal–semiconductor, later on named Schottky barriers, for the separation of the generated electron and hole photocarriers [16]. Since the 1950s, we have witnessed the introduction of a new approach for the separation of the generated photocarriers [15, 17]. The metal–semiconductor junction was replaced with a junction of semiconductor with two different doping polarities (p-n junction) formed using two different dopants (see Chapter 1). Here the material used was crystalline silicon, which although itself not a suitable

material for large area electronics, the developed p-n junction approach is the basis of many subsequent TFSCs demonstrated using a wide range of methods and technologies.

TFSCs consist of a number of layers deposited sequentially under well-controlled process conditions and stacked on one another with particular attention to the interfaces between each later. Due to their thinness and mechanical fragility, the TFSC film stacks require a substrate to act as carrier. Typically glass is used for this purpose owing to its relatively low cost, optical transparency and high stability when exposed to high temperatures processes used in conventional solar cells. In addition to glass, TFSCs on paper, plastic, and metal foils have been demonstrated, yielding devices with flexible form factor, light weight and sometimes lower cost.

The active layers in TFSC devices consist of high conductivity electrodes, one of which should have high reflectivity (back reflector electrode) and the other high transparency (window electrode), electron and hole collector layers (typically doped semiconductors) and a light absorber layer where photocarriers are generated. The absorber layer can either be an additional layer in the TFSC stack, or be one of the doped layers. Main absorber films are silicon, cadmium telluride (CdTe), copper indium gallium diselenide (CIGS), and carbon-based layers. In addition to these, DSSC and quantum dot cells form two important categories of emerging solar cells compatible for large area applications [12, 18].

> **Thin-Film Solar Cell Fabrication**
>
> A range of fabrication tools have been used in the production of thin-film solar cells (TFSCs).
>
> TFSCs can only be economically viable if production methods are scalable to large area and have high throughput manufacturing capability.
>
> Material systems and design of the TFSCs dictates the choice of fabrication tools; chemical vapor deposition (CVD) and spray coating are two commonly used methods.
>
> Thanks to their large area compatibility and high throughput, these methods are also used for the fabrication of thin-film photodiodes and thin-film transistors at low cost.

17.1.2 Amorphous, Nano- and Micro-Crystalline Silicon

Hydrogenated amorphous silicon (a-Si:H) solar cell was first reported in 1974 by Carlson and Wronski from RCA Laboratory with efficiency of 1.1% [19]. Spear in parallel with Chittick demonstrated glow discharge deposited a-Si:H films which yielded sufficiently low defect density [20] due to the passivation of defective dangling bonds: inevitable in disordered semiconductors. Defects can act as recombination centers for photogenerated electron–hole pairs [21], leading to solar cells with low collection efficiency, making films with low defect densities essential for efficient operation of solar cells. Using the glow discharge method in 1975, Spear and LeComber reported deposition of a-Si:H using mixtures of silane and either phosphine or diborine, making them n-type and p-type, respectively [20, 22]. This allowed the use of p-type and n-type materials for the fabrication of p-i-n solar cells, making a

departure from Schottky based a-Si:H solar cells, and thus leading to significant improvements in the conversion efficiently of these type of devices.

Light induced degradation in the photoconductivity of a-Si:H thin-films and solar cells were first reported by Staebler and Wronski in 1977 [23]. This effect, commonly known as Staebler–Wronski effect, leads to subsequent degradation of a-Si:H solar cell efficiency after prolong exposure to illumination [24]. Through adjustments in deposition conditions such that the a-Si:H film is grown at the onset of crystal growth, it is possible to minimize the degradation of a-Si:H solar cells [25].

Alternatively multi-junction tandem solar cells, based on stacking two or more solar cells each with thin absorber layers with different bandgaps, have been shown to reduce the degradation of the a-Si:H TFSC [26, 27]. In addition to higher stability, the multi-junction approach also increases the maximum efficiency of the solar cell, compared with single junction device [26, 27]. By tuning the absorption spectrum of each layer, the captured energy from each photon can be maximized. In silicon TFSC this is typically achieved through alloying silicon with germanium a-SiGe:H, or though introducing crystallinity in the amorphous silicon matrix (i.e., nanocrystalline or microcrystalline silicon).

17.1.3 Cadmium Telluride (CdTe)

CdTe solar cells trace their history to 1950 [28]. From the early days it was recognized that due to the direct bandgap of 1.5 eV, theoretically a high solar cell efficiency ($\sim$28%) can be achieved using a thin absorber layer [29, 30]. A typical active device structure is a p-n heterojunction using p-type CdTe and n-type CdS [30]. CdTe solar cells can be fabricated using low cost and highly scalable methods such as electrodeposition, closed-space sublimation and screen plating [31]. However, these steps are typically followed by a high temperature annealing for removal of defects within the solar cell [31], placing limitations on the choice of substrate materials for flexible solar cells. Furthermore the presence of cadmium raises environmental concerns on the wide-scale use this technology [31, 32], requiring governmental control measures to deal with its recycling.

The highest efficiency device reported base on CdTe/CdS structure with the value of 16.5% [33]. Some progress has also been made in the use of CdTe nanoparticles in quantum dot solar cells [34]. Here the quantum size allows tuning of the bandgap by controlling the size of the nanoparticle, opening the possibility of low cost multi-junction solar cells. Research continues in the quest for improved material systems and higher device efficiencies.

17.1.4 Calcogenides

Copper indium gallium diselenide (CIGS) and its gallium free counterpart (CIS) are two of the main calcogenide semiconducting films used for fabrication of solar cells. The first reported device in this category was by Wagner et al. in 1974 based on crystalline CuInSe2/CdS p-n heterojunction device [35]. CIGS and CIS are both direct bandgap compound semiconductors [30], typically used as p-type material in heterojunction p-n solar cells with the CdS as the n-type material [30]. Owing to their direct bandgap, which can be as low as 1.1 eV, they have a high absorption coefficient and are one of the highest efficiency TFSC devices reported thus far [31]. Furthermore they offer the possibility of tailoring the semiconductor's bandgap by adjusting the composition of the compound [36]. The highest efficiency reported for CIGS

is 19.9% [37]. However, the high process temperatures required for the fabrication this type of solar cells places limitations on the choice of substrate materials [32]. Furthermore use of highly toxic gases and materials, such as selenide and cadmium as the precursors in the production of calcogenide solar cells [30, 32], places stringent requirements on the manufacturing, installation and disposal of the solar cell modules.

17.1.5 Organics and Polymers

This category includes both organic and polymeric solar cells, themselves divided into a large number of subcategories. The first solar cell was demonstrated in 1950s, where a number of organic dyes, particularly chlorophyll and related compounds, were investigated [38]. In 1986 the first heterojunction based organic solar cell which used both donor and acceptor material was demonstrated [39], resulting in a significant increase in the device efficiency. This was followed by development of bulk heterojunctions in 1995 [40] leading to significant increase in the generation of photocarriers, and therefore short circuit current and device efficiency in the range of ~5%.

Since then organic solar cells have improved further through the use of nanoparticles and multi-junction tandem structures, with most recent reported efficiency of 10.0% [41]. However, it should be noted that the reported efficiencies are typically for small devices (area of ~1 cm^2) with a considerably lower efficiency for larger devices [42]. Furthermore, organic solar cells suffer from short lifetime when exposed to illumination due to the changes in the material properties [42], making them less commercially viable. However, despite these current short comings, the low deposition temperature, large area and low cost fabrication processes are some of the major advantages of organic TFSCs, fuelling research in this field and making organic solar cells an attractive option a number of applications.

17.2 Large Area Imaging

17.2.1 Introduction and Applications

Large area digital imaging development has flourished since the widespread adoption of flat panel displays such as active matrix liquid crystal displays (AMLCDs) in the consumer marketplace. The technological and manufacturing development on AMLCDs, especially related to semi-conductor backplane and optical integration, serve as the

Flat Panel Imagers

Thin-film electronics also facilitates the seamless heterogeneous integration of system components. In the case of active-matrix flat panel imagers (AMFPIs), an array of thin-film transistor circuits are integrated with light sensing elements, such as photodiodes or phototransistors to produce an array of pixelated imaging elements over large area at low cost.

Furthermore use of thin-film technology allows use of a diverse range of substrates with the possibility of integrating of AMFPIs with other systems such as lab-on-chip for biomedical applications.

| High voltage ICs emerge | **2004** | ELECTRON DEVICES SOCIETY® | IBM, UCLA and Delft University in the Netherlands demonstrate reading of a single spin | CNT memories demonstrated |

enabling ground work in fabricating low-cost, high performance, and dimensionally large active matrix flat panel imagers (AMFPIs). AMFPIs address a unique set of requirements such as large area coverage (>8" diagonal), low cost, adaptability to different imaging modalities, and mechanical robustness. These technology attributes opens the door to many digital imaging applications that are challenging to be fulfilled technologically and cost effectively by conventional imaging solutions.

Table 17.1 compares AMFPIs with conventional imaging solutions. It is worth highlighting that AMFPIs provide a host of imaging possibilities that is non-overlapping with the well-developed crystalline silicon-based sensor application. In most scenarios, the dimensional coverage of Charged-coupled-device (CCD)

Table 17.1 Examples and comparison of AMFPIs and conventional imaging solutions

	AMFPIs	CCD Sensor	CMOS Sensor
Technology	Based on a-Si:H or low temperature polysilicon (LTPS) backplane with front plane customizable for different imaging needs such as x-ray, visible light, etc.	Customized fabrication procedures for crystalline silicon process to ensure high performance.	Using standard crystalline silicon process.
Performance (SNR, readout rate, quantum efficiency)	Low to mid	High	Mid to high
Coverage area	>8″ diagonal	<1.7″ area scan <2.4″ line scan	<1.7″
Integration	Mid- to high-, depends on fabrication technology	Mid to high	High
Cost	Low	High	Low- to Mid-
Application	X-ray imaging, large area cargo scanner, high-throughput pharmaceutical imager, low-cost finger print scanner.	High-end digital SLRs, scanners, and industrial based precision controlled device.	Mid- to low-end SLRs, cellphone camera, web-cam, etc.

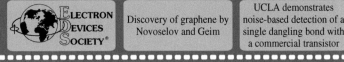

or Complementary-metal-oxide-semiconductor (CMOS) sensors is too small for the application to be cost effective (see Chapter 12). For example, static X-ray radiogram covers an imaging area with diagonal length range of 18″ to 24″ (45.72 cm to 60.96 cm). Addressing such area coverage either requires focusing X-ray source, or tiling multiple sensors that are both functionally and economically ineffective. AMFPIs coupling with indirect or indirect X-ray conversion material provide a low cost, large area, and scalable technology that is being heavily focused and researched in recent years [43, 44].

17.2.2 Imaging Architecture

There are two architectures currently employed in large area AMFPIs: the linear architecture, used in photocopiers, fax machines, and scanners, and the two-dimensional array architecture, employed in digital (including video) lens-less cameras as well as X-ray imaging systems [45, 47]. In both architectures, the basic imaging unit is the pixel, which consists of an image sensor and in-pixel circuit. The pixel is accessed by a matrix of gate and data lines, and operated in storage (or integration) mode. Here, during the off period of the pixel, the sensor charge is integrated in the sensor element, and when the pixel is addressed, the charge in the sensor is transferred to the data line where it is then detected by a charge sensitive amplifier. There are various metal interconnects used to control the readout of the imaging information from the arrays of pixel. The imaging system is completed with peripheral circuitry that amplifies, digitizes by analog-to-digital (A/D) converter and synchronizes the readout of the image, and a computer that manipulates and distributes the final image to the appropriate soft- or hard-copy device. A sample array of imaging pixels with column parallel readout architecture is illustrated in Figure 17.2.

The photosensitive area in each pixel as illustrated in Figure 17.2 can consist of various material or structure that enables the detection of incident signal of different forms. Along the lines of the same example used earlier, an X-ray imaging AMFPI will employ an X-ray sensitive stack to convert incident X-ray photons into electrical signals. Figure 17.3 illustrates a cross-sectional diagram for an indirect X-ray detector pixel. A phosphor scintillator layer is coupled on top of the AMFPI to convert X-ray photons into visible light photons, which is subsequently detected by light sensitive device on every pixel.

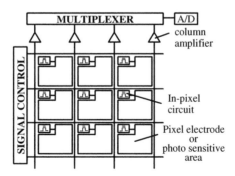

Figure 17.2   Schematic view of an active matrix detector array with peripheral electronics

First demonstration of 5-nm MOSFET

Vardeny *et al.* demonstrate spin-valve effect in the organic molecule Alq3

ELECTRON DEVICES SOCIETY®

C.H. Lee *et al.* publish NWL (Negative Word Line) -/FINFET-/RCAT-merged cell introduction to DRAM

J. Park *et al.* Present 8 Gb MLC (Multi-level Cell) Flash memory with 63-nm technology

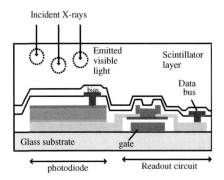

Figure 17.3   Indirect X-ray detection method based on ITO/a-Si:H Schottky photodiode integrated with phosphor layer

The thin-film-transistors (TFTs) along with the light sensitive devices (n-i-p or Schottky diode) constitute the backplane, and are fabricated on a glass substrate [48–50].

Other forms of imaging modality (visible light, IR, etc.) can be realized using similar architecture by employing a different combination of coupling layer and signal detection device (photodiode, IR sensor, etc.). This allows AMFPIs to be highly adaptive to design specifications and target applications.

17.2.3 Challenges

The key technological challenges of AMFPIs can be attributed to the electrical and optical performance of the system implemented in large area semiconductor technologies. The following outlines several major hurdles that are faced by AMFPI design regardless of imaging modality:

1. Backplane technology: TFTs fabricated using a-Si:H or LTPS typically lack behind in key performance metrics comparing to crystalline silicon. This includes in areas such as device mobility, minimum feature size, and spatial and/or temporal. These limitations exhibit themselves in challenges in achieving small pixel pitch, high scanning frequency/frame-rate, and low power operation. Improvements both in fabrication and imager architectures attempt to solve some of the aforementioned concerns.
2. Low noise requirement: AMFPIs strength in large area coverage also results in large line capacitance in the signal detection path. This unavoidably results in high noise when the signal lines are switched during readout. Development in active amplifiers within the pixel level and high gain readout circuitry are underway to address the low noise requirement.
3. Temporal and spatial performance shift: AMFPIs fabricated using a-Si:H benefit from the large area uniformity, however, a-Si:H TFTs' meta-stability cause the performance to shift over time upon prolonged usage. On the other hand, while LTPS TFTs' temporal performance change is minimal, its spatial uniformity can cause fixed-pattern noise (FPN). Both temporal and spatial non-idealities of AMFPIs require development of in-pixel, off-pixel, image correction, or a combination of these techniques.

| 2005 | Molecular many-body transport signatures explored - Muralidharan/ Ghosh/Datta | **E**LECTRON **D**EVICES **S**OCIETY® | SiC power rectifier demonstrated | Initiation of the Education Award of the IEEE Electron Devices Society |

17.2.4 Future Prospects and Development

AMFPIs' unique technological attributes enable large area digital imaging to meet specifications of many applications, from medical X-ray imaging to commercial photocopier/scanner. The future development in semiconductor backplane and integration techniques will further allow AMFPIs to improve performance and broaden its reach into other applications. Several key areas that AMFPIs are receiving significant developments from are:

1. Alternative substrate fabrication: Many large area imaging applications, such as portable scanner, portable X-ray imagers involve the electronic devices to be transported frequently; hence portability is important. Typical AMFPIs on glass substrates require cushioning to prevent glass crack when dropped, and such breakage often renders the module non-functional. Alternative substrates such as stainless steel foil or plastic not only provide mechanical robustness, and will further reduce the module thickness and weight. However, fabricating TFTs on alternative substrates require processing temperatures less than 250°C, and is typically lower than the semiconductor processing temperatures. This will greatly reduce the quality of TFTs. Methods of fabricating low-temperature and high quality TFTs is one of the heavily focused research subjects that can bring forth mechanically robust, lightweight and high performance AMFPIs.
2. Fully integrated imagers: Modern day camera modules used on cellphones, digital camera, and digital SLRs offers fully integrated module solution. Contrary to those applications, current AMFPIs rely on many external electronics such as chip-on-flex (CoF), chip-on-glass (COG), and/or flexible printed circuit (FPC) to interface with other modules. These interfaces unavoidably introduce noise when signals are passed through without amplification. Hence, integrating readout and addressing circuit on-panel is important. Research and development in realizing high speed, and low-noise gate drivers, column based amplifiers, charge-pump, and even analog-to-digital converter (ADC) are popular research and development topics. Successfully integrating signal processing and amplification on panel not only improves performance in signal to noise ratio, also aid in reducing module cost.

17.3 Flat Panel Displays

A display is a transducer device to convert information into a format suitable for comprehension by the human brain through the visual system. One of the most successful display technologies in the past is the cathode ray tube (CRT), which had been widely used in televisions (TVs) and desktop computer monitors. However, disadvantages of bulky size and heavy weight limit its applications.

Flat panel displays (FPDs), as the name implies and shown in Figure 17.4, have a much thinner profile; that is, several centimeters or less. Various types of FPD technologies have been developed in the past and can be generally classified as emissive and non-emissive [51], as shown in Figure 17.5. Emissive displays, including plasma display panel (PDPs), organic light-emitting diodes (OLEDs), and field emission displays (FEDs), emit light from each pixel with different intensity and colors to form images. Non-emissive displays, such as liquid crystal displays (LCDs), electronic-paper displays (EPDs), do not emit light themselves, but modulate light in each pixel to display images. Therefore, a light source is needed for a non-emissive display. In a transmissive display, the light source is put behind the display panel (named as backlight),

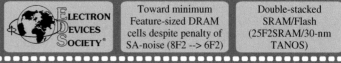

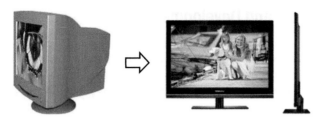

Figure 17.4 Transition from the bulk cathode ray tube (CRT) display to the thin flat panel displays (FPDs)

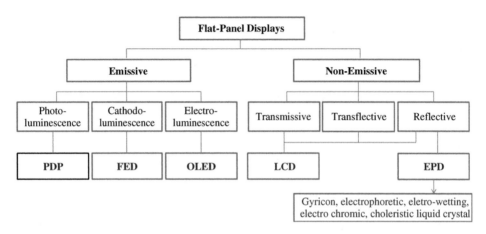

Figure 17.5 Classifications of various flat panel display technologies: plasma display panel (PDP), field emission display (FED), organic light emitting diode (OLED), liquid crystal display (LCD), and electronic paper display (EPD)

while, in a reflective display, the ambient light is used as the light source. Since no backlight is needed in a reflective display, its power consumption is relatively low. However, the reflective displays generally have inferior performances compared to the transmissive ones, and are not readable at dark ambient. Hence, transflective displays, which try to combine the advantages of both technologies, have also been developed.

The LCD is currently the dominant FPD technology. The developments of LCDs started in 1964 when a dynamic scattering mode was discovered [52]. In the early stage, direct and passive matrix addressing LCDs were applied in pocket calculators and digital watches, which gave birth to the LCD industry in the late 1970s. The subsequent development of the thin-film transistor (TFT) active matrix addressing technology provided a significant boost to the industry in the late 1980s. In the 1990s, active-matrix LCDs (AMLCDs) occupied an important

Flat Panel Displays

Thin-film technology is a key enabler for flat panel displays. It constitutes a technology platform for integrating electronics on glass, or other transparent substrates. For instance, using large area thin-film electronics, it is possible to produce flat panel displays at a cost of approximately one-hundredth of that associated with conversional silicon chip technology for similar size.

| 2006 | Record power density in GaN MODFET | ELECTRON DEVICES SOCIETY® | First graphene FETs with very low on-off ratios by DeHeer | Introduction of 65-nm silicon technology node in manufacturing |

position in the display market primarily through notebook computer applications. However, since the AMLCD technology suffered from serious limitations in terms of off-axis image quality, moving picture quality, and manufacturing difficulties for large-area TFT panels, it was believed that the LCD technology was suited only to smaller-size TV applications. In that time, the PDP, which was invented in 1964 and appeared in the commercial market in late 1990s, pioneered the large-size FPD market [53]. Until the early 2000s, PDPs were the most popular choice for large-size TVs as they had many benefits over LCDs, including higher contrast ratio, faster response, greater color, wider viewing angle, and a low-cost and simple manufacturing process. However, with improvements of TFT large-area fabrication technologies, LC modes, color filters and backlights, the increased size, lower weight, falling prices, and often lower power consumption of AMLCDs made them competitive with PDPs [54]. AMLCDs have thus gradually penetrate into the market of TVs and now even very-large-area digital information displays.

The OLED, regarded as the most promising technology to be able to compete with the LCDs, has evolved rapidly since the first working OLED developed at Eastman Kodak in 1987 [55]. Compared to LCDs, OLEDs own the advantages of faster response, greater color, and wide viewing angle, as an emissive type display. An additional special feature of OLEDs is being able to achieve thin and flexible panels [56]. OLEDs also require active matrix addressing with TFTs to achieve high resolution displays. Its current driven feature puts more strict requirements of the TFTs in terms of high carrier mobility, and uniform and stable current driving. With the improvement of material/device efficiency, reliability and manufacturing processes, AMOLEDs have been used in mobile phones, media players and digital cameras since the early 2000s, and now continues to make progress toward low-power, low-cost and large-size (>40-inch) FPD applications.

Motivated by the requirements of replicating paper-like reading experience with electronic displays, EPD technologies were developed to mimic the appearance of ordinary ink on paper [57, 58]. Previously introduced displays require that pixels be periodically refreshed to retain any states, even for a static image. EPDs reflect light like ordinary paper. Many of the technologies can hold static text and images indefinitely without using electricity, and thus greatly reduce the power consumption. The applications include electronic pricing labels, digital signage, electronic billboards, mobile phone displays, and e-readers for books or magazines.

References

[1] M. Danesh and J. R. Long, "An Autonomous Wireless Sensor Node Using a Solar Cell Antenna for Solar Energy Harvesting", *Microwave, MTT-S International Symposium-MTT*, pp. 1–4, 2011.

[2] A. Ahnood and A. Nathan, "Flat-Panel Compatible Photovoltaic Energy Harvesting System", *Journal of Display Technology*, vol. 8, no. 4, pp. 204–211, Apr. 2012.

[3] H. Zhou, A. Colli, A. Ahnood, Y. Yang, N. Rupesinghe, T. Butler, I. Haneef, P. Hiralal, A. Nathan, and G. A. J. Amaratunga, "Arrays of Parallel Connected Coaxial Multiwall-Carbon- Nanotube–Amorphous-Silicon Solar Cells", *Advanced Materials*, vol. 21, no. 38–39, pp. 3919–3923, Oct. 2009.

[4] O. K. Varghese, M. Paulose, and C. A. Grimes, "Long vertically aligned titania nanotubes on transparent conducting oxide for highly efficient solar cells", *Nature Nanotechnology*, vol. 4, no. 9, pp. 592–597, Aug. 2009.

[5] V. Sivakov, G. Andrä, A. Gawlik, A. Berger, J. Plentz, F. Falk, and S. H. Christiansen, "Silicon Nanowire-Based Solar Cells on Glass: Synthesis, Optical Properties, and Cell Parameters", *Nano Letters*, vol. 9, no. 4, pp. 1549–1554, Apr. 2009.

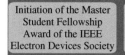

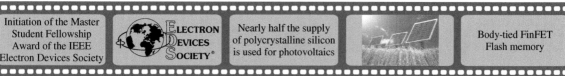

Initiation of the Master Student Fellowship Award of the IEEE Electron Devices Society

ELECTRON DEVICES SOCIETY®

Nearly half the supply of polycrystalline silicon is used for photovoltaics

Body-tied FinFET Flash memory

[6] L. Dou, J. You, J. Yang, C.-C. Chen, Y. He, S. Murase, T. Moriarty, K. Emery, G. Li, and Y. Yang, "Tandem polymer solar cells featuring a spectrally matched low-bandgap polymer", *Nature Photonics*, vol. 6, no. 3, pp. 180–185, Feb. 2012.

[7] H. A. Atwater and A. Polman, "Plasmonics for improved photovoltaic devices", *Nature Materials*, vol. 9, no. 3, pp. 205–213, Feb. 2010.

[8] M. A. Green, "Third generation photovoltaics: Ultra-high conversion efficiency at low cost", *Progress in Photovoltaics: Research and Applications*, vol. 9, no. 2, pp. 123–135, Mar. 2001.

[9] S. Price, R. Margolis, G. Barbose, *et al.*, "*2008 Solar Technologies Market Report*", LBNL-3490E, 983330, Jan. 2010.

[10] J. Kalowekamo and E. Baker, "Estimating the manufacturing cost of purely organic solar cells", *Solar Energy*, vol. 83, no. 8, pp. 1224–1231, Aug. 2009.

[11] V. Petrova-Koch, "Milestones of Solar Conversion and Photovoltaics", in *High-Efficient Low-Cost Photovoltaics*, vol. 140, V. Petrova-Koch, R. Hezel, and A. Goetzberger, (eds). Berlin, Heidelberg: Springer Berlin Heidelberg, pp. 1–5.

[12] M. Grätzel, "Photoelectrochemical cells", *Nature*, vol. 414, no. 6861, pp. 338–344, Nov. 2001.

[13] D. Wei, "Dye Sensitized Solar Cells", *International Journal of Molecular Sciences*, vol. 11, no. 3, pp. 1103–1113, Mar. 2010.

[14] W. G. Adams and R. E. Day, "The action of light on selenium", *Proc. Roy. Soc. London*, vol. 113, p. A 25, 1877.

[15] M. A. Green, "Photovoltaic principles", *Physica E: Low-dimensional Systems and Nanostructures*, vol. 14, no. 1–2, pp. 11–17, Apr. 2002.

[16] C. R. Wronski, D. E. Carlson, and R. E. Daniel, "Schottky-barrier characteristics of metal–amorphous-silicon diodes", *Applied Physics Letters*, vol. 29, no. 9, p. 602, 1976.

[17] R. S. Ohl, "*Light-sensitive electric device*", U.S. Patent 24026621946.

[18] A. Nozik, "Quantum dot solar cells", *Physica E: Low-dimensional Systems and Nanostructures*, vol. 14, no. 1–2, pp. 115–120, Apr. 2002.

[19] D. E. Carlson, "Semiconductor device having a body of amorphous silicon", U.S. Patent 40645211977.

[20] E. A. Davis, "Tribute", *Philosophical Magazine*, vol. 89, no. 28–30, pp. 2427–2429, Oct. 2009.

[21] R. E. I. Schropp and M. Zeman, *Amorphous and Microcrystalline Silicon Solar Cells: Modeling, Materials and Device Technology*. Boston [u.a.]: Kluwer Academic, 1998.

[22] R. Street, "Doping and the Fermi Energy in Amorphous Silicon", *Physical Review Letters*, vol. 49, no. 16, pp. 1187–1190, Oct. 1982.

[23] D. L. Staebler and C. R. Wronski, "Reversible conductivity changes in discharge-produced amorphous Si", *Applied Physics Letters*, vol. 31, no. 4, p. 292, 1977.

[24] D. E. Carlson, C. R. Wronski, J. I. Pankove, P. J. Zanzucchi, and D. L. Staebler, "Properties of amorphous silicon and a-Si solar cells", *RCA Review*, vol. 38, pp. 211–225, 1977.

[25] D. V. Tsu, B. S. Chao, S. R. Ovshinsky, S. Guha, and J. Yang, "Effect of hydrogen dilution on the structure of amorphous silicon alloys", *Applied Physics Letters*, vol. 71, no. 10, p. 1317, 1997.

[26] B. Yan, G. Yue, J. M. Owens, J. Yang, and S. Guha, "Light-induced metastability in hydrogenated nanocrystalline silicon solar cells", *Applied Physics Letters*, vol. 85, no. 11, p. 1925, 2004.

[27] H. Keppner, J. Meier, P. Torres, D. Fischer, and A. Shah, "Microcrystalline silicon and micromorph tandem solar cells", *Applied Physics A: Materials Science & Processing*, vol. 69, no. 2, pp. 169–177, Aug. 1999.

[28] B. Goldstein and L. Pensak, "High-Voltage Photovoltaic Effect", *Journal of Applied Physics*, vol. 30, no. 2, p. 155, 1959.

[29] P. Rappaport, "Photovoltaic effect and its utilization", *Solar Energy*, vol. 3, no. 4, pp. 8–18, 1959.

Tape is dead, Disk is Tape, Flash is Disk, RAM locality is King". (Soft-ware Revolution in mass memories) | 2007 | ELECTRON DEVICES SOCIETY® | Supriyo Bandyopadhyay and Marc Cahay groups demonstrate possibly 1 second spin relaxation time in organic molecules at 77 K | Use of high-k metal gate stack at Intel Co.

[30] A. Shah, "Photovoltaic Technology: The Case for Thin-Film Solar Cells", *Science*, vol. 285, no. 5428, pp. 692–698, Jul. 1999.

[31] K. L. Chopra, P. D. Paulson, and V. Dutta, "Thin-film solar cells: an overview", *Progress in Photovoltaics: Research and Applications*, vol. 12, no. 23, pp. 69–92, Mar. 2004.

[32] T. B. Johansson and L. Burnham, *Renewable Energy: Sources for Fuels and Electricity*. Washington, D.C.: Island Press, 1993.

[33] X. Wu, J. C. Keane, R. G. Dhere, C. DeHart, D. S. Albin, A. Duda, T. A. Gessert, S. Asher, D. H. Levi, and P. Sheldon, "16.5%-efficient CdS/ CdTe polycrystalline thin-film solar cell", presented at the *17th European Photovoltaic Solar Energy Conference*, 2001, pp. 995–1000, 2001.

[34] P. V. Kamat, "Quantum Dot Solar Cells. Semiconductor Nanocrystals as Light Harvesters", *Journal of Physical Chemistry B*, vol. 112, no. 48, p. 18737–18753, Oct. 2008.

[35] S. Wagner, "CuInSe2/CdS heterojunction photovoltaic detectors", *Applied Physics Letters*, vol. 25, no. 8, p. 434, 1974.

[36] T. Dullweber, G. Anna, U. Rau, and H. Schock, "A new approach to high-efficiency solar cells by bandgap grading in Cu(In,Ga)Se2 chalcopyrite semiconductors", *Solar Energy Materials and Solar Cells*, vol. 67, no. 1–4, pp. 145–150, Mar. 2001.

[37] I. Repins, M. A. Contreras, B. Egaas, C. DeHart, J. Scharf, C. L. Perkins, B. To, and R. Noufi, "19.9%-efficient ZnO/CdS/CuInGaSe 2 solar cell with 81.2% fill factor", *Progress in Photovoltaics: Research and Applications*, vol. 16, no. 3, pp. 235–239, May 2008.

[38] H. Spanggaard and F. C. Krebs, "A brief history of the development of organic and polymeric photovoltaics", *Solar Energy Materials and Solar Cells*, vol. 83, no. 2–3, pp. 125–146, Jun. 2004.

[39] C. W. Tang, "Two-layer organic photovoltaic cell", *Applied Physics Letters*, vol. 48, no. 2, p. 183, 1986.

[40] G. Yu, J. Gao, J. C. Hummelen, F. Wudl, and A. J. Heeger, "Polymer Photovoltaic Cells: Enhanced Efficiencies via a Network of Internal Donor-Acceptor Heterojunctions", *Science*, vol. 270, no. 5243, pp. 1789–1791, Dec. 1995.

[41] R. F. Service, "Outlook Brightens for Plastic Solar Cells", *Science*, vol. 332, no. 6027, p. 293–293, Apr. 2011.

[42] E. A. Chandross, "Not-So-Sunny Outlook for Organic Photovoltaics", *Science*, vol. 333, no. 6038, pp. 35–36, Jun. 2011.

[43] M. J. Yaffe and J. A. Rowlands, "X-Ray Detectors for Digital Radiology". *Phys. Med. Biol.*, vol. 42, pp. 1–39, 1997.

[44] A. R. Cowen, "Digital X-Ray Imaging". *Meas. Sci. Technol.*, vol. 2, pp. 691–707, 1991.

[45] P. G. LeComber, W. E. Spear, and A. Ghaith, "Amorphous Silicon Field Effect Device and possible Application", *Electronic Letters*, vol. 15, pp. 179–181, 1979.

[46] R. L. Weisfield, "Amorphous Silicon Linear-Array Device Technology: Applications in Electronic Copying", *IEEE Transaction on Electron Devices*, vol. 36, pp. 2935–2939, 1989.

[47] W. Zhao and J. A. Rowlands, "X-ray imaging using amorphous selenium: Feasibility of a flat panel self-scanned detector for digital radiology", *Medical Physics*, vol. 22, no. 10, pp. 1595–1604, 1995.

[48] R. A Street and L. E. Antonuk, "Amorphous silicon arrays develop a medical image", *IEEE Circuits and Devices*, vol. 9, no. 4, pp. 38–42, 1993.

[49] R. A. Street, X. D. Wu, R. Weisfield, S. Ready, R. Apte, M. Mgyuen, and P. Nylen, "Two dimensional amorphous silicon image sensor arrays", *Amorphous Silicon Technology – 1995: Materials Research Society Symposia Proceedings*, vol. 377, pp. 757–766, 1995.

[50] W. Zhao and J. A. Rowlands, "A large area solid-state detector for radiology using amorphous selenium", *Medical Imaging VI: Instrumentation: Proceedings of SPIE*, 1651, pp. 134–143, 1992.

[51] J.-H. Lee, D. N. Liu and S.-T. Wu, Introduction to Flat Panel Displays, Chichester: John Wiley & Sons, Ltd., 2008.

[52] H. Kawamoto, "The history of liquid-crystal displays", *Proc. of the IEEE*, vol. 90, no. 4, pp. 460–500, April 2002.

2008 | Chemically modified graphene nanoribbons with 200 mV bandgaps by Hongjie Dai | ELECTRON DEVICES SOCIETY | HP identifies observed pinched hysteresis curves as "memristors"

[53] Larry F. Weber, "History of the Plasma Display Panel", *IEEE Tran. Plasma Science*, vol. 34, no. 2, pp. 268–278, April 2006.

[54] K.-H. Kim and J.-K. Song, "Technical evolution of liquid crystal displays", *NPG Asia Mat.*, vol. 1, no. 1, pp. 29–36, Oct. 2009.

[55] J. N. Bardsley, "International OLED technology roadmap", *IEEE J. Selected Topics in Quantum Electronics*, vol. 10, no. 1, pp. 3–9, Jan.-Feb. 2004.

[56] A. B. Chwang, M. Hack and J. J. Brown, "Flexible OLED display development: strategy and status", *J. the SID*, vol. 13, no. 6, pp. 481–486, 2005.

[57] B. Comiskey, J. D. Albert, H. Yoshizawa, and J. Jacobson, "An electrophoretic ink for all-printed reflective electronic displays", *Nature*, vol. 394, pp. 253–255, Jul. 1998.

[58] R. A. Hayes, and B. J. Feenstra, "Video-speed electronic paper based on electrowetting", *Nature*, vol. 425, pp. 383–385, Sept. 2003.

| Introduction of 45-nm silicon technology node in manufacturing | Barack Obama elected first African American president of the US | ELECTRON DEVICES SOCIETY® | Introduction of the Early Career Award of the IEEE Electron Devices Society | Record of monolithic integration for avalanche photodiodes (Y. Kang *et al.*) |

Chapter 18

Microelectromechanical Systems (MEMS)

Darrin J. Young and Hanseup Kim

18.1 Introduction

Microelectromechanical Systems, generally referred to as MEMS, has been researched and developed over the past decades. Besides the traditional microfabricated sensors and actuators, the field covers micromechanical components and systems integrated or micro-assembled with electronics on the same substrate or package, achieving high-performance functional systems. This chapter presents an overview of MEMS innovations starting from the 1960s. A large number of MEMS devices have been developed for specific applications. Therefore, it is difficult to provide an overview covering every aspect of the technology innovation. It is the authors' intent to illustrate a few key MEMS innovations and their demonstrative impacts.

18.2 The 1960s – First Micromachined Structures Envisioned

The 1960s marks the birth of micromachined integrated sensors. In 1965, M. Lepselter at Bell Labs reported "beam-lead technology" and formation of "air-isolated integrated circuits" [1]. Individual transistor islands were held together with electroplated gold beams, as depicted in Figure 18.1(a), with an intent to eliminate parasitic capacitance and leakage current associated with junction-isolated transistors. In that same

Guide to State-of-the-Art Electron Devices, First Edition. Edited by Joachim N. Burghartz.
© 2013 John Wiley & Sons, Ltd. Published 2013 by John Wiley & Sons, Ltd.

Philip Kim and co-workers show 230,000 cm²/Vs mobility ... ... at room temperature with suspended sheets ELECTRON DEVICES SOCIETY® 2009 Introduction of 40-nm silicon technology node in manufacturing

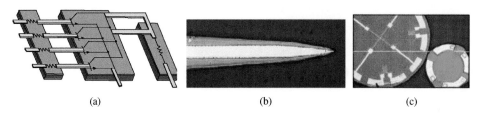

(a)   (b)   (c)

Figure 18.1   (a) Individual transistor islands held together by electroplated gold beams, (b) Photo of a fabricated neural recording electrode, (c) Photo of diaphragm-based piezoresistive pressure sensor (Courtesy of K. D. Wise)

year, J. Moll at Stanford University suggested to use beam-lead technology to construct recording electrodes to monitor individual neurons activities *in vivo*. Selective silicon etching techniques were developed by K. D. Wise, J. B. Angell and A. Starr around 1968 and 1969 to produce recording electrodes shown in Figure 18.1(b) [2].

This work led to a number of critical research efforts devoted to microsensors and actuators in the 1970s, forming the foundation of the field known as integrated sensors, MEMS, and microsystems.

18.3 The 1970s – Integrated Sensors Started

Using integrated circuit technology to fabricate integrated sensors was a natural outgrowth of activities in microelectronics. In 1973, millimeter-sized silicon diaphragm-based piezoresistive pressure sensors fabricated by using micromachining techniques was demonstrated by Gieles, *et al.* [3, 4]. Figure 18.1(c) shows a photo of fabricated pressure sensor. Micromachined pressure sensors were critical for a large number of applications. Significant research and commercialization efforts were devoted from the early 1970s onward, including the development of silicon diaphragm-based capacitive pressure sensors by Ko *et al.* at Case Western Reserve University [5] and Lee and Wise at the University of Michigan in 1982 [6], and Nova Sensors as a pioneer company successfully bringing the pressure sensor technology into commercial products.

In 1979, Terry, Jerman, and Angell demonstrated a single-wafer micro gas chromatography system for air analysis for the Viking Mars Lander mission [7]. The system integrated a spiral microcolumn, where analytes were separated, and a microdetector on a single two-inch silicon wafer as shown in Figure 18.2. The prototype device detected eight different hydrocarbons within 10 s. Such a comprehensive sensing scheme enhanced the dynamic range and the selectivity of a chemical sensor beyond the conventional limitation of a stand-alone device.

Bulk Micromachining

The development of MEMS requires appropriate fabrication technologies that enable the definition of small geometries, precise dimension control, design flexibility, repeatability, reliability, high yield, and low cost. In early stage, mechanical sensors and actuators were fabricated from silicon wafer by using bulk micromachining process, which employed chemical etchants to selectively remove silicon materials to form desirable microstructures. The process could produce large and thick structures but exhibited a limited process control capability. Integration with standard IC process remained as a challenge.

| Multi-stacked and -layered 3D NAND Flash (R. Katsumata, Toshiba) | George Smith, William Boyle and Charles K. Kao receives the Nobel Price in Physics ... | ELECTRON DEVICES SOCIETY® | ... for the invention of the charge-coupled device (CCD) |  |

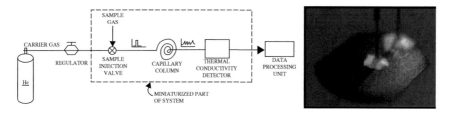

Figure 18.2   A gas chromatograph system architecture with prototype device photo (Courtesy of K. D. Wise)

Notably it provided the foundation for the emerging concepts later in the 1990s of so-called Bio-MEMS, Lab-on-Chips, and Micro Total Analysis Systems (μTAS).

18.4 The 1980s – Surface Micromachining Emerged

A number of technology innovations including integration of micro-machined structures with MOS electronics, surface-micromachined sensors and actuators, wafer-level vacuum packaging marked the key milestones in the 1980s. K. E. Petersen and A. Shartel first demonstrated a silicon micromechanical accelerometer integrated with MOS detection circuitry in 1980 [8]. This ground-breaking research work opened the field of integrated silicon mechanical structures with electronics. During the same timeframe, polycrystalline silicon surface micromachining process was being developed by R. T. Howe and R. S. Muller at the University of California at Berkeley. Surface micro-machining technology would greatly simplify the integration process with electronics. In 1984, R. T. Howe and R. S. Muller reported a surface micromachined resonant-microbridge vapor sensor integrated with NMOS detection circuit shown in Figure 18.3(a) [9]. This critical milestone motivated a significant follow-up research and development effort in integrated microelectro-mechanical devices and systems.

Researchers envisioned moving mechanical parts made of silicon, for example gears and motors, could be fabricated on silicon substrates with controlled motion by integrated electronics. Around 1987 and 1988, M. Mehregany, K. Gabriel, W. S. N. Trimmer, S. F. Bart, L. S. Tavrow, J. H. Lange, and S. D. Senturia from M.I.T. and AT&T Bell Laboratories, and L.-S. Fan, Y. C. Tai and R. S. Muller at U.C. Berkeley demonstrated surface micromachined polycrystalline silicon micro-gears and micro-motors [10, 11, 12]. Figures 18.3(b) and (c) present scanning electron micrograph (SEM) photos of fabricated microgears and micromotor, achieving a diameter on the order of 100 μm with a minimum vertical feature size on the order of a micrometer.

Surface Micromachining

Surface micromachining enables the fabrication of MEMS above a substrate surface. Microstructures are formed by multiple layers of polycrystalline silicon deposited by LPCVD technique. A sacrificial layer, typically silicon dioxide, is deposited between the structural layers and is ultimately removed by HF solution or vapor to form an air gap, thus creating a freely suspended MEMS structure. Surface-micromachined MEMS typically exhibit a vertical and lateral dimension on the order a few micrometers and a few hundred micrometers, respectively. The dimensions can be accurately controlled in fabrication. The process is amenable to standard CMOS process integration, thus enabling the realization of MEMS on top of integrated electronics.

2010's   Era of More Moore and More than Moore   **2010's**

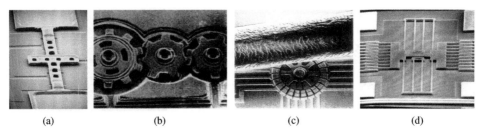

(a)   (b)   (c)   (d)

Figure 18.3   (a) Surface micromachined resonant-microbridge vapor sensor (© 1984 IEEE. Reprinted, with permission, from Howe, R.T.; Muller, R.S.; Integrated resonant-microbridge vapor sensor; *IEEE Proceedings* 30; 1984), (b) Surface micromachined microgears (© 1988 IEEE. Reprinted, with permission, from Mehregany, M.; Gabriel, K.J.; Trimmer, W.S.N.; Integrated fabrication of polysilicon mechanisms; *Electron Devices, IEEE Transactions* 35(6); 1988), (c) Surface micromachined micromotor (© 1988 IEEE. Reprinted, with permission, from Long-Sheng Fan; Yu-Chong Tai; R.S. Muller; IC-processed electrostatic micro-motors; *IEEE Proceedings*; 1988), (d) Surface micromachined laterally driven resonator (© 1989 IEEE. Reprinted, with permission, from Tang, W.C.; Nguyen, T.-C.H.; Howe, R.T.; Laterally driven polysilicon resonant microstructures; *IEEE Proceedings*; 1989)

The micromotors can be electrostatically actuated to obtain a rotational speed of approximately 1000 rpm. Besides rotational motion, a large lateral motion is desirable for resonant sensors and actuators. In 1989, Tang, Nguyen and Howe at U.C. Berkeley demonstrated a laterally driven polysilicon resonator by using electrostatic comb drive structures [13, 14]. Figure 18.3(d) presents an SEM photo of a fabricated resonant microstructure occupying an area on the order of 500 μm × 300 μm. Prototype devices can be electrostatically driven into mechanical resonance, developing a lateral motion with resonant frequencies in range from 8–80 kHz and a quality factor of approximately 50,000 in vacuum, thanks to the inherent large inductance values in MEMS structures. This research work set a milestone for the next two decades of MEMS resonators development, which ultimately achieved resonant frequency above 1 GHz.

With the emerging commercialization effort of MEMS devices in the 1980s, MEMS packaging became another major research focus in the era. Notable was the vacuum packaging for individual tiny micro-devices that reduced damping around a movable micro-actuator, such as a microresonator, as well as prevented particles. In 1988, Yokokawa Electric Company in Japan demonstrated a vacuum-sealed resonant pressure gauge shown in Figure 18.4(a) [15]. Subsequently, Lin, *et al.* from U.C. Berkeley demonstrated vacuum-encapsulation of large-area comb-drive mechanical resonators shown in Figure 18.4(b) [16].

In parallel to the development of integrated mechanical sensors and actuators, researchers at the University of Michigan were pioneering integrated neural probes, which brought sensors and electronic interface to the central nervous system. In 1984, K. Najafi and K. D. Wise demonstrated a multichannel bi-directional recording and stimulating probe, consisting of ten integrated preamplifiers, a time-division multiplexer and self-test circuitry shown in Figure 18.5(a) [17]. This achievement inspired many researchers to devote the next three decades to develop the area known as neural recording, neural interface, and brain-machine interface. Another significant achievement was the three-dimensional electrode array developed by Jones, Campbell and Normann from the University of Utah in 1990, as shown in Figure 18.5(b) [18].

| **2010** | Wide bandgap MOSFETs emerge | **E**LECTRON **D**EVICES **S**OCIETY® | Introduction of 32-nm silicon technology node in manufacturing | Metal-gated cells in DRAM (More array-efficiency: More cells/WL) |

(a)   (b)

Figure 18.4   (a) Vacuum-sealed resonant pressure sensor (© 1998 IEEE. Reprinted, with permission, from Esashi, M.; Sugiyama, S.; Ikeda, K.; Wang, Y.; Miyashita, H.; Vacuum-sealed silicon micromachined pressure sensors; *Proceedings of the IEEE* 86(8) 1998), (b) Vacuum-encapsulated comb-drive mechanical resonator (© 1998 IEEE. Reprinted, with permission, from Liwei Lin; Howe, R.T.; Pisano, A.P.; Microelectromechanical filters for signal processing; *IEEE/ASME Journal of Microelectromechanical Systems*; 7(3) 1998)

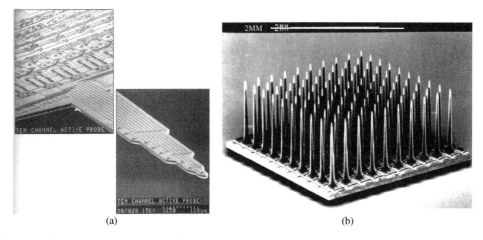

(a)   (b)

Figure 18.5   (a) Multichannel neural recording probe (© 1986 IEEE. Reprinted, with permission, from Najafi, K.; Wise, K.D.; An implantable multielectrode array with on-chip signal processing; *IEEE Journal of Solid-State Circuits*; 21(6), 1986), (b) 3D Utah electrode array (© 1990 IEEE. Reprinted, with permission, from Jones, K.E.; Campbell, P.K.; Normann, R.A.; Interelectrode Isolation In A Penetrating Intracortical Electrode Array; *IEEE Proceedings*; 1990)

Around the same time frame, MEMS technology started being utilized for the detection of biological and chemical subjects, which in turn ignited an increasing attention to microscale fluidic manipulation techniques, such as micropumps and valves. van Lintel, van de Pol, and Bouwstra in Europe [19], Esashi, Shoji, and Nakano in Asia [20], and Smits in North America [21], respectively, demonstrated chip-scale micropumps.

| 3D NAND flash memory vertically arrayed | Islandic volcanic ash disrupts flights in much of Europe | ELECTRON DEVICES SOCIETY® | IBM demonstrates 100 GHz graphene devices | ZSW in Germany breaks 20% efficiency barrier for CIGS solar cells |

18.5 The 1990s – MEMS Impacted Various Fields

MEMS started a rapid development into various fields in the 1990s. Significant progresses and achievement were made in the area of inertial sensors. In 1991, Charles Stark Draper Laboratory demonstrated one of the first batch fabricated silicon micromachined rate gyroscopes [22]. Subsequent to this pioneering effort, vibrating ring gyroscopes were developed by Putty, Ayzai and Najafi at the University of Michigan in 1994, shown in Figure 18.6(a) [23, 24]. Around the same time frame, dual-axis rate gyroscopes were also demonstrated by T. Juneau and A. P. Pisano at U.C. Berkeley as shown Figure 18.6(b) [25]. Prior to this achievement, Yun, Gray and Howe from U.C. Berkeley demonstrated a surface micromachined polycrystalline z-axis accelerometer integrated with CMOS detection circuitry in 1992 [26].

Another notable milestone in the 1990s is the invention of digital micromirror devices (DMD) by L. J. Hornbeck from Taxes Instruments in 1993 [27]. A DMD consists of a large array of small micromachined aluminum mirrors. Figure 18.7 shows an SEM micrograph of a close-up view of a DMD pixel array. Each mirror is capable of rotating $+/-10°$ by an electrostatic actuation force to reflect light through a projection lens, thus creating images on a large screen. The DMD technology can achieve higher performance than conventional cathode ray tube and has become a successful commercial display product.

Around the same time-frame, O. Solgaard and a few other researchers at Stanford University were developing MEMS-based grating light value [28]. Subsequently, the technology was successfully commercialized by a number of companies for HDTV projector applications.

High-Aspect Ratio Molding

Molding technique shapes pliable materials by forging it over a hard pre-patterned mold and produces a high-aspect ratio microstructure that provides large surface-to-volume ratio. It enables the use of non-silicon materials, such as various polymers, to create high-aspect ratio pillars, trenches, and channels that are transparent, compliant, bio-friendly, and disposable. The aspect ratio, typically larger than 10, is restricted by the dimensions of the mold, material properties, and process conditions.

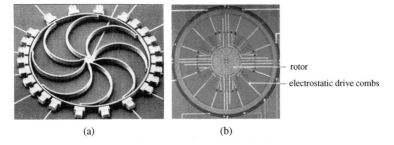

— rotor
— electrostatic drive combs

(a)   (b)

Figure 18.6   (a) Vibrating ring gyroscope (© 1998 IEEE. Reprinted, with permission, from Ayazi, F.; Najafi, K.; Design and fabrication of high-performance polysilicon vibrating ring gyroscope; *IEEE Proceedings*; 1998), (b) Dual-axis rate gyroscope. (© 1997 IEEE. Reprinted, with permission, from Juneau, T.; Pisano, A.P.; Smith, J.H.; Dual axis operation of a micromachined rate gyroscope; *IEEE Proceedings*; 2; 1997.)

| Supriyo Bandyopadhyay and Jayasimha Atulasimha … | … propose hybrid spintronics and straintronics for energy efficient computing | **E**LECTRON **D**EVICES **S**OCIETY® | **2011** | Introduction of 22/28-nm silicon technology node in manufacturing |

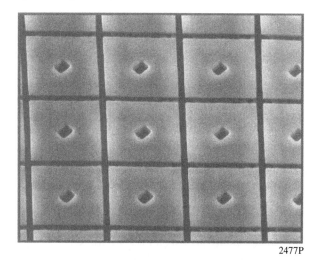

Figure 18.7 Close-up view of a DMD pixel array. (© 1993 IEEE. Reprinted, with permission, from Hornbeck, L.J.; Current status of the digital micromirror device (DMD) for projection television applications; *IEEE Proceedings*; 1993.)

The development of MEMS devices for wireless communication applications, known as RF MEMS, was a major research effort in the 1990s. D. J. Young and B. E. Boser at U.C. Berkeley demonstrated a MEMS high-Q tunable capacitor in 1996 [29]. Similar efforts were devoted to implement MEMS switches by Yao and Chang at the Rockwell Science Center [30], and Goldsmith, *et al.* from Raytheon [31].

Figure 18.8(a) and (b) show SEM micrographs of a fabricated MEMS tunable capacitor and capacitive switch, respectively. RF MEMS components based on the demonstrated operating principles became available in commercial processes and products recently.

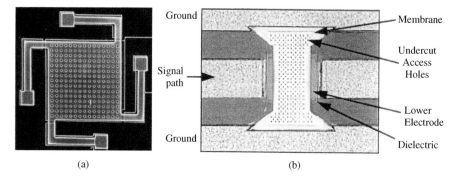

(a)   (b)

Figure 18.8 (a) RF MEMS tunable capacitor (Courtesy of D. J. Young.), (b) RF MEMS capacitive switch. (© 1998 IEEE. Reprinted, with permission, from Goldsmith, C.L.; Zhimin Yao; Eshelman, S.; Denniston, D.; Performance of low-loss RF MEMS capacitive switches; *Microwave and Guided Wave Letters*, IEEE; 8:8, 1998.)

Besides using traditional materials to fabricate MEMS devices, researchers were also developing silicon carbide (SiC) thin-film technology for harsh environment applications. Starting from the early 1990s, M. Mehregany at Case Western Reserve University led the research effort to develop SiC-based surface micromachining technology, which ultimately demonstrated MEMS devices operated above 500°C [32].

The 1990s also observed explosive demands of MEMS in identifying and analyzing various biological and chemical objects, forming a research area of bio-MEMS or μTAS. The term, μTAS, was coined by Manz, Graber, and Widmer in 1990 [33]. Bio-MEMS pursued a new vision of shrinking labor-intensive analysis process into an automated microchip performance, including various MEMS technologies for separation, detection, and pumping of nanoscale biological and chemical subjects, for portable and low-cost assays. Burns and Mastrangelo at the University of Michigan demonstrated the first integrated DNA analysis micromachined chip in 1996 as shown Figure 18.9 [34].

The Si-based DNA analyzer utilized local heating to move discrete nanoliter drops of samples and integrated thermal-cycling chamber, gel electrophoresis channels, and radio-labeled DNA detectors in a single substrate. Subsequently numerous MEMS technologies for separation, detection, and pumping of various biological and chemical subjects were developed, including polymer-based disposal chips for point-of-care clinical diagnostics by Chong Ahn's group at University of Cincinnati [35].

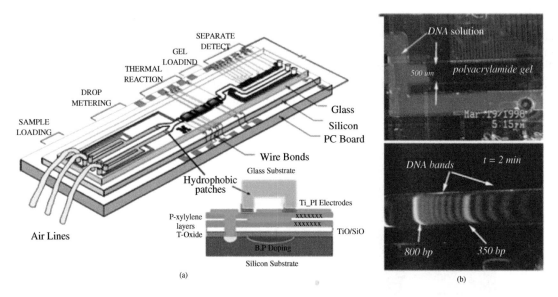

Figure 18.9 (a) Illustration of the microchip for DNA sequencing containing liquid pumping channels, a thermally controlled reaction chamber, electrophoresis separation channels, DNA band detectors, and controlled circuitry with the cross-section view of the structure (Courtesy of Carlos Mastrangelo.), (b) SEM views of the gel loading and separated DNA bands. (© 1998 IEEE. Reprinted, with permission, from Mastrangelo, C.H.; Burns, M.A.; Burke, D.T.; Microfabricated devices for genetic diagnostics; *Proceedings of the IEEE*; 86:8, 1998.)

| Solarjunction Inc. demonstrates ... | ... 43.5% efficiency four-junction III-V solar cell | **E**LECTRON **D**EVICES **S**OCIETY® | UC Santa Barbara demonstrates HBTs ... | ... with fT>500 GHz and f_{max}>1 THz |

18.6 The 2000s – Diversified Sophisticated Systems Enabled by MEMS

During the 2000s MEMS grew explosively in quantity and spread across the world. Building MEMS structures from a pre-fabricated CMOS wafer can potentially achieve highly miniaturized and low-cost microsystems. In 2002, Xie, *et al.* from Carnegie Mellon University demonstrated a CMOS-MEMS process, which enabled post-CMOS processing for obtaining high-aspect ratio integrated silicon microstructures [36]. Three-dimensional assembly and fabrication technologies became another notable technological trend, and implementation of sophisticated inertial sensing devices provided an excellent example. In 2011, Zotov, *et al.* from University of California at Irvine demonstrated a 3D compact Inertial Measurement Unit by using a planar multisensory structure, which was fabricated and subsequently folded in a pyramidal shape [37]. Around this time frame, Prikhodko, S *et al.* from the same university demonstrated a microscale glass-blown 3D spherical shell resonator [38]. Visvanathan, Li, and Gianchandani from the University of Michigan also developed spherical structures based on microelectro discharge machining technology [39]. Spherical structures are critical for high-performance miniature gyroscopes. Figure 18.10 shows representative device photos.

Significant efforts were also devoted in MEMS packaging technology in the early 2000s. M. Schmidt from M.I.T. demonstrated

Substrate Transfer

Substrate transfer is a key MEMS fabrication technology that involves substrate etching, bonding, and thinning in order to vertically stack two or more substrates atop. The stacked substrates are capable of providing tall, deep or embedded structural profiles with dimensions spanning over several hundreds of μm, which is not possible by conventional thin-film-based microfabrication. Substrate transfer, thus, enables high-aspect ratio microstructures, such as a suspended diaphragm and a closed channel, unique and prevalent in MEMS and now in 3D packaging of microelectronics.

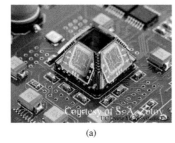 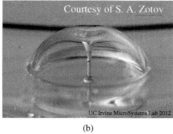 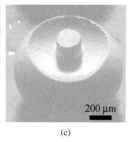

(a)   (b)   (c)

Figure 18.10   (a) 3D compact IMU (© 2011 IEEE. Reprinted, with permission, from Zotov, S.A.; Rivers, M.C.; Trusov, A.A.; Shkel, A.M.; Folded MEMS Pyramid Inertial Measurement Unit; *IEEE Sensors Journal*; 11(11); 2011), (b) Glass-blown 3D spherical shell resonator(© 2011 IEEE. Reprinted, with permission, from Prikhodko, I.P.; Zotov, S.A.; Trusov, A.A.; Shkel, A.M.; Microscale Glass-Blown Three-Dimensional Spherical Shell Resonators; *IEEE/ASME Journal of Microelectromechanical Systems*; 20(3); 2011), (c) Micromachined spherical structure (© 2011 IEEE. Reprinted, with permission, from Visvanathan, K.; Tao Li; Gianchandani, Y.B.; 3D-soule: A fabrication process for large scale integration and micromachining of spherical structures; *IEEE Proceedings*; 2011)

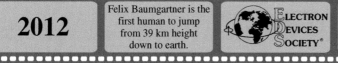

| 2012 | Felix Baumgartner is the first human to jump from 39 km height down to earth. | ELECTRON DEVICES SOCIETY® | In contrast, work on electron devices will continue at the highest level ... | ... and further advance in miniturization, performance and diversification |

bonding methods to stack two or multiple wafers [40]. Between 2005 and 2009, researchers from the University of Michigan demonstrated a vacuum packaging platform with a comprehensive set of electrical feed-through, anti-vibration-shock, and thermal isolation capabilities [41, 42]. Around the same time, monolithic vacuum packaging for mass production was demonstrated by R. N. Candler and T. W. Kenny from Stanford University [43].

Manipulating fluids based on mechanical actuation was another emerging research topic in this era, known as microfluidics. Glezer, Smith and Allen from Georgia Institute of Technology first introduced a synthetic jet concept in 1995 [44]. In 2007, Kim, *et al.* from the University of Michigan demonstrated the first high-pressure (>18 kPa), high flow-rate (>4 sccm), low-power (<56 mW) microgas pump in an 18-stage design, where 18 pumps and 19 valves were integrated into a compact chip of 2.5 × 1.9 × 0.1 cm^3 [45].

Non-mechanical fluidic handling method was also investigated. One notable was a purely-electrical control of liquid droplets on a flat substrate: Electro Wetting on Dielectric (EWOD), demonstrated by C. J. Kim's research group at UCLA in 2001 [46]. Self-assembly techniques were also reported by K. Bohringer from the University of Washington in 2001, demonstrating multi-step bonding of micro component batches by manipulating the surface functionalities of a substrate [47].

18.7 Future Outlook

Continuing the success over the past four decades, MEMS technologies are envisioned to extend their impacts to broader societal aspects, including environment, health care, energy, transportation, security, and so on. Some of the technical challenges are being investigated and others remain to be solved. Some recent examples of the MEMS neuronal interface system for insects demonstrated by A. Lal's group at Cornell University [48] and M. Maharbiz's group at U.C. Berkeley [49], various energy harvesting devices from ambient vibration and thermal sources, and bio-packaging against foreign body effects could be the apparent indicators. Resolution of these challenges requires interdisciplinary knowledge and strategic efforts by the community and the next-generation of researchers. Clearly, the field of MEMS is, and will be, full of promise and excitement.

Acknowledgment

The authors of Chapter 18 gratefully acknowledge Professor Kensall D. Wise of the EECS Department of the University of Michigan at Ann Arbor for his valuable advice and perspectives on MEMS history.

References

[1] M. Lepselter, "Beam-Lead Devices and Integrated Circuits", *Proceedings of the IEEE*, 53 (4), p. 405, 1965.

[2] K. D. Wise, J. B. Angell, and A. Starr, "An Integrated Circuit Approach to Extracellular Microelectrodes", *IEEE Transactions on Biomedical Engineering*, vol. BME-17, no. 3, pp. 238–247, 1970.

[3] S. Samaun, K. D. Wise, and J. B. Angell, "An IC Piezoresistive Pressure Sensor for Biomedical Instrumentation", *IEEE Transactions on Biomedical Engineering*, vol. BME-20, no. 2, pp. 238–247, 1973.

[4] A. C. M. Gieles and G. H. J. Somers, "Miniature pressure transducers with a silicon diaphragm", *Philips Tech. Rev.*, 33, pp. 14–20, 1973.

[5] W. Ko, M. Bao and Y. Hong, "A high sensitivity integrated circuit capacitive pressure transducer", *IEEE Trans. Electron Devices*, ED-29, pp. 48–56, 1982.

[6] Y. Lee and K. Wise, "A batch-fabricated silicon capacitive pressure transducer with low temperature sensitivity", *IEEE Trans. Electron Devices*, ED-29, pp. 42–48, 1982.

[7] S. C. Terry, J. H. Jerman, and J. B. Angell, "A gas chromatographic air analyzer fabricated on a silicon wafer", *IEEE Trans. on Electron Devices*, Vol. 26, No. 12, pp. 1880–1886, 1979.

[8] K. E. Petersen and A. Shartel, "Micromechanical accelerometer integrated with MOS detection circuitry", *IEEE International Electron Devices Meeting*, pp. 673–675, 1980.

[9] R. T. Howe and R. S. Muller, "Integrated Resonant-Microbridge Vapor Sensor", *Techn. dig. IEEE International Electron Devices Meeting*, pp, 213–216, 1984.

[10] M. Mehregany, K. J. Gabriel, W. S. N. Trimmer, "Integrated Fabrication of Polysilicon Mechanisms", *IEEE Transactions on Electron Devices*, vol. 35, No. 6, pp. 719–723, June 1988.

[11] L.-S. Fan, Y.-C. Tai and R. S. Muller, "IC-Processed Electrostatic Micro-motors", *Techn. Dig. IEEE International Electron Devices Meeting*, pp. 666–669, 1988.

[12] M. Mehregany, S. F. Bart, L. S. Tavrow, J. H. Lang, and S. D. Senturia, "Principles in Design and Microfabrication of Variable-Capacitance Side-Drive Motors", *J. Vac. Sci. Tech. A*, vol. 8, pp. 3614–3624, 1990.

[13] W. C. Tang, T.-C. H. Nguyen, and R. T. Howe, "Laterally drive polysilicon resonant microstructures", *Sensors and Actuators*, vol. 20, pp. 25–32, 1989.

[14] W. C. Tang, *Electrostatic Comb Drive for Resonant Sensor and Actuator Applications*, PhD Thesis, Electrical Engineering and Computer Sciences, University of California at Berkeley, 1990.

[15] K. Ikeda, H. Kuwayama, T. Kobayashi, T. Watanabe, T. Nishikawa, T. Yoshida, and J. Harada, "Silicon pressure sensor with resonant strain gages built into diaphragm", *Tech. Dig. 7th Sensor Symp.*, pp. 55–58, 1988.

[16] L. Lin, K. M. McNair, R. T. Howe, and A. P. Pisano, "Vacuum-encapsulated lateral microresonators", in *Proc. Transducers' 93*, 1993, pp. 270–273.

[17] K. D. Wise and K. Najafi, "A Micromachined Integrated Sensor with On-Chip Self-Test Capability", *IEEE International Solid-State Circuits Conference*, pp. 12–16, 1984.

[18] K. E. Jones, P. K. Campbell, R. A. Normann, "Interelectrode isolation in a penetrating intracortical electrode array", *Proc. Ann. Int. Conf. IEEE Eng. Med. Biol. Soc.*, vol. 12, pp. 496–497, 1990.

[19] H. T. G. van Lintel, F. C. M. van De Pol, and S. Bouwstra, "A piezoelectric micropump based on micromachining of silicon", *Sensors and Actuators A*, vol. 15, no. 2, pp. 153–167, 1988.

[20] M. Esashi, S. Shoji, and A. Nakano, "Normally piezoelectric micropump based on micromachining of silicon", *Sensors and Actuators A*, vol. 15, no. 2, pp. 153–167, 1988.

[21] J. G. Smits, "Piezoelectric micropump with three valves working peristaltically", *Sensors and Actuators A*, vol. 21, no. 1–3, pp. 203–206, 1990.

[22] P. Greiff, B. Boxenhorn, T. King, and L. Niles, "Silicon Monolithic Micromechanical Gyroscope", *Tech. Dig. 6th Int. Conf. Solid-State Sensors and Actuators*, pp. 966–968, 1991.

[23] M. W. Putty and K. Najafi, "A Micromachined Vibrating Ring Gyroscope", *Tech. Dig. Solid-State Sensor and Actuator Workshop*, pp. 213–220, 1994.

[24] F. Ayazi and K. Najafi, "Design and Fabrication of a High-Performance Polysilicon Vibrating Ring Gyroscope", *Proc. IEEE Micro Electro Mechanical Systems Workshop*, pp. 621–626, 1998.

[25] T. Juneau and A. P. Pisano, "Micromachined Dual Input Axis Angular Rate Sensor", *Tech. Dig. Solid-State Sensor and Actuator Workshop*, pp. 299–302, 1996.

[26] W. Yun, P. R. Gray and R. T. Howe, "A Surface Micromachined, Digitally Force-balanced Accelerometer with Integrated CMOS Detection Circuitry", *IEEE Solid-State Sensor and Actuator Workshop*, pp. 122–125, 1992.

[27] L. J. Hornbeck, "Current Status of The Digital Micromirror Device (DMD) for Projection Television Applications", *Techn. Dig. IEEE International Electron Devices Meeting*, pp. 381–384, 1993.

[28] O. Solgaard, F. S. A. Sandejas, D. M. Bloom, "A deformable grating optical modulator", pp. 688–690, 1992.

to be continued   to be continued   ELECTRON DEVICES SOCIETY®   to be continued   to be continued

[29] D. J. Young and B. E. Boser, "A Micromachined Variable Capacitor for Monolithic Low-Noise VCOs", *Technical Digest, IEEE Solid-State Sensor and Actuator Workshop*, pp. 86–89, 1996.

[30] J. J. Yao and M. F. Chang, "A Surface Micromachined Miniature Switch for Telecommunication Applications with Signal Frequencies from DC up to 40 GHz", *Technical Digest 8th International Conference on Solid-State Sensors and Actuators*, pp. 384–387, 1995.

[31] C. L. Goldsmith, Z. Yao, S. Eshelman, and D. Denniston, "Performance of Low-Loss RF MEMS Capacitive Switches", *IEEE Microwave and Guided Wave Letters*, vol. 8, no. 8, pp. 269–271, 1998.

[32] M. Mehregany, C. A. Zorman, N. Rajan, and C. H. Wu, "Silicon Carbine MEMS for Harsh Environment", *Proceedings of the IEEE*, pp. 1594–1610, August, 1998.

[33] A. Manz, N. Graber, and H. M. Widmer, "Miniaturized total chemical analysis systems: A novel concept for chemical sensing", *Sensors and Actuators B*, vol.1, no. 1–6, pp. 244–248, Jan. 1990.

[34] M. A. Burns, C. H. Mastrangelo, T. S. Sammarco, F. P. Man, J. R. Webster, B. N. Johnsons, B. Foerster, D. Jones, Y. Fields, A. R. Kaiser, and D. T. Burke, "Microfabricated structures for integrated DNA analysis", *PNAS*, vol. 93, no. 11, pp. 5556–5561, May 1996.

[35] C. H. Ahn, J.-W. Choi, G. Beaucage, J. H. Nevin, J.-B. Lee, A. Puntambekar, and J. Y. Lee, "Disposal smart lab-on-a-chip for point-of-care clinical diagnostics", *Proceedings of the IEEE*, vol. 92, no. 1, pp. 154–173, 2004.

[36] H. Xie, L. Erdmann, K. Gabriel, and G. Fedder, "Post-CMOS Processing for High-Aspect-Ratio Integrated Silicon Microstructures", *Journal of Microeletro-Mechanical Systems*, vol. 11, no. 2, pp. 93–101, April 2002.

[37] S. A. Zotov, M. C. Rivers, A. A. Trusov, and A. M. Shkel, "Folded-MEMS-Pyramid Inertial Measurement Unit", *IEEE Sensors Journal*, vol. 11, no. 11, pp. 2780–2789, Nov. 2011.

[38] I. P. Prikhodko, S. A. Zotov, A. A. Trusov, and A. M. Shkel, "Microscale Glass-Blown Three-Dimensional Spherical Shell Resonators", *IEEE/ASME Journal of Microelectromechanical Systems*, vol. 20, no. 3, pp. 691–701, Jun. 2011.

[39] K. Visvanathan, T. LI, and Y. B. Gianchandani, "3D-SOLE: A Fabrication Process for Large Scale Integration And Micromachining of Spherical Structures", *The 24th IEEE International Conference on Micro Electro Mechanical Systems*, pp. 45–48, 2011.

[40] M. A. Schmidt, "Wafer-to-wafer bonding for microstructure formation", *Journal of Microelectromechanical Systems*, vol. 86, no. 8, pp. 1575–1585, 1998.

[41] J. Mitchell, G. R. Lahiji, and K. Najafi, "Encapsulation of vacuum sensors in a wafer level package using a gold-silicon eutectic", *Proc. Int. Conf. Solid-State Sensors, Actuators, and Microsystems (Transducers '05)*, pp. 928–931, 2005.

[42] S.-H. Lee, S. W. Lee, and K. Najafi, "A generic environment-resistant packaging technology for MEMS", *Proc. Int. Conf. Solid-State Sensors, Actuators, and Microsystems (Transducers '07)*, pp. 335–338, 2007.

[43] R. N. Candler, W.-T. Park, H. Li, G. Yama, A. Partridge, M. Lutz, and T. W. Kenny, "Single wafer encapsulation of MEMS devices", *IEEE Transactions on Advanced Packaging*, vol. 26, no. 3, pp. 227–232, Aug. 2003.

[44] M. G. Allen, B. K. Smith, and A. Glezer, "Addressable micromachined jet arrays", *Proc. Int. Conf. Solid-State Sensors, Actuators, and Microsystems (Transducers '95)*, pp. 329–332, 1995.

[45] H. Kim, A. Astle, K. Najafi, L. Bernal, and P. Washabaugh, "A fully integrated high-efficiency peristaltic 18-stage gas micropump with active microvalves", *Proc. IEEE Int. Conf. on Micro Electro Mechanical Systems (MEMS '07)*, pp. 131–134, Jan. 2007.

[46] F. Saeki, J. Baum, H. Moon, J. Y. Yoon, C. J. Kim, and R. L. Garrell, "Electrowetting on dielectrics (EWOD): reducing voltage requirements for microfluidics", *Polym. Mater. Sci. Eng.*, vol. 85, pp. 12–13, 2001.

[47] K. F. Bohringer, U. Srinivasan, and R. T. Howe, "Modeling of capillary forces and binding sites for fluidic self-assembly", *Proc. IEEE Int. Conf. on Micro Electro Mechanical Systems (MEMS '01)*, pp. 369–374, 2001.

[48] A. Paul, A. Bozkurt, J. Ewer, B. Blossey, and A. Lal, "Surgically implanted micro-platforms in Manduca-sexta", *Proc. Solid State Sens. Actuator Workshop*, pp. 209–211, 2006.

[49] H. Sato, C. W. Berry, B. E. Casey, G. Lavella, Y. Yao, J. M. VandenBrooks, and M. Maharbiz, "A cyborg beetle: Insect flight control through an implantable tetherless microsystem", *the 21st IEEE Intl. Conf. on Micro Electro Mechanical Systems (MEMS 2008) Technical Digest*, pp. 164–167, 2008.

Chapter 19

Vacuum Device Applications

David K. Abe, Baruch Levush, Carter M. Armstrong, Thomas Grant and
William L. Menninger

19.1 Introduction

Vacuum electronic (VE) devices produce coherent electromagnetic radiation through the interaction of a beam of electrons with an electromagnetic structure. In response to the needs of the defense, scientific, and commercial communities, VE technology has continually provided breakthrough performance in power, bandwidth, and efficiency from UHF to THz frequencies (see also Chapter 14). The technological benefits have been far reaching, ranging from radar, terrestrial and space-based communications, high energy physics, fusion, industrial processing, and medicine down to the countertop microwave oven. In the twenty-first century, applications such as high resolution radar and digital television continue to stimulate the development of new devices to reach new highs in linearity and power-to-mass ratio performance. Even "mature" devices such as traveling-wave tubes and klystrons that had their beginnings in the 1930–1940s have yet to reach their ultimate performance limits, particularly at millimeter and sub-millimeter wavelengths.

Vacuum electronics is a multidisciplinary field. Advances are driven not only by innovations in electromagnetic design and innovative beam-wave interaction structures, but also by the development of new materials, thermal management, fabrication technologies, and computational techniques. From the outset, computers and computational analysis have played an important role in the advancement and understanding of VE devices. Digital computers were first used in the early 1950s to perform large-signal analyses of

Guide to State-of-the-Art Electron Devices, First Edition. Edited by Joachim N. Burghartz.
© 2013 John Wiley & Sons, Ltd. Published 2013 by John Wiley & Sons, Ltd.

to be continued   to be continued   ELECTRON DEVICES SOCIETY®   to be continued   to be continued

traveling-wave tubes. For further background on the history and science of computational modeling of VE devices, the reader is directed to chapter 10 of Ref. [1] and to Ref. [2].

The following sections provide brief descriptions of VE applications and the evolution of device performance for five main classes of devices. More detailed accounts of the history and physics of particular devices can be found in [1–4] and the references contained therein.

19.2 Traveling-Wave Devices

The traveling-wave tube (TWT) is a linear electron beam amplifier comprised of a periodic electromagnetic structure with a beam of electrons traveling down the axis. The periodic structure slows the axial phase velocity of electromagnetic waves to be in synchronism with the velocity of the beam electrons, facilitating a transfer of kinetic energy in the beam to electromagnetic energy in the waves. Depending on the power, bandwidth, and frequency of operation, different periodic circuit elements are used including helices, coupled-cavities, and ring-bar structures. Today, TWTs have broad applications as amplifiers for terrestrial and space-based communications, radar, and electronic warfare at frequencies ranging from hundreds of megahertz to over one hundred gigahertz. Figure 19.1 is a schematic of a helix TWT depicting

What is Velocity Tapering?

In a TWT, the pitch (period) of the circuit is designed to match the phase velocity of the electromagnetic wave to the electron velocity to maximize the transfer of energy. As the beam electrons give up their energy to the growing wave, the electrons lose velocity. To optimize the efficiency of the interaction, the circuit pitch can be modified as a function of axial position to better maintain synchronism between the slowing electrons and the wave. This technique is known as "velocity tapering." When used in combination with collector depression (Section 19.3), device efficiencies of >73% can be realized.

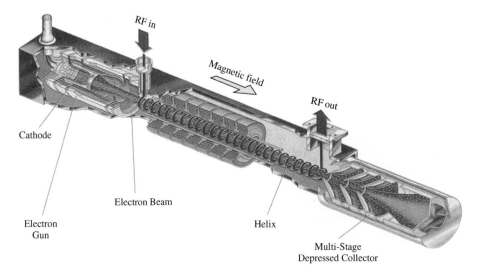

Figure 19.1   Schematic model of a helix TWT (Courtesy of Thales Electron Devices)

to be continued   to be continued   to be continued   to be continued

the main features of the amplifier including the cathode, electron beam, helix interaction circuit, and multi-stage depressed collector.

The TWT was invented in 1942 by R. Kompfner. In the 1950s, guided by the wave theory developed by J. R. Pierce, intense research stimulated an exponential rise in TWT performance starting from a few watts at the beginning of the decade to over one megawatt with 10% bandwidth in the 10-cm wavelength band in 1959. Since the 1960s, helix TWTs have played a major role in space communications. They were used in both ground terminals and space-based transponders in the first telecommunications satellite (Telstar I, 1962) and in the first geo-synchronous satellite (Syncom II, 1963). To illustrate the growth of TWT applications in space, consider that in 1977 there were about 500 tubes in space with a total of slightly more than six million operational hours. Thirty-five years later, there are over 10 000 TWTs operating in approximately 1000 satellites. As an example, Figure 19.2 shows two banks of 130-W CW (continuous-wave), 10-cm band TWTs built by L-3 Communications Electron Technologies Inc. installed on the 6800-kg DBSD satellite. As of 2012, the total number of operational hours in space exceeds one billion, marking the TWT as one of the most long-lived and reliable electronic devices ever manufactured.

The TWTs on the *NASA Pioneer 6* spacecraft are perhaps the finest testament to VE device reliability in one of the harshest environments around. *Pioneer 6* was launched into a heliocentric orbit on December 16, 1965 and is the oldest operating spacecraft in existence. Sometime after December 1995, the primary data-link TWT failed. Fortunately, the backup TWT responded to command (after 30 years of inactivity) and continued to send data. It was still functioning in December 2000 when contact with *Pioneer 6* was established one last time to commemorate its 35 years in orbit [5]. With their ability to provide high bandwidth and linear power with high efficiencies ($>73\%$ DC-to-RF efficiencies are now readily available),

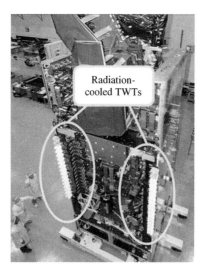

Radiation-cooled TWTs

Figure 19.2 Radiation-cooled TWT amplifiers on the DBSD (formerly ICO G1) satellite (Courtesy of Space Systems/Loral)

to be continued   to be continued   ELECTRON DEVICES SOCIETY®   to be continued   to be continued

TWTs will continue to be the predominant amplifier technology for space applications into the foreseeable future.

In addition to communications applications, the helix TWT has widespread use in electronic counter-measure (ECM) systems, where its capacity for multi-octave bandwidths combined with tens to hundreds of watts of output power enables a single system to function against a wide variety of threats. In the early 1970s, new high energy product rare-earth permanent magnet materials such as samarium cobalt and neodymium-iron-boron made it possible to focus and transport significantly higher current electron beams at relatively modest voltages, enabling the creation of higher power devices with lower weight and volume. The TWT "mini-tube" was an important byproduct enabled by these new materials and would become a key enabling technology for the microwave and millimeter-wave power modules of the 1990s.

Coupled-cavity TWTs span the performance space between helix TWTs and klystrons. With their all-metal construction, they are capable of supporting higher peak and average powers than the dielectrically-supported helix TWTs, and their traveling-wave circuits are capable of higher bandwidths compared with resonant cavity-based klystrons. An example of high power performance is the 100-kW Millimeter-Wave radar at the Kwajalein Atoll driven by a pair of 50-kW peak, 5-kW average power, Ka-band coupled-cavity TWTs [6]. In the 3-mm wavelength band, Communications and Power Industries has demonstrated permanent-magnet focused devices with 90-W CW power and solenoid-focused devices with 6-kW pulsed power. Also in the 3-mm band, Thales Electron Devices has demonstrated >200 W peak pulsed power, >500 MHz bandwidth in a permanent magnet focused amplifier weighing <1 kg using a double-comb delay line circuit.

In the 1990s, TWT mini-tube technology was combined with an integrated power conditioner and emerging monolithic microwave integrated circuit (MMIC) solid-state amplifier technology (see Chapter 14) to produce a self-contained, compact, high power transmitter package. Termed a "microwave power module" or MPM, the configuration leverages the best of both vacuum and solid-state technologies by partitioning the overall transmitter gain between a low-noise solid-state RF driver and a high power, high efficiency miniature helix TWT [2]. Broadband MPMs have been configured to generate 50–200 W of CW RF output power with gains in excess of 50 dB and bandwidths of more than two octaves. Narrowband MPMs have demonstrated overall efficiencies of 50% and multi-octave versions have demonstrated efficiencies >35%. With their compact size and low weight, coupled with their high power and efficiency, MPMs are ideally suited for use on power- and volume-constrained airborne platforms such as unmanned aerial vehicles (UAVs), where they are used in radar and high-speed data links.

In 2011, an MPM was used in the "mini-RF" synthetic aperture radar on the NASA Lunar Reconnaissance Orbiter (LRO) that recorded the first full polarimetric images of the Moon's surface – the first time that an MPM was used in space. The MPM was ideal fit to the stringent platform constraints and its broad bandwidth enabled a single amplifier to cover both the 12.6- and 4.2-cm wavelength bands required by the mission [7].

As suitable solid-state driver amplifiers have become available, MPM technology has been extended to millimeter-wave applications such as high-speed data links in the 1-cm and 7-mm wavelength bands. Figure 19.3 is an example of a 1-cm band millimeter-wave power module (MMPM) manufactured by L-3 Communications Electron Devices Division. The unit can generate a saturated output power of 100 W from 30–31 GHz or 50 W of linear power with a pre-distortion linearizer in a < 3760 cm^3, air-cooled package weighing < 4.4 kg.

Figure 19.3   A 1-cm wavelength band MMPM capable of generating 100 W of saturated output power (Courtesy of L-3 Communications Electron Devices Division)

19.3 Klystrons

The klystron is a linear electron beam amplifier comprised of a series of re-entrant cavity circuits with an electron beam traveling down the axis. In the input cavity, a low-level RF drive signal creates a time-varying longitudinal electric field across the input cavity gap. Depending on the phase of the electric field in the gap, the beam electrons will either be accelerated or decelerated, producing a "velocity modulation" on the beam. Downstream of the input cavity, the beam propagates in a tunnel that is cut-off to external electric fields. With distance, the electrons begin to bunch producing a "density modulation" on the beam; the wave propagation on the beam is at the reduced plasma wavelength. Subsequent resonant cavities are used to increase the intensity of the bunching and/or broaden the bandwidth of interaction by de-tuning the cavity frequencies. Power is extracted in an output cavity that is excited by the time-varying electric field induced by the bunched beam as it passes the output cavity gap (gains of 40–60 dB are typical).

The first modern klystron amplifier was demonstrated in 1939 by R. and S. Varian. During World War II, the klystron was eclipsed by the magnetron as the device of choice to generate coherent high power microwave radiation for radar. However, low-power reflex klystrons were widely used as local oscillators in superheterodyne receivers, where their ability to be tuned in frequency by changing the reflector electrode voltage was a valuable feature [8].

Immediately after the war, klystron development focused on applications in linear accelerators to support experiments in high-energy particle physics. Much of the early work was performed at Stanford University in California for the 10-cm band Mark III electron accelerator. In 1953, the group reported the successful operation of the first multi-megawatt microwave source of any kind whose 30 MW of peak output power at a frequency of 2.857 GHz (43% efficiency) represented a three order-of-magnitude increase over the previous state-of-the-art [3]. Other applications of linear accelerators ("linacs") quickly followed, including medicine (the first medical linac was used in 1953 in Hammersmith Hospital, UK). Klystron development has kept pace with the growing needs of the accelerator community through the use of new materials and improved manufacturing techniques. By the late 1990s, for example, researchers at the Stanford Linear

Accelerator Center developed the solenoid-focused 11.4-GHz XL-4 klystron that reliably produced 75 MW of output power in a 1.2 μs pulse with an efficiency of 50% [1].

The results from the early Stanford 10-cm band tube and others like it demonstrated that klystrons could achieve higher peak and average RF powers compared with the magnetrons used in World War II. Furthermore, the klystron could generate a highly stable signal and, as an amplifier, could support complex, time-varying waveforms. These features were highly attractive for radar transmitters where the higher power enabled greater range, the higher stability contributed to improved performance against moving targets in clutter, and the frequency agility allowed the use of techniques such as pulse compression to enhance range resolution. By the late 1950s, "super-power" klystrons (1.25 MW peak power, 75 kW average power) were being used in the three large radar stations that comprised the US Ballistic Missile Early Warning System (BMEWS). This period also saw the first uses of klystrons in radar systems used to track satellites (beginning with Sputnik in 1957) and for extra-terrestrial planetary measurements such as the first 2D radar mapping of the moon in 1960 [9].

The 1960s saw continued improvement in klystron power, efficiency, and bandwidth, driven by the requirements of both radar and communications systems. A notable accomplishment was the demonstration of a 50-dB gain, 1-MW CW 3-cm band klystron with a 1-dB bandwidth of 1% and an efficiency of 45% [3] – a 50-fold increase in average power over the state-of-the-art at the beginning of the decade and still an impressive achievement today.

Figure 19.4 is a photograph of the Klystron Gallery in the accelerator-driven Spallation Neutron Source (SNS) at the Oak Ridge National Laboratory (Oak Ridge, TN USA). The SNS uses 70 VKP-8291A

Figure 19.4 The Klystron Gallery at the Oak Ridge National Laboratory Spallation Neutron Source (Photo courtesy of Communications and Power Industries LLC)

to be continued   to be continued   **ELECTRON DEVICES SOCIETY**®   to be continued   to be continued

klystrons manufactured by Communications and Power Industries, each operating at a frequency of 805 MHz and generating 550 kW of peak output power at 9% duty with an overall efficiency of 67%. As of February 2008, the 70 SNS klystrons had collectively accumulated over one million hours of high voltage operation with only one failure.

The extended interaction cavity concept that was first used in the multi-megawatt Stanford klystrons of the 1950s has subsequently been used in a wide variety of extended interaction klystrons (EIKs) to improve power performance and to achieve higher bandwidths. EIKs have been operated at frequencies up to 280 GHz. The first space-qualified 94.05-GHz EIK was developed in the 1990s for the CloudSat project. The 2-kW peak power, 32% efficient, 56-dB gain, 250-MHz bandwidth EIK weighed only 6.2 kg and enabled the first millimeter-wave radar measurements of the vertical profile of clouds [10].

Ultra-high frequency (UHF) television was one of the earliest commercial applications for the klystron, where the requirements of high power and efficiency at relatively narrow bandwidths made it a natural choice. Communications applications for klystrons have continued into the twenty-first century with satellite uplink transmitters and terrestrial digital radio repeaters that can reach areas where satellite coverage is poor or non-existent. High power over-the-horizon ("troposcatter") systems use 5-cm band klystrons to bounce signals off of the troposphere to provide medium-to-high data rate coverage (up to 20 Mbps) to fixed and mobile sites over ranges of tens to hundreds of miles. Similar to TWTs, multi-stage depressed collectors (MSDCs) are used in klystrons to significantly improve DC-to-RF efficiency. The addition of an MSDC into a 2.4-kW klystron designed for a Direct Broadcast Satellite (DBS) band uplink raised its efficiency from 24% to 40%, enabling a substantial savings in annual energy costs. Klystrons are also at the heart of space communications. For example, the NASA Deep Space Network 70-m diameter antennas employ a variety of high power klystrons for orbital communications, tracking, and spacecraft command/control.

Shortly after the demonstration of the first klystron in 1939, tube designers began to consider multiple electron beam versions of the device as a means to circumvent the space-charge limitations of a single intense electron beam in order to achieve higher power at relatively low operating voltages. In a multiple-beam klystron (MBK), parallel electron beams propagate in independent beam tunnels and interact only in the short gaps of the resonant cavities. While the individual beams have low perveance ($I_{beam}V^{-3/2}$), which is conducive to low-loss transport and efficient electron bunching, the total current can be

Collector Depression

Collector depression is a unique feature of VE devices. After interaction with the circuit, the electron beam must be "collected" and its energy dissipated. By spreading and electrostatically decelerating the "spent" electron beam using negative ("depressed") voltages, the residual kinetic energy in the beam can be converted to potential energy, dramatically improving the device efficiency while at the same time providing a large thermal conductance path for the remaining dissipated power. Overall amplifier efficiencies >73% have been demonstrated with collectors using multiple negatively biased electrodes.

Figure 19.5 18-beam cathode designed for the US Naval Research Laboratory's 3-cm band MBK. Inset: The cathode and thermionic emitters undergoing a high-temperature test (Courtesy of the US Naval Research Laboratory)

high, facilitating high power. Figure 19.5 is a photograph of an 18-beam cathode designed for an MBK operating in the 3-cm wavelength band [11].

From approximately 1960–1990, the majority of MBK development took place in the former Soviet Union, where researchers developed an impressive array of high-power multiple-beam amplifiers primarily for radar applications. Operating wavelengths ranged from 30 cm down to 2 cm with power gains from 40–45 dB, electronic efficiencies from 30–65%, and bandwidths up to 18% [12]. Highly compact "Miniature MBKs" (MMBKs) with power-to-mass ratios as high as 500 W kg^{-1} were also developed [13]. MBKs have also found applications in high-energy accelerators. For example, the European X-ray Free Electron Laser (XFEL) uses 1.3-GHz MBKs with 10-MW pulsed output power, 1.3-millisecond pulse lengths, and 10-Hz repetition-rates operating at efficiencies >65% [14].

19.4 Inductive Output Tubes

The inductive output tube (IOT) uses a closely-spaced floating grid to gate the emission from a thermionic cathode to form density modulated electron bunches at UHF to lower microwave band frequencies. The bunched beam is post-accelerated to high potential and the amplified signal is extracted from an output cavity or broadband circuit. Conceived by A. Haeff in 1938, the IOT began to see widespread commercial use in the over-the-air broadcast TV industry with the introduction of laser-cut pyrolytic graphite for the

grid structure of the RF electron gun in the early 1980s [15]. At present, commercial UHF IOTs operate at CW power levels of up to 90 kW; higher frequency versions can provide over 30 kW up to ~1 GHz. High basic efficiency and excellent linearity are other hallmarks of the IOT. A typical broadcast TV IOT is run as a class AB amplifier and has an electronic efficiency of over 40%. The compact (~80-cm in length) device generates 120 kW of peak output power and 30 kW of average power while amplifying an 8VSB digital broadcast signal.

In the late 1990s, the invention of the constant efficiency amplifier (CEA) represented yet another significant advance in IOT technology [16]. The CEA replaced the grounded collector of a standard IOT with a multi-stage depressed collector enabling energy recovery from the electron beam. The result is a power amplifier with near constant efficiency from saturation to well into output power backoff. Such performance is critical when processing input signals with large amplitude modulation content, such as those found in Digital Broadcast TV. The CEA provides an improvement in efficiency of approximately 50% over a standard IOT (corresponding to an overall CEA efficiency of 60%) when operated with an 8VSB signal.

IOT Linearity

The input structure of the IOT behaves like a tetrode in that, to first order, the beam current is limited only by the maximum allowable current interception on the grid due to thermal considerations. As a result, the transfer characteristics of the IOT do not saturate at the point of maximum efficiency and retain a highly linear characteristic.

19.5 Crossed-Field Devices

Crossed-field devices are so named for their use of orthogonal ("crossed") electric and magnetic fields to guide the motion of the electron beam. The magnetron is the most familiar of this type of device due to its critical role in providing the Allies with effective radar in World War II and its widespread use today in countertop microwave ovens. A key technological innovation occurred in Britain in 1939–1940 with the development of the resonant cavity magnetron [17] which enabled breakthrough high power performance at microwave frequencies.[1] The war years of 1940–1945 saw an explosive growth in magnetron performance and the implementation of the magnetron in numerous centimeter-wavelength radar systems (Ref. [18] cites the development of over one hundred different "radar equipments" at the Massachusetts Institute of Technology Radiation Laboratory alone in less than five years). By the end of the war, typical peak (average) powers had reached 500 kW (500 W) in the 10-cm band and 100 kW (100 W) in the 3-cm band [17].

In the 1950s, advances in supporting technologies such as high purity alumina ceramics (high voltage insulation, vacuum windows),

---

[1] Both Japan [18] and Germany [19] appear to have invented a version of the microwave cavity magnetron before the British but did not exploit the technology.

metal-ceramic joining technology (ultra-high vacuum), and thoriated tungsten and molybdenum cathodes able to withstand high energy electron bombardment, enabled another order of magnitude increase in magnetron performance. For example, the Raytheon QK338 developed during this period produced 5 MW and 4 kW of peak and average power, respectively, in the 10-cm band [17]. The "circular electric mode" (CEM) magnetron was another breakthrough invention from this period [17]. The CEM or coaxial magnetron featured increased stability and the capability for fast mechanical frequency tuning and was subsequently incorporated into many instrumentation radar systems and missile seekers. During the 1950s, magnetron-based radar systems were also being applied to scientific measurements, enabling, for example, a more accurate measure of the distance from the earth to the moon.

In the 1960s, magnetron performance pushed into the upper millimeter-wave range, with devices reported at 75 GHz (5–40 kW peak power) and 120 GHz (2.5 kW peak, 0.5 W average power) [20]. Further improvements, particularly in average power performance at millimeter-wavelengths, have continued into the twenty-first century.

In parallel with the development of pulsed magnetrons for radar, continuous-wave magnetrons have been developed for industrial drying and heating and other commercial uses as well as for medical applications [21]. Common operating frequencies are 915 MHz (5–100 kW, 80–87% efficiency) and 2.45 GHz (1–30 kW, 50–72% efficiency). Invented in 1945, the microwave oven is probably the most ubiquitous and well-known application of magnetron technology. The first "Radarange" designed for kitchen countertop use was introduced by Amana in 1967. A series of innovations has reduced the size and cost of "cooker" magnetrons and, by 2002, automated manufacturing reduced the unit cost of the magnetron to as low as US$7.00 [22].

Medical microwave diathermy machines for deep tissue heating were first marketed in 1947 [16]. Therapeutic use of magnetron-based systems expanded in the 1970s with RF hyperthermia and the treatment of cancer and has continued into the present with new treatments such as microwave ablation of human tissue and nerves [23].

In addition to oscillators, amplifying devices based on crossed-field interaction have also been developed. In response to growing demands by new radar concepts that required high power, broadband amplifiers, the crossed-field amplifier (originally called a "platinotron" and trademarked as the "Amplitron" by Raytheon) was invented in 1953 by W. Brown. These devices have found numerous applications in high power search and surveillance radar, including passive phased array systems, and their use has continued into the present [3].

19.6 Gyro-Devices

In linear electron beam devices such as TWTs and klystrons, the critical dimension of the beam-wave interaction structure is smaller than the operating electromagnetic wavelength. As the operating frequency increases, linear beam tubes encounter increasing challenges due to tight mechanical tolerances, electric field breakdown, and thermal management. In contrast, "cyclotron resonance maser" (CRM) devices operate on the principle of coherent bremsstrahlung radiation resulting from the motion of gyrating electrons in a constant magnetic field. Electron bunching can occur azimuthally and the critical dimensions of the interaction structures can be many times an electromagnetic wavelength, facilitating high power operation at very high frequencies compared with linear beam devices. Gyrotron oscillators and gyro-amplifiers are

a type of CRM whose development began in the 1950s and were first demonstrated in the early 1960s. In the succeeding decades, the field has undergone considerable growth in both theoretical understanding and device performance resulting in the demonstration of remarkable peak and average power at frequencies into the low terahertz regime. Much of the early pioneering work took place in the former Soviet Union; see [24] for a unique and detailed account of both the history and physics of these devices.

The first gyrotron was successfully demonstrated by M. I. Petelin and his associates at the Radiophysical Research Institute in Gorki (now Nizhny Novgorod) in 1964. This device operated at the fundamental cyclotron resonance mode at a frequency of ~8.82 GHz and produced a CW power of 6 W.

The 1970s saw continued gyrotron development with a focus on applications in electron-cyclotron resonance heating (ECRH) to support magnetic fusion experiments. The TM-3 Tokamak at the Kurchatov Institute in Moscow and the TUMAN-2 at the Ioffe Institute in Leningrad (now St. Petersburg) both used 4.5-mm wavelength, 80-kW peak power gyrotrons with pulse lengths of 0.6 ms, a significant step toward achieving the long pulse lengths necessary for plasma confinement. By the end of the 1970s, higher order $TE_{m,n,p}$ whispering gallery mode structures (with large azimuthal index $m \gg 1$ for $p \sim 1$) were introduced to address the mode competition issues present in the symmetric $TE_{0,p}$ mode cavities in use at the time. Using such a structure, a megawatt-level, 100-GHz gyrotron was demonstrated in 1978 [24].

In the 1990s, the development of chemical vapor deposition (CVD; see Chapter 10) diamond vacuum windows enabled even longer pulse durations and higher power CW operation. Requirements for large-scale plasma reactors such as the Wendelstein 7-X stellarator and the International Thermonuclear Experimental Reactor (ITER), an advanced tokamak with the goal of demonstrating a net energy gain of 10 (500 MW of output power produced from an input power of 50 MW), would drive the development of fusion gyrotrons for the next two decades. By 2009, the US, European Union, Russia, and Japan had all demonstrated megawatt-class fusion gyrotrons at frequencies of 140 and 170 GHz with pulse durations ranging from 1.7–30 min [25].

As gyrotron technology has matured, the devices have found application in other non-fusion technological, industrial, and medical areas. Devices operating up to the low THz frequency regimes have been developed for plasma diagnostics and spectroscopy. In 2008, a gyrotron using a pulsed magnetic field of up to 40 T was reported to produce 1.5 kW at 1.022 THz [26]. Gyrotrons have also been developed for a variety of industrial uses including the sintering of nanostructured ceramic and metal powders, surface hardening, joining, and dielectric coating of metals and alloys [25]. Gyrotrons operating at 263, 395, and 527 GHz are used in medical applications such as dynamic nuclear polarization (DNP), a technique that uses electron spin resonance excitation with a transference to nuclear spins to enhance nuclear magnetic resonance (NMR) imaging [27].

The development of gyro-amplifiers occurred simultaneously with that of the gyrotron oscillator. The first gyro-klystron was demonstrated in 1967, a 3-cm band device that produced several kilowatts of output power at an efficiency of 70%. Starting in the 1970s, gyro-klystrons have been developed in Russia for radar applications, primarily at the atmospheric transmission frequency windows of 34–35 GHz and 93–95 GHz. In 1987–1989, "Ruza", a megawatt mechanically-steered phased array radar was built at the Sary-Shagan test range in Kazakhstan, powered by two 3-cm band, 750 kW, 100 μs pulse duration, 2% duty gyro-klystrons. The radar can track up to 30 targets up to a distance of 4000 km and its large power-aperture product enables it to detect objects in space with a radar cross-section as small as 0.01 m^2 at ranges up to 420 km [28]. In the 1990s, a government-industry-university team led by researchers at the US Naval Research Laboratory developed a 3-mm band five-cavity gyro-klystron with an average power of 10.2 kW (10% duty), 700-MHz bandwidth, and an efficiency of 31% [1] that was subsequently integrated into the WARLOC W-band radar system.

to be continued   to be continued   ELECTRON DEVICES SOCIETY®   to be continued   to be continued

References

[1] R. J. Barker, J. H. Booske, N. C. Luhmann Jr., and G. S. Nusinovich (eds), *Modern Microwave and Millimeter-Wave Power Electronics*, Piscataway, NJ, USA, IEEE Press, 2005.

[2] V. L. Granatstein and C. M. Armstrong (eds), "Special Issue on New Vistas for Vacuum Electronics", *Proc. IEEE*, vol. 87, no. 5, May 1999.

[3] L. L. Clampitt (ed.), "Special Issue on High-Power Microwave Tubes", *Proc. IEEE*, vol. 61, no. 3, Mar. 1973.

[4] N. H. Pond, *The Tube Guys*, West Plains, MO, USA, Russ Cochran Publisher, 2008.

[5] NASA, *NASA – The Pioneer Missions*, available online: http://www.nasa.gov/centers/ames/missions/archive/pioneer.html (accessed October 10, 2012), Mar. 2007.

[6] M. D. Abouzahra and R. K. Avent, "The 100-kW millimeter-wave radar at the Kwajalein Atoll", *IEEE Ant. and Prop. Mag.*, vol. 36, no. 2, pp. 7–19, Apr. 1994.

[7] R. K. Raney, P. D. Spudis, B. Bussey, J. Crusan, J. R. Jensen, W. Marinelli, P. McKerracher, C. Neish, M. Palsetia, R. Schulze, H. B. Sequeira, and H. Winters, "The lunar mini-rf radars: Hybrid polarimetric architecture and initial results", *Proc. IEEE*, vol. 99, no. 5, pp. 808–823, May 2011.

[8] J. R. Pierce, "History of the microwave-tube art", *Proc. IRE*, vol. 50, no. 5, pp. 978–984, May 1962.

[9] H. G. Weiss, "The Millstone and Haystack radars", *IEEE Trans. Aerospace and Electronic Sys.*, vol. 37, no. 1, pp. 365–379, Jan. 2001.

[10] A. Roitman, D. Berry, and B. Steer, "State-of-the-art W-band extended interaction klystron for the CloudSat program", *IEEE Trans. Electron Dev.*, vol. 52, no. 5, pp. 895–898, May 2005.

[11] K. T. Nguyen, E. L. Wright, D. E. Pershing, D. K. Abe, J. J. Petillo, and B. Levush, "Broadband high-power 18-beam S-band klystron amplifier design", *IEEE Trans. Electron Dev.*, vol. 56, no. 5, pp. 883–890, May 2009.

[12] A. N. Korolev, S. A. Zaitsev, I. I. Golenitskij, Y. V. Zhary, A. D. Zakurdayev, M. I. Lopin, P. M. Meleshkevich, E. A. Gelvich, A. A. Negirev, A. S. Pobedonostsev, V. I. Poognin, V. B. Homich, and A. N. Kargin, "Traditional and novel vacuum electron devices", *IEEE Trans. Electron Dev.*, vol. 48, no. 12, pp. 2929–2937, Dec. 2001.

[13] A. S. Kotov, E. A. Gelvich, and A. D. Zakurdayev, "Small-size complex microwave devices (CMD) for onboard applications", *IEEE Trans. Electron Dev.*, vol. 54, no. 5, pp. 1049–1053, May 2007.

[14] V. Vogel, S. Choroba, T. Froelich, T. Grevsmuehl, F.-R. Kaiser, V. Katalev, I. Sokolov, H. Timm, and A. Cherepenko, "Testing of 10 MW multibeam klystrons for the European X-ray FEL at DESY", *Proc. of the 2007 IEEE Particle Accelerator Conf.*, pp. 2077–2079, Jun. 2007.

[15] D. H. Preist and M. B. Shrader, "The klystrode – An unusual transmitting tube with potential for UHF-TV", *Proc. IEEE*, vol. 70, no. 11, pp. 1318–1325, Nov. 1982.

[16] R. Symons, M. Boyle, T. Bemis, J. Cipolla, H. Schult, and R. True, "Prototype constant-efficiency amplifiers", *IEEE Trans. Broadcasting*, vol. 47, no. 2, pp. 147–152, Jun. 2001.

[17] W. C. Brown, "The microwave magnetron and its derivatives", *IEEE Trans. Electron Dev.*, vol. 31, no. 11, pp. 1595–1605, Nov. 1984.

[18] M. Skolnik, "Role of radar in microwaves", *IEEE Trans. Microwave Theory Tech.*, vol. 50, no. 3, pp. 625–632, Mar. 2002.

[19] H. E. Hollmann, "Magnetron", U.S. Patent no. 2,123,728, Jul. 1938.

[20] G. H. Plantinga, "Conventional rising-sun magnetron for 120 Gc", *IEEE Trans. Electron Dev.*, vol. 11, no. 2, pp. 76–77, Feb. 1964.

[21] J. Thuéry, *Microwaves: Industrial, Scientific and Medical Applications*, Boston, MA, USA, Artech House, 1992.

[22] K.-S. Chang, D.-K. Suh, C.-B. Lee, W.-W. Eom, J.-J. Cha, D.-W. Kim, and H.-J. Ha, "Mass production of magnetron for microwave oven", *2002 IEEE Int. Vacuum Electronics Conf*, pp. 7–8, Apr. 2002.

[23] A. Rosen, M. A. Stuchly, and A. Vander Vorst, "Applications of rf/microwaves in medicine", *IEEE Trans. Microwave Theory Tech.*, vol. 50, no. 3, pp. 963–974, Mar. 2002.

[24] G. S. Nusinovich, *Introduction to the Physics of Gyrotrons*, Baltimore, MD, USA, The Johns Hopkins University Press, 2004.

[25] M. Thumm, "History, presence, and future of gyrotrons", *2009 IEEE Int. Vacuum Electronics Conf.*, pp. 37–40, Apr. 2009.

[26] M. Yu. Glyavin, A. G. Luchinin, and G. Yu. Golubiatnikov, "Generation of 1.5-kW, 1-THz coherent radiation from a gyrotron with a pulsed magnetic field", *Phys. Rev. Lett.*, vol. 100, 015101, Jan. 2008.

[27] M. Blank, P. Borchard, S. Cauffman, and K. Felch, "High-frequency CW gyrotrons for NMR/DNP applications", *2012 IEEE Int. Vacuum Electronics and Vacuum Electron Sources Conf.*, pp. 327–328, Apr. 2012.

[28] A. A. Tolkachev, B. A. Levitan, G. K. Solovjev, V. V. Veytsel, and V. E. Farber, "A megawatt power millimeter-wave phased-array radar", *IEEE Aerospace and Electronic Sys. Magazine*, vol. 15, no. 7, pp. 25–31, Jul. 2000.

to be continued   to be continued   ELECTRON DEVICES SOCIETY®   to be continued   to be continued

Chapter 20

Optoelectronic Devices

Leda Lunardi, Sudha Mokkapati and Chennupati Jagadish

20.1 Introduction

Optoelectronic devices have been made possible only due to simultaneous developments in device concepts and designs, crystal growth techniques with ultra-fine control over the material compositions and thickness, and also solving some fundamental materials-related issues like doping. Noteworthy are light emitting diodes and lasers while operating on different principles (spontaneous emission and stimulated emission, respectively) the historical developments associated with these two devices are closely related and cannot be decoupled. Lasers are used for telecommunications, solid-state laser pumping, barcode scanning, materials processing, optical data storage, research, and medical diagnostics while light emitting diodes are preferred for consumer electronics, traffic signals, automotive, solid state and liquid crystal displays, and especially attractive for 'lighting effects' that need different levels of colour-mixing.

Electroluminescence

Light emitting diodes and lasers convert electrical energy into optical radiation. The frequency or wavelength of this radiation depends on the intrinsic properties of the material they are made from. The optical energy released from a laser, i.e. from its optical resonant cavity is monochromatic with spatial and temporal coherence.

In contrast incoherent light emitting diodes have no optical resonant cavities for wavelength selectivity.

Guide to State-of-the-Art Electron Devices, First Edition. Edited by Joachim N. Burghartz.
© 2013 John Wiley & Sons, Ltd. Published 2013 by John Wiley & Sons, Ltd.

to be continued   to be continued   ELECTRON DEVICES SOCIETY®   to be continued   to be continued

20.2 Light Emission in Semiconductors

The first report on light emitting diodes (LEDs) dates back to 1907, when Henry J. Round published *A Note on Carborundum* [1], however the discovery and development of the theory and understanding of p-n junctions in semiconductor materials encompassing the basis of present devices did not happen until few decades later [2]. Semiconductor p-n junctions were first (accidentally) formed and discovered by R. Ohl in 1940 and reported in 1947 [3]. Luminescence from Ge and Si junction diodes was observed as early as 1952. However, Ge and Si were poor light emitters because of the indirect bandgap [4, 5]. The importance of direct bandgap semiconductors for efficient radiative recombination was first identified by Bardeen in 1956 [6]. In less than a decade, four research groups independently reported the first p-n junction lasers [7–10].

H. Kroemer and Z. I. Alferov won the 2000 Noble Prize in Physics for their work on double heterojunctions (DHs) [11, 12], for carrier and optical confinement to achieve continuous operation at elevated temperatures. Demonstration of lasing at room temperature using DHs was reported simultaneously by two groups [13, 14]. Dupuis *et al.* demonstrated the first heterostructure laser based on superlattices (GaAs/AlGaAs) with characteristics comparable to those of conventional DHs lasers in 1978 [15].

The next technology breakthrough for semiconductor-based lasers was the conceptual development and experimental realization of pseudomorphic or strained materials that would decrease the threshold current value consequently resulting in higher output power [15–19]. In 1982, Arakawa *et al.* theoretically studied the advantages of using quantum boxes and/or quantum dots in lasers [20]. While quantum dot lasers in research laboratories have demonstrated properties superior to those of quantum well lasers, commercial LEDs and lasers are still mostly based on quantum well structures [20–22]. Figure 20.1 illustrates a quantum dot laser.

Another laser concept, the quantum cascade laser (QCL) [23] proposed by Capasso *et al.* in the mid-1990s was made possible by the fine control obtained with MBE of thin heterostructure layers. It relies on confined electrons for light emission inside a sequence of multiple thin quantum wells instead of band to band recombination in the bulk material, with a wider range of wavelength tuning (4–19 μm).

20.2.1 Crystal Growth Techniques

AlGaAs-GaAs was one of the initial systems with direct bandgap investigated for the emitting layer of heterostructure devices because of the possible high efficiency but growth with defect free boundaries was initially obtained only by liquid phase epitaxy [24, 25]. The development of molecular beam epitaxy and metal organic chemical vapour deposition (MOCVD) techniques in 1970s provided sharp heterointerfaces with better accuracy in composition and thickness control [26, 27].

Using MOCVD Dupuis *et al.* reported the first room temperature pulsed operation of AlGaAs DHs laser as well as the first quantum well laser with performance matching that of a standard DH laser [28]. After the seminal work of Arakawa *et al.*, MOCVD and MBE (see Chapter 14) were also used to grow quantum dots in InAs/GaAs lattice mismatched system using the Stranski–Krastanow growth mode. Nonetheless MOCVD is the growth technique suitable for large area substrates, overall cost, and currently widely employed by the optoelectronics industry in general.

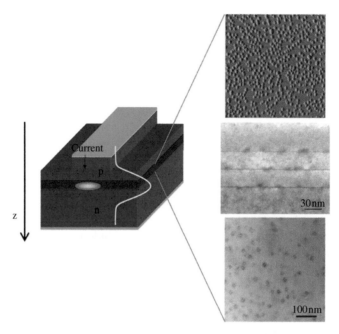

Figure 20.1 (*left*) Schematic of a quantum dot laser indicating the different regions. (*right*) Atomic force microscope image (*top*); cross-section transmission electron microscope (X-TEM) micrograph of three stacks of quantum dot layers (*middle*); plan-view TEM (*bottom*) of quantum dots used for making lasers

20.2.2 Materials System Development

The first visible LED was demonstrated using GaAsP compound semiconductor by Holonyak in 1962 [9]. In the early 1960s his group developed the vapour phase epitaxy (VPE) growth of GaAsP and GaAsP heteroepitaxy on GaAs substrates, the first step toward commercial LEDs. For visible LEDs, GaAsP alloys would be replaced by AlGaAs because of the direct bandgap in the visible region of the electromagnetic spectrum and lattice matched to GaAs substrate. The wider bandgap enabled efficient heteroepitaxial device designs. With high purity AlGaAs layers, AlGaAs-based LEDs dominated the visible LED market to become the first high brightness red LEDs to be used in automotive industry. While research shifted towards heterostructures for efficient LEDs and lasers, there was a need for identification of lattice matched heterostructures over a range of emission wavelengths.

In 1970 the first semiconductor quaternary device using AlGaAsP-GaAsP heterojunctions was demonstrated, indicating that quaternary III-V compounds could offer bandgap tuning in lattice matched heterojunctions [29]. InAlGaP became the next material system of interest for high efficiency visible LEDs. Some of solid state commercial LEDs are in fact based on this material system.

In the early 1980s material quality and p-type doping in GaN were unsolved issues for fabrication of efficient light emitting p-n junctions. In 1986, Akasaki *et al.* showed a two-step MOCVD growth process

for low defect density GaN on a low temperature AlN buffer [30]. Meanwhile, successful p-type doping of GaN was demonstrated using Mg [31, 32]. These developments led Nakamura [33] to demonstrate the first bright GaN-based blue LEDs after two decades of the first report [34] to initiate the commercial solid state lighting era. The discoveries on optical devices and advances on active compound semiconductor materials have covered a large portion of the electromagnetic spectrum from the visible to the far infrared.

20.3 Photodetectors

Photodetectors are optoelectronic devices that convert the optical energy into an electrical signal. They depend on the intrinsic material properties (referred as absorption edge) and can have internal gain. The simplest type is the photoconductor where the active conductivity increases by the incident photons, but its performance is limited by thermal noise (large dark current).

The p-n junction photodiode is the most common type used for all applications in all material systems, where the energy of photons absorbed is equal to the material bandgap energy. The principle is based on absorbed photons in the reverse bias region of the p-n junction creating electron–hole pairs, and therefore contributing to the electrical current. If there is internal gain the photocurrent can be amplified.

Photodiodes are essential for all applications that need light detection with low noise: from engine monitoring to astronomy. In communication systems photodiodes have applications in high and low power detection, radio-over-fibre transmission, phase array radar, photonic analogue to digital converters (ADCs), and stable microwave signal generation.

Observation of photovoltaic effect in p-n junctions was first reported by R. Ohl in 1941 [2]. The first p-n photodiodes were made of silicon and germanium, because of their intrinsic material properties [35, 36]. Germanium crystals were cooled (at 77 K) to reduce dark currents with limited use in spectroscopy. After Anderson's report of Ge-GaAs heteroepitaxy, p(Ge)-n(GaAs) heterojunction diodes were demonstrated for infrared detection [37, 38].

At that time heteroepitaxy of Ge on Si substrates was not considered an alternative because the 4.2% lattice mismatch between the two materials resulted in high dislocation density combined with surface roughness and prevented further development as required for integration with electronics. Instead II-VI or III-V materials in ternary or

Integrated Optics

Advantages:

- Immunity to electromagnetic interference (EMI)
- Low loss transmission
- Wide bandwidth
- No short circuits
- No ground loops
- Secure from eavesdropping
- Scalable
- Low crosstalk
- High throughput capacity

Disadvantages:

- Insertion loss
- Temperature dependence
- Need of external power supply

to be continued   to be continued   ELECTRON DEVICES SOCIETY®   to be continued   to be continued

quaternary compositions for heterojunction p-n photodiodes have been developed for different wavelengths and applications [39].

For very low levels of optical signal photodetectors with internal gain are preferred. Avalanche photodetectors (APDs) have large gain due to the impact ionization process at breakdown voltage [40] while phototransistors exhibit internal gain without bias dependence [41].

20.4 Integrated Optoelectronics

The first concept of optoelectronic integrated circuits was proposed in the late 1960s but practical demonstrations took decades to realize because individually optimized components when assembled together for a specific application could still outperform a monolithically integrated chip [42]. Only after common substrates, mostly of insulating type for reduced parasitic reactance associated with shorter bonding wire inductances and smaller capacitors, computer aided design for circuit, and advanced processing techniques, monolithic optoelectronic integrated circuits started performing or outperforming 'hybrid optical integrated circuits'. When the integration included elements such as optical waveguides, optical couplers, with light sources and detectors for signal integration and distribution without further processing throughout electronic circuits, this monolithic integration was referred as photonic integrated circuits (PICs).

20.4.1 Integrated Light Sources

During the 1970s and 1980s the constant progress made in optical fibre communications technology also encouraged the pursuit of semiconductor devices that could improve transmission, reception and processing of optical signals including integrated solutions. The first demonstrations of semiconductor lasers on semi-insulating substrates compatible to single mode optic fibre were soon followed by reports of integration of semiconductor lasers with simplified electronics (transistors) as driver circuit [43, 44]. However, in 1986 the monolithic integration of an electroabsorption modulator and a quantum well laser operating at 1.55 μm proved to be the successful approach to date [45]. Electroabsorption modulators are based on either the Franz-Keldysh effect in case of bulk material or the quantum confined stark effect (QCSE) for structures with quantum confinement [46]. The first electroabsorption optical modulator based on Franz-Keldysh effect was demonstrated in 1976 [47]. Electroabsorption (EA) modulators are simple, compact, and efficient semiconductor devices in managing data transmission in addition to having a structure compatible for integration with DH semiconductor lasers. Their technology has been refined in terms of processing with epitaxial regrowth, packaging, and development for commercial products in the twenty-first century.

20.4.2 Integrated Detectors

The design of an integrated detector is controlled by external system-level parameters depending on specific applications, for example, in data communications obtaining the lowest electrical noise while keeping the widest electrical bandwidth of the post electronics, fast acquisition time, and transparency to the data format. With a variety of different choices for detectors and transistor circuitry, the basic design for integration

relies on tradeoffs between the detector performance (optical sensitivity and electrical bandwidth) and the electronic processing circuitry (preamplifier, post amplifier) as well as light coupling into the device. For optical fibre communications in 1970 the receiver design aspects were addressed by Personick [48], followed by reports of integration on the same substrate for different components and photodetectors with different transistors [49–51]. Only in the early 1990s InP-based long wavelength monolithically integrated optical receivers with p-i-n detectors and heterojunction bipolar transistors excelled the performance of hybrid receivers [52, 53]. To date due to the advanced technology development for CMOS and electronics in general, the density of integration between photonics and electronics differ by several orders of magnitude. One exception is the demonstration of InP-based photonic integrated circuits for large aggregated data in optical communications applications [54]. Another is for imaging CMOS-based integrated detectors based on charge coupled devices (CCDs) as shown in Figure 20.2 (see also Chapter 13). The concept at Bell Labs started in 1969 [55], and later developed into prototypes for imagers [56]. CCDs demonstrated high sensitivity and low noise for applications in high resolution astronomy and high resolution instrumentation. Subsequently its applications popularized into scanners, digital cameras, and bar code readers.

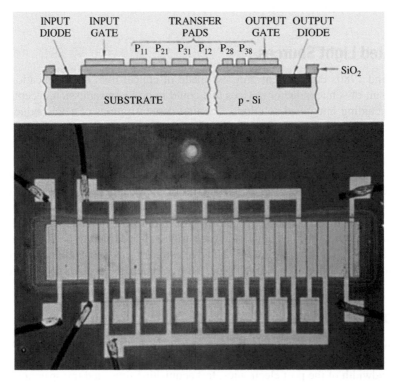

Figure 20.2 Schematic (*top*) and photograph of the first chip with charge coupled device (Reprinted with permission from George E. Smith, ''The invention and early history of the CCD,'' *J. of Applied Physics*, May 31, 2011. Copyright 2011)

20.5 Optical Interconnects

In the late 1960s after work published by Smith some studies started using optics for digital logic gates [58]. The perception was that the speed of CMOS-based circuits would be mainly limited by interconnection delays instead of gate delay with the decreasing feature sizes. With the demand for wider optical bandwidth and increase in electronic chip area, circuits and systems designers started looking for different alternatives. For the following two decades (1970–1980s) the quest for optical computing and optical digital gates continued as well as rapid development in optical fibre communication capacity and epitaxial materials [59, 60]. The concept of vertical-cavity surface-emitting laser (VCSEL) was developed in 1977 by Iga and co-workers and reported in 1979 [61]. This original class of semiconductor lasers can have very low value of threshold current, in a shorter cavity, smaller active area, and smaller volume than conventional strip lasers. The monolithic fabrication and easy device separation without the need for perfect cleaving (against edge emitting) also indicated that VCSELs could be a promising technology for coupling to parallel optical fibres or vertical integration with photodetectors in an array mode.

Both VCSELs and electroabsorption modulators in optical communications offered new options as multidimensional arrays of optical devices: VCSELs in optical data links (short distances) and modulators integrated with lasers (long distances). Especially GaAs-based VCSELs enabled parallel processing as optical interconnects offering an alternative for wider aggregated bit rate capacity than electrical interconnects without the limitations of impedance matching, isolation, and attenuation in addition to smaller form factor and lower total electrical power dissipation. In 2012, IBM announced 1 Tb/s data parallel transfer capacity using an integrated GaAs-based VCSEL chip with 24 transmitters and 90 nm CMOS technology receivers in 150 m short link [62, 63].

20.6 Closing Remarks

Optoelectronic devices have continuously evolved with new materials, technologies, process optimization, and integration for new uses. Their role in society has become ubiquitous by improving quality of life and expanding into new applications for medical, automotive and consumer electronics.

References

[1] H.J. Round, *"A Note on Carborundum"*, Electrical World, 49(6), (1907).
[2] W. Shockley, "The Theory of p-n junctions in semiconductors and p-n junction transistors", *Bell Syst. Tech. J.*, 28(3), 435 (1951).
[3] R. S. Ohl, "Light-sensitive electric device", *US Patent Appl.* 2,402, 662 (1941).
[4] J. R. Haynes, "New Radiation Resulting from Recombination of Holes and Electrons in Germanium", *Phys. Rev.*, 98(6), 1866 (1955).
[5] J. R. Haynes and W.C. Westphal, "Radiation Resulting from Recombination of Holes and Electrons in Silicon", *Phys. Rev.*, 101(6), 1676 (1956).
[6] J. Bardeen, F.J. Blatt, and L. H. Hall, "Indirect Transitions from the valence to the conduction bands", in Photoconductivity Conference, Atlantic City (1956), John Wiley & Sons, Inc.

[7] R. N. Hall, G. E. Fenner, J. D. Kingsley, *et al.*, "Coherent Light Emission from GaAs Junctions", *Phys. Rev. Lett.*, 9(9), 366 (1962).

[8] M. I. Nathan, W. P. Dumke, G. Burns, *et al.*, "Stimulated Emission of Radiation from GaAs p-n Junctions", *Appl. Phys. Lett.*, 1(3), 62 (1962).

[9] N. Holonyak and S.F. Bevacqua, "Coherent (Visible) Light Emission from Ga(As$_{1-x}$P$_x$) Junctions", *Appl. Phys. Lett.*, 1(4), 82 (1962).

[10] T.M. Quist, R.H. Rediker, R.J. Keyes, *et al.*, "Semiconductor Maser of GaAs", *Appl. Phys. Lett.*, 1(4), 91 (1962).

[11] H. Kroemer, "A proposed class of hetero-junction injection lasers", *Proc. of the IEEE*, 51(12), 1782 (1963).

[12] Z. I. Alferov, "Nobel Lecture: The double heterostructure concept and its applications in physics, electronics, and technology", *Rev. of Modern Phys.*, 73(3), 767 (2001).

[13] M. B. Panish and S. Sumsk, "Double-Heterostructure Injection Lasers with Room Temperature Thresholds as low as 2300 A/cm^2", *Appl. Phys. Lett.*, 16(8), 326 (1970).

[14] J. E. Ripper, J. C. Dyment, L. A. Dasaro, *et al.*, "Stripe-geometry double heterostructure junction lasers - mode structure and CW operation above room temperature", *Appl. Phys. Lett.* 18(4), 155 (1971).

[15] R. D. Dupuis and P. D. Dapkus, *et al.* "operation of Ga(1-x)AlxAs/GaAs double heterostructure lasers grown by metalorganic chemical vapor deposition (late paper)", *IEEE Trans. on Electron Devices*, 24(9), 1195 (1977).

[16] R. G. Waters, P. K. York, K. J. Beernink, and J. J. Coleman, "Viable strained-layer laser at $\lambda = 1100$ nm", *J. Appl. Phys.*, 67(2), 1132 (1990).

[17] H. K. Choi, C. A. Wang, D .F. Kolesar, *et al.*, "High-power, high-temperature operation of AlInGaAs-AlGaAs strained single-quantum-well diode lasers", *IEEE Photon. Technol. Lett.*, 3(10), 857 (1991).

[18] D. P. Bour, D. B. Gilbert, L. Elbaum, and M. G. Harvey, "Continuous, high-power operation of a strained InGaAs/AlGaAs quantum well laser", *Appl. Phys. Lett.*, 53(24), 2371 (1988).

[19] D.F. Welch, W. Streifer, C. F. Schaus, *et al.*, "Gain characteristics of strained quantum well lasers", *Appl. Phys. Lett.*, 56(1), 10 (1990).

[20] Y. Arakawa and H. Sakaki "Multidimensional quantum well laser and temperature dependence of its threshold current", *Appl. Phys. Lett.*, 40(11), 939 (1982).

[21] G. Park, O. B. Shchekin, D. L. Huffaker, and D.G. Deppe, "Low-threshold oxide-confined 1.3-μm quantum-dot laser", *IEEE Photon. Technol. Lett.* 12(3) 230 (2000).

[22] A. E. Zhukov, A. R. Kovsh, S. S. Mikhrin, *et al.*, "3.9 W CW power from sub-monolayer quantum dot diode laser", *Electron Lett.*, 35(21), 1845 (1999).

[23] J. Faist, F. Capasso, D. L. Sivco, *et al.*, "Quantum Cascade Laser", *Science*, 264(5158), 553 (1994).

[24] H. Rupprecht, J. M. Woodall, and G. D. Petit, "Efficient Visible Electroluminescence at 300 K from Ga$_{1-x}$Al$_x$As p-n junctions grown by liquid epitaxy phase", *Appl. Phys. Lett.*, 11(81), (1967).

[25] H.M. Manasevit, "Single-Crystal Gallium Arsenide on Insulating Substrates", *Appl. Phys. Lett.*, 12(4), 156 (1968).

[26] I. Hayashi, M. B. Panish, P. W. Foy, and S. Sumski, "Junction Lasers which operate continuously at room temperature", *Appl. Phys. Lett.*, 17(3), 109 (1970).

[27] A.Y. Cho, "Film Deposition by Molecular-Beam Techniques", *J. Vacuum Sci. and Tech.*, 8(5), S31 (1971).

[28] R. Dupuis, P. D. Dapkus, "Room-temperature operation of Ga$_{1-x}$Al$_x$As/GaAs double-heterostructure lasers grown by metalorganic chemical vapor deposition", *Appl. Phys. Lett.*, 31(7), 466 (1977).

[29] R. D. Burnham, N. Holonyak, and D. R. Scifres, "Al$_x$Ga$_{1-x}$As$_{1-y}$P$_y$/GaAs$_{1-y}$P$_y$ Heterostructure Laser and Lamp Junctions", *Appl. Phys. Lett.*, 17(10), 455 (1970).

[30] H. Amano, N. Sawaki, I. Akasaki, and Y. Toyoda, "Metalorganic vapor phase epitaxial growth of a high quality GaN film using an AlN buffer layer", *Appl. Phys. Lett.*, 48(5), 353 (1986).

[31] S. Nakamura, T. Mukai, M. Senoh and N. Iwasa, "Thermal annealing effects on p-type Mg-doped GaN films", *Jap. J. Appl. Phys.*, 31(2B), L139 (1992).

[32] S. Nakamura, M. Senoh, and T. Mukai, "Highly p-type Mg-doped GaN films grown with GaN buffer layers", *Jap. J. Appl. Phys.*, 30(10A), (1991).

to be continued   to be continued   ELECTRON DEVICES SOCIETY   to be continued   to be continued

[33] S. Nakamura, T. Mukai, and M. Senoh, "High-power GaN p-n junction blue light emitting diodes", *Jap. J. Appl. Phys.*, 30(12A), (1991).

[34] J. I. Pankove, E. A. Miller, D. Richman, and J. E. Berkeyheiser, "GaN Blue Light-Emitting Diodes", *J. of Luminescence*, 4, 63 (1971).

[35] J. N. Shive, "A new germanium photoresistance cell", *Phys. Rev.*, 76, 575 (1949).

[36] F. S. Goucher, G. L. Pearson, M. Sparks, *et al.*, "Theory and an experiment for a germanium p-n junction", *Phys. Rev.*, 42, 1267, (1947).

[37] R. L. Anderson, "Germanium-Gallium Arsenide Heterojunctions", *IBM J. Res. & Develop.*, 4(3), 283 (1960).

[38] R. H. Rediker, T. M. Quist, and B. Lax, "High Speed Heterojunction Photodiodes and beam-of-light transistors", *Proc. of the IEEE*, 51(5), 219 (1963).

[39] L. K. Anderson and B. J. McMurtry, "High speed photodetectors", *Proc. IEEE*, 54(10), 1335, (1966).

[40] H. Melchior and W. T. Lynch, "Signal and Noise Response of High Speed Germanium Avalanche Photodiodes", *IEEE Trans. Elect. Dev.*, **ED13**(12), 829 (1966).

[41] H. Beneking, G. Schul, G. Mischel, *et al.*, "High-gain wide-gap emitter $Ga_{1-x}Al_xAs$-GaAs phototransistor", *Elect. Lett.*, 12, 395, (1976).

[42] S. E. Miller, "Integrated optics: An introduction", *Bell Syst. Tech. J.*, 48, 2059, (1969).

[43] C. P. Lee, "Double heterostructure GaAs-GaAlAs injection lasers on semi-insulating substrate using carrier crowding", *Appl. Phys. Lett.*, 31, 281, (1977).

[44] T. Matsuoka, K. Takahei, Y. Noguchi, and H. Nagai, "1.5 micron region InP/GaInAsP buried heterostructure lasers on semi-insulating substrates", *Electron. Lett.*, 17, 12, (1981).

[45] T. H. Wood, C. A. Burrus, D. A. B. Miller, *et al.*, "High-speed optical modulation with GaAs/GaAlAs quantum wells in a p-i-n diode structure", *Appl. Phys. Lett.*, 44(1), 16 (1984).

[46] G.E. Stillman, C. M. Wolfe, C. O. Bozler, and J. A. Rossi, "Electroabsorption in GaAs and its application to waveguide detectors and modulators", *Appl. Phys. Lett.*, 28(9), 544 (1976).

[47] Y. Kawamura, K. Wakita, Y. Itaya, *et al.*, "Monolithic integration of InGaAs/InP DFB lasers and InGaAs/InAlAs MQW optical modulators", *Elect. Lett.*, 22(5), 242 (1986).

[48] S. D. Personick, "Receiver design for digital fiber optic communication systems I and II", *Bell Syst. Tech. J.*, 52(6), 843 (1973).

[49] C. P. Lee, S. Margalit, I. Ury, and A. Yariv, "Integration of an injection laser with a Gunn oscillator on a semi-insulating GaAs substrate", *Appl. Phys. Lett.*, 32, 806 (1978).

[50] R. F. Leheny, R.E. Nahory, M. A. Pollack, *et al.*, "Integrated $In_{0.53}Ga_{0.47}As$ P-I-N FET Photoreceiver", *Elec. Lett.*, 16(10), 353 (1980).

[51] O. Wada, "Recent progress in optoelectronic integrated circuits (OEICs)", *IEEE J. Quant. Electron.*, **QE-22** (6), 805 (1986).

[52] S. Chandrasekhar, L. M. Lunardi, A.H. Gnauck, *et al.*, "High Speed Monolithic p-i-n/HBT and HPT/HBT Photoreceivers implemented with simple phototransistor structure", *IEEE Photonics Tech. Lett.*, 5 (11) 1316 (1993).

[53] S. Chandrasekhar, M. Zirngibl, A. G. Dentai, *et al.*, "Monolithic 8-wavelength demultiplexed receiver for dense WDM Applications", *IEEE Photon. Tech. Lett.* 7(11), 1342 (1995).

[54] R. Nagarajan, M. Kato, J. Pleumeekers, *et al.* "InP photonic integrated circuits", *IEEE J. Sel. Top. Quantum Electron.* 16(5) 1113, (2010).

[55] W. S. Boyle and G. E. Smith, "Charge Coupled Semiconductor Devices", *Bell Syst. Tech. J.*, 49, 587 (1970).

[56] G. F. Amelio, W. J. Bertram, and M. F. Tompsett, "Charge Couple Devices: Design Considerations", *IEEE Trans. on Elect. Dev.*, 23(2), 183 (1976).

[57] G. E. Smith, "The invention and early history of the CCD", *J. Appl. Phys.*, 109(10), 102421 (2011).

[58] P. W. Smith, "On the physical limits of digital optical switching and logic elements", *Bell Syst. Tech. J.*, 61, 1975 (1982).

[59] D. A. B. Miller, "Rationale and challenges for optical interconnects to electronic chips", *Proc. IEEE*, 88, 728 (2000).

[60] F. B. McCormick, T. J. Cloonan, F. A. P. Tooley, *et al.*, "Six-stage digital free-space optical switching network using symmetric self-electro-optic-effect devices", *Appl. Opt.*, 32, 5153 (1993).

[61] H. Soda, K. Iga, C. Kitahara and Y. Suematsu, "GaInAsP/InP Surface Emitting Injection Lasers", *Jap. J. Appl. Phys.*, 18, 2329 (1979).

[62] L. Schares, J. A. Kash, F. E. Doany, *et al.*, "Terabus: Terabit/second-class card-level optical interconnect technologies", *IEEE J. of Selected Topics in Quantum Electronics*, 12(5), 1032 (2006).

[63] J. Proesel, C. Schow, and A. Rylyakov, "Ultra Low Power 10- to 25-Gb/s CMOS Driven VCSEL Links", *OW4I.3 OSA Optical Fiber Conference*, March 4–8, 2012, Los Angeles (2012).

to be continued   to be continued   ELECTRON DEVICES SOCIETY®   to be continued   to be continued

Chapter 21

Devices for the Post CMOS Era

Wilfried Haensch

21.1 Introduction

The slowdown of scaling intensified the search for the "next switch" [1], (see Chapters 2 and 11). The dream is, of course, to find a new switching element that can replace the conventional transistor. Preferably without any change of the existing infrastructure – new materials and fabrication methods would be tolerated. Before we will discuss some of the new device concepts there are some fundamental properties that need to be considered for these devices fit into the existing space of combinatorial logic:

1. Performance is measured by how fast the device can drive signals on the connectivity network and how fast the device drive signals through a network comprised of its own (logic gates) (Figure 21.1). For MOSFETs a measure for this was I_{on} or an effective current I_{eff} which is proportional to I_{on} [2]. Performance is also impacted by device structure-related parasitic elements in particular for the case of gate-loaded circuits, like series resistance and various capacitance components. The potential for density scaling will be determined by role of contacts and these parasitics, which are well understood in

Device Requirements?

A device is defined by an intrinsic part and the extrinsic component that connects it to the environment. With increasing integration density the immediate environment in which the device is placed can no longer be separated from the intrinsic device. The extrinsic components are collectively summarized as parasitic elements. To quantify these parasitics correctly is important for the evaluation of device performance and switching power.

Guide to State-of-the-Art Electron Devices, First Edition. Edited by Joachim N. Burghartz.
© 2013 John Wiley & Sons, Ltd. Published 2013 by John Wiley & Sons, Ltd.

to be continued   to be continued   ELECTRON DEVICES SOCIETY®   to be continued   to be continued

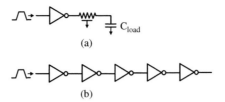

(a)

(b)

Figure 21.1 (a) wire loaded circuits for a wire lorded and a (b) gate lorded circuit. The triangle represents an inverter switching from 0 to supply voltage V_dd

the case of conventional MOSFET. When density scaling potential for a new device is considered very often the role of the contacts and their influence on device performance is usually forgotten.

2. There has to be significant distinction between the active and the quiescent state of the device. In MOSFET this is expressed by the I_{on}/I_{off} ratio. A value of several orders of magnitude is probably necessary to support high density integrated circuits. Typical values for MOSFET technology are 10^4 and 10^6 for high performance and low power space, respectively. This ratio can get smaller if supply voltage can be significantly reduced.

3. The device has to have gain. Gain is the property that the input signal can be amplified and regenerated. In a MOSFET this quantity is related to the ratio of transconductance $g_m = \delta I_d/\delta V_{gs}$ and output conductance $g_d = \delta I_d/\delta V_{ds}$. Gain, $G = g_m/g_d$, is needed for signal restoration in logic chains or signal propagation along long wire connections. For logic gain of around 10 is sufficient. Analog circuits need higher values.

4. Input and out put have to be isolated. If there is a dissipative feedback between input and output signal propagation is degraded and the noise margin of the circuits will decrease significantly.

5. Connectivity between devices has to be assured without significant dissipative losses. This is especially important if the switch is operating on a different set of state variables than the connectivity solution. In this case conversion losses between switch and wire can be significant.

Devices have to live with in the magic triangle: Power, performance and density (see Chapter 2). In recent years it became clear that power constraints will limit further performance gain of CMOS technology, as measured by the clock frequency. Clock frequency scaling has virtually stopped (see also Chapter 11). However, economy will drive continued density scaling, possibly at the cost of

Conventional Materials?

The natural device development will move from a doping controlled device towards the multi-gated fully depleted devices to improve short channel effects at shorter gate length. Performance is addressed by high mobility channel materials. Conventional materials are channel materials that are established in the semiconductor industry. They are Si, Ge and some of the III/V compounds. The properties of these materials are understood (see Chapter 6).

device performance in the future. The key to density scaling is not measured by the feature size of the device but has to include also the contacts and wires that connect the devices. In Si technology the figure of merit is the contacted gate pitch and the first level wire pitch. The first is the sum of gate length, contact length and other process related dimensions and the second the sum of line width of the wire and space between them.

Classical scaling rules [3, 4], for the device, rely on continuous reductions of the dielectric thickness, junction depth and increased channel doping. Even today there is no solution in sight that can push conventional doping controlled devices significantly beyond the 20 nm gate-length mark. Fully depleted devices like, FinFETs and extremely thin SOI, will present an opportunity to extend gate length scaling without stressing dielectric or junction scaling [5]. It is expected that these new devices will be widely introduced for the high end computing space as early as the 14-nm node. Low power technologies might adopt these devices earlier. Alternate device concepts or channel materials are currently investigated. They are neither demonstrated at relevant dimensions nor are they proven to deliver what they promise at this point in time.

21.2 Devices for the 8-nm Node with Conventional Materials

It is believed that conventional CMOS technology could be operated at around 0.5 V without loss of functionality and design robustness [6]. It is, however, clear that this reduction of supply voltage will cause a significant reduction clock frequency f. On the chip level we find that clock frequency f and active power P_{active} are related according to:

$$f = \alpha \left(V - V_0 \right) \tag{21.1}$$

$$P_{active} = \alpha \, C_{eff} V^2 \left(V - V_0 \right) + I_{leak} V \tag{21.2}$$

where C_{eff} is the total effective load capacitance of a chip, V is the operating voltage V_0 is the voltage at which frequency approaches zero ($\sim$0.25 V for modern technologies) and α is a constant that depends on the circuit and technology, and I_{leak} is the total leakage current of the chip. It should be noted that P_{active} is the power for an operating chip that has always active and quiescent components. Equations 21.1 and 21.2 give a rough idea of the cost in performance if power needs to be reduced. At the device level performance scaling can be captured by:

$$I_{on} = w C_{inv} v_{sat} \left(V - V_{th} - \frac{1}{2} V_{sat} \right) \tag{21.3}$$

$$I_{off} = w I_0 10^{-\frac{V_{th}}{S}} \tag{21.4}$$

where w is the average device width, C_{inv} the inversion capacitance, v_{sat} the saturation velocity of the device, V_{th} the threshold voltage, S the sub threshold slope, I_0 the target current at which the threshold voltage is defined, and V_{sat} is the pinch-off voltage, which is the drain voltage at which the carrier supply to the drain is saturated. The threshold voltage V_{th} and V_0 are closely correlated. Effective chip capacitance C_{eff} is related to the device width that will also determine wire length and pitch, and is therefore directly proportional to density of integration. C_{eff} will have a front end or device related, and a back-end or wiring related contribution. Since wire delay limitations can be mitigated by the insertion of repeaters, both parts are

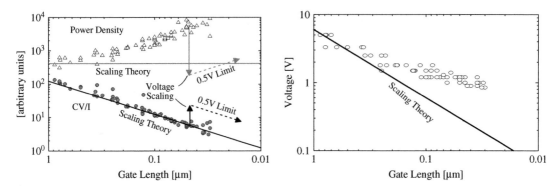

Figure 21.2   Voltage and performance scaling (© 2009 IEEE. Reprinted with permission from L. Chang, *et al.*, "Practical Strategies for Power-Efficient Computing Technologies," submitted to Proc. IEEE, 2009)

interrelated. It is notable that from the device perspective we have the material parameters C_{inv} ($\sim$ dielectric scaling) and v_{sat} ($\sim$ transport enhancement) and the device architecture parameters V_{th} and S to work with. To meet performance targets the industry has seen a slowdown in voltage scaling as shown in Figure 21.2, at the cost of power density [7]. This slow down in voltage scaling is primarily related to the fact that the sub-threshold slope is limited to 60 mV/dec at room temperature. To maintain a tolerable device leakage, according to Equation 21.4 the threshold voltage can no longer be reduced, which, under the constraint of maximal drive current, will according to Equation 21.3 neither allow the supply voltage to be reduced.

21.2.1 Si Channel Devices

To increase performance dielectric scaling was improved with the introduction of hi-k materials to increase charge and strain engineering was employed to boost carrier velocity [5]. With these a higher drive current was achieved at constant V_{th} and supply voltage, according to Equations 21.3 and 21.4. It is however noticeable that gate length scaling has slowed down due to the fact that the basic device architecture did not change. In Figure 21.3 we show the possible progression of device architecture for the coming technology nodes. The transition from doping controlled devices to fully depleted devices is expected to take place earliest in the 14 nm and at the latest in the 11 nm node technology for high performance technologies. In the near future doping controlled devices will be replaced by the fully depleted body devices. In contrast to the doping controlled devices where threshold voltage and sub-threshold slope are controlled by gate dielectric thickness, junction scaling, and channel doping, in the fully depleted devices they are controlled by the body thickness [8]. Typically the body thickness will determine the minimum supported gate length L_{min} at which electrostatic integrity is guaranteed. For the ETSOI device that is built on a very thin Si film of thickness T_{si} on insulating buried oxide (BOX) we have $L_{min} \sim 4\ T_{si}$ and for the FinFET, which is a 3D double gated device structure $L_{min} \sim 2\ D_{fin}$, with D_{fin} the thickness of the Fin structure. The FinFET has a superior scaling behavior compared to the ETSOI device because of its double gate nature. However, processing issues related to the 3D nature of the structure will make it more difficult to build. The FinFET would be able to support gate length of about 10 nm, which is expected at the 8 nm

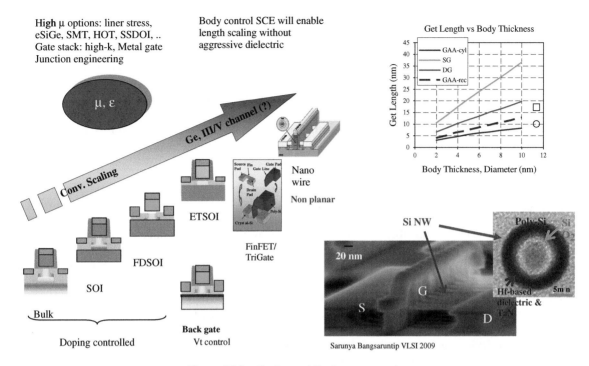

Figure 21.3 Device architecture progression

technology node, at a Fin width of 5 nm. Reducing the Fin width further would result in a dramatic mobility loss and unacceptable threshold variation due to Fin width tolerances in the manufacturing process [9, 10]. An evolution of the ETSOI device architecture on thick BOX ($\sim$ 120 nm) towards a device on thin BOX ($\sim L_g$) will enable gate length scaling at a relaxed body thickness T_{si}. However, in the extreme case of a very thin BOX at the critical body thickness of $T_{si} = 5$ nm it cannot scale better than a FinFET. Therefore both FinFETs and ETSOI devices will not be able to be scaled below 10 nm gate length without significant performance loss.

21.2.2 High-Mobility Channel Materials

In the transition for 14-nm to 8-nm technology node high mobility channel materials like Ge for PFETs and III/V compounds for NFETs might emerge to boost device performance, similar to strain engineering in Si. However, the existing device architectures will prevent scaling beyond 10 nm gate length or limited to even larger gate length. In particular for III/V materials quantum confinement effects are expected to dominate much earlier than in Si or Ge due to their small effective masses [11]. Their light effective mass will make III/V material channels more sensitive to direct S/D tunneling which will limit gate length scaling. A possible advantage, due to the high mobility of the III/V compounds, is expected at low voltage

operation in a range of about 0.5 V. Taking projections from existing devices and benchmarking these to the best Si devices at the 32 nm node at 0.5 V operation, it is reasonable to assume that a performance gain of a factor two seems possible. Besides the unknowns in the series resistance scaling, the dielectric/channel interface and the junction formation are still subject of intensive research and it is not clear at present how these devices will perform in a properly scaled environment. A fully fledged III/V technology is at this point inconceivable because of the lack of available large wafer materials and the lack of a good PFET solution. A possible PFET solution might be a device that has a Ge channel. However, no Ge device has been produced as of today that could compete with a properly scaled Si FET with strain engineering. Like the III/V materials device leakage is aggravated by the small band gap and would require very thin body devices to reduce the device leakage.

21.2.3 Wire Device Architecture

The best possible scaled device that can be built with conventional semiconducting materials has wire geometry, as indicated in the evolution sequence in Figure 21.3. The ultimate wire geometry would be a wire with circular cross section. In principle this device would allow to push gate length beyond 10 nm. For wires a scaling behavior of $L_{min} \sim 1.2 \ldots 1.5 * d_{wire}$ is expected, depending on the cross section control. For the 5-nm technology node a gate length of 8 nm is expected which would require a wire diameter of about 5–7 nm. III/V materials will not be feasible at this wire dimension because of the large quantum confinement effects due to their low effective mass. Variability impact on device parameters would be unacceptable. Si devices and circuits have been demonstrated in that diameter range [12, 13]; however, it is not clear if they can deliver the needed performance at the desired integration densities. The net is that with the existing known materials and architectures there is no solution to build a high performance device beyond the 8-nm technology node.

21.2.4 Tunnel FET

The device solutions we have discussed so far have a common operating principle. A potential barrier (measured with reference to the source Fermi potential) that separates the source from the drain is modulated by the gate voltage. Carriers form the source are injected into the channel that have an energy higher that the potential barrier. This means that a change of the potential barrier will sample the Boltzmann tail of the Fermi distribution of the carriers in the source. This is the ultimate reason why the sub thresholds slope is limited by 60 mV/dec at room temperature.

To beat this restriction a new switching mechanism needs to be employed. Band-to-band tunneling offers such a mechanism. The probability of carriers tunneling from the valence band to the conduction band of a semiconductor depends on the alignment of the band edges to provide states to tunnel from and to provide states to tunnel into. Furthermore the occupation of these states needs to be such that the former are occupied and the latter are empty. In a tunnel FET the gate voltage will modulate the alignments of the band edges as shown in Figure 21.4(b). In contrast to the conventional FET shown in Figure 21.4(a), the tunnel FET will not sample the Boltzmann tail of the distribution function it rather will sharply turn on when the band edges are aligned properly and the tunnel process can kick in. Characteristic for the tunnel FET is a continuously increasing slope with increasing current as indicated in Figure 21.4(b). The measure

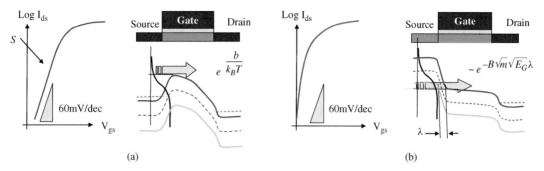

Figure 21.4 (a) MOSFET: carriers are injected over a barrier into the channel (b) Tunnel FET: carriers are injected into the channel by band-to-band tunneling at a proper alignment of conduction and valence band edges

for the quality of the tunnel FET is that the device turns on at a rate smaller than 60 mV/dec over a wide gate voltage range between I_{off} and I_{on}, The gate voltage range connecting these two currents would set the supply voltage, which potentially could be significantly smaller than that for a conventional MOSFET at the same I_{off} and I_{on} specifications.

So far, tunnel FETs are not able to compete with Si devices in the high performance space. For that space I_{on}/I_{off} ratios are approximately 10^4 with I_{on} at about 1.5 mA/μm at a 1 V supply voltage. The low power space requires an I_{on}/I_{off} ratio of 10^6 with an I_{on} current spec of several 100 μA/μm. For ultralow power applications with very relaxed performance requirements tunnel FETs could play a role. In Table 21.1 we list the primary parameter space for the optimization of the tunnel FET: bandgap E_g, effective mass m, and the sharpness λ of the potential profile. Preferred device materials are III/V compounds as they provide the opportunity for bandgap engineering and offer a light effective mass. The sharpness of the electric field

Table 21.1 Trade-off for tunnel FET design parameters

	Need	Con	Solution
E$_G$	Small band gap for high Ion	Large band gap for low Ioff	Hetero Structures with appropriate band off sets
m	Small tunnel mass for high Ion (band-to-band tunneling)	Large tunnel mass for low Ioff (S/D tunneling)	Hetero structures with large channel mass
λ	Small tunnel distance for high Ion		Planar → multigate → Wire, CNT

profile is directly related to the electrostatic integrity of the device architecture. Like the scaling of the conventional FETs, for the tunnel FET the wire geometry is also the device architecture with the best gate control. The literature on tunnel FETs is full of examples and new devices are added continuously [14]. So far no one has produced a device that simultaneously can provide a low off current and a high on current with in a supply voltage window smaller than 1V.

Before we leave the tunnel FET we need a short discussion on assessing the performance of a tunnel FET. Since it controls the flow of charge the CV/I metric introduced for conventional MOSFET is appropriate. When the inverter switches gate (input) and drain (output) move in the opposite way. The capacitance that couples the gate to the drain is counted double in this case, which is also known as the *Miller effect*. For a conventional FET in saturation this capacitance is the gate-to-drain overlap capacitance. In a tunnel FET the drain is not decoupled from the channel in saturation and therefore the gate capacitance has to be added to the gate-to-drain coupling. This will add a significant capacitive load that will degrade performance unless it is compensated by a higher drive current.

21.3 New Channel Materials and Devices

Device and material solutions that were discussed in Chapter 6 would fit into the existing microelectronic ecosystem without major disruption. Circuits will work the same way as they do today and in general business can continue as usual. At the lowest level, in the fab, however, there will be significant changes driven by the new materials and modified process flows. The question is: is there a solution available that will fit into the existing infrastructure, maybe with small modifications, and allow continuous growth of performance? The ideal candidate would be a MOSFET-like device that could work at sub 500 mV operating voltage and could provide continuous performance growth with in today's power budget. This solution should be available for implementation into a technology in the 2020 timeframe when the device solutions discussed so far run out of steam. Among all explorative devices the carbon nanotube (CNT) transistor would fit the bill. There are other options under investigation that look promising [15] (see Section 21.2). They would, however, not be consistent with the existing infrastructure and would not be available before the middle of the next decade.

21.3.1 Carbon Nanotubes (CNTs)

CNTs combine good transport properties and excellent electrostatic scaling. Recently scalability of a CNT device below the 10 nm gate length was shown with good drive current and no degradation of short channel behavior over a wide range of gate lengths [16, 17]. In addition it would allow an operation voltage significantly lower than 500 mV without performance loss compared to a Si device operated at the same gate length. However, many issues need to be solved to get this technology ready for the 5 nm node time frame. In addition to the usual device scaling issues CNT material solutions are critical. Naturally grown CNTs will have a population that is 70% semiconducting and 30% metallic. Growth conditions can be found to shift this ratio somewhat but not significantly. For the 5 nm node a gate length in the sub 10 nm range is expected with a contacted gate pitch in the range of 25–30 nm. A logic device will have 5–6 CNTs at 6–8 nm apart. From these numbers it is easy to see that requirements for CNT placement and purity will

be extremely aggressive. Ordered growth has been shown on slightly miscut quartz wafers [18]. However, neither the uniformity nor the purity is sufficient for the desired end goal of ~nm placement accuracy. The second method is to grow the CNTs in solution, apply a purification process, and then selectively place the tubes at pre-determined positions. In this approach purity and placement are to a high degree decoupled and an independent optimization of both processes is possible. At this point in time it is not clear which method will provide a placement accuracy and purity that is required for high density integrated circuits [19]. From statistical considerations one finds that for a placement accuracy of several nm and a purity in access of 10^{-5} has to be reached to yield 1 000 000 000 transistors. This statistical behavior has to be incorporated in the circuit design to provide operating components.

The problems associated with making CNTs a viable option for post 8-nm node technology are well established and are on the way to be solved. The device concept is clear and to a high degree it will be consistent with the existing Si CMOS ecosystem.

New Materials and Devices

Explorative research looks at materials and device options that can replace the conventional device at the 5-nm node. The choice of existing options ranges for seamless fits into the CMOS ecosystems to disruptive solutions that will impact the whole CMOS infrastructure. The device that will mature in the 2020s will have a chance to intercept the 5-nm node.

21.3.2 Other Devices and Materials

Spin manipulation is considered a promising alternative to a charge based switch. In the most general case a spin FET can come in three realizations: Voltage controlled, current controlled, and all spin logic [20] In a voltage controlled spin FET carriers are injected into the channel that are filtered with respect to a preferred spin direction, they pass through the gate controlled channel and are than detected with a similar spin filter at the drain side. The action of the gate will be to rotate the spin direction such that for the off-state the resulting spin orientation is opposite to the preferred direction of the filter and the on state is aligned with the preferred direction of the filter. The physical effect that is utilized in these devices is the Rash-bar effect; it allows manipulation of the spin orientation in an electric field through a strong spin-orbit-interaction [21]. It is to be expected that spin filtering at source and drain will create significant losses, currently efficiencies of 20–40% can be achieved at room temperature [22]. The second class of spin FET is a current controlled device. A current will drive domain walls from source to drain or vice versa to set the magnetic polarization of the channel. A spin polarized gate current will than probe the resistance of a magnetic tunnel junction connected to the channel [23]. To be useful for high density application the drive current to move the domain walls needs to be every low (<1 μA) and the

motion of the domain walls needs to be fast to meet the performance requirement. Like the voltage controlled device this device will also suffer from conversion losses between the wire and the device. The third option of spin FET would be an all-spin logic [24, 25]: switching and connectivity is accomplished in spin space. The logic state of the device is altered by a charge current that changes the magnetization of an output magnet by spin torque and signal propagation is accomplished by coupling of adjacent magnets. Conversion losses are eliminated. Finally manipulation of spin waves would avoid charge transport all to gather. There are several all spin proposals under investigation. It is not clear, however, if these can be used to build complex circuits: Speed, power, signal restoration and clocking schemes are issues that need to be addressed.

Recently a device was suggested in which the conventional gate dielectric is replaced with a combination of conventional dielectric and a ferroelectric material [26]. Ferrolectricity leads to a hysteretic behavior which can be under special circumstances lead to a negative capacitance ($\delta Q / \delta V < 0$). In a simple model the charge stored in a ferroelectric dielectric material is a function of the charge itself.

$$Q = C_0 \left(V + \alpha Q \right) \tag{21.5}$$

Here C_0 and α are material parameters. If the gate dielectric is a combination of a conventional dielectric with capacitance C_{ins} and a depletion capacitance C_s we will have for the sub-threshold slope:

$$S = 2.3 k_B T \left(1 + \frac{C_s}{C_{ins}} \left(1 + \frac{C_{ins}}{C_0} \left(1 - a C_0 \right) \right) \right) \tag{21.6}$$

A proper choice of material parameters would allow making the slope significant smaller than 60 mV/dec. This simplified picture has several caveats: First it assumes that the semiconductor capacitance is constant, which it is not. It changes, in particular, very rapidly when the device transitions from depletion into inversion. Equation 21.5 oversimplifies the situation in the ferroelectric material and a more detailed description is needed. At the end it needs to be shown if the available space of material parameters will permit to build a device that can take advantage of the steeper slope and at the same time provide enough drive current at the on state. A material combination that satisfies both requirements is still elusive.

The last group of devices we will discuss will use graphene as the channel material. Graphene is a perfectly two-dimensional array of carbon atoms arranged in a hexagonal honeycomb lattice. It has transport properties similar to CNTs; however, it is missing a band gap. The lack of a band gap has consequences. FET like devices do not saturation since pinch off does not occur and the I_{on}/I_{off} ratio at room temperature is usually less than 10. Its usefulness as FET is limited to RF application where its good transport properties could give an advantage. A band gap can be opened in very narrow stripes of several nm width [27]. The band gap approximately scales approximately as 2.4 eVnm/width. Very narrow stripes of graphene most likely will suffer a degradation of the transport properties from the impact of none passivated edges. Another possibility to open a band gap exists in bilayer graphene subjected to a transversal field. Although interesting from the physics point of view it is not clear how this can be used in a device configuration. To open up the band gap and to maintain good transport properties in graphene is an active field of research. The manipulation of the bandgap with the width of the graphene stripe is of particular interest for the design of tunnel FETs, a device that we discussed earlier [28].

The two-dimensional honeycomb crystal structure of graphene is not a Bravais lattice. It is built up from two triangular lattices with base atoms A and B. As a consequence the wave function in graphene can be described with two components that take the form of a bi-spinor. This can be interpreted such that the electron carries, in addition to its real spin, a pseudo spin 1/2. The pseudo spin character of the graphene

wave function opens up new switching mechanisms that are beyond what is possible with the classical known transistors.

The BisFET (**Bi** Layer Pseudo **S**pin **F**ield **E**ffect **T**ransistor) is a device that operates on a new collective coherent state between electrons and holes separated by a thin dielectric layer. This phenomenon is observed at low temperatures in GaAs/InGaAs bilayer systems [29]. It manifests itself as strong reduction of the tunnel resistance across the thin insulating layer at very low bias conditions, 50 mV or less. It is speculated that such a collective phenomenon should be observed between two graphene layers that are separated by a thin insulator of the order 1 nm, even at room temperature (Figure 21.5). Spontaneous coherences, or the formation of the new collective state, is most likely if hole and electron density in the upper and lower layer, respectively, are the same. This situation can be triggered by applying the appropriate gate bias at the device. A small voltage between source drain will than draw a current that is only limited by the source drain junction resistance. High current levels will destroy the n and p balance and the current will cease. Therefore the IV characteristic of this device will be significantly different than a conventional FET. While the latter will show current saturation at increased drain voltage in the BiSFET the drain current will go to zero (Figure 21.5) Therefore it requires a different circuit solution than conventional MOSFET. However, the operating voltage of the BisFET will be significantly smaller than that of a conventional FET. It is expected that this device can work at supply voltages of only several 10 mV at a clock frequency of 100 GHz [29]. At this point in time it is not clear how such a clock rate can be generated without impeding the power budget of the system. The BisFET can follow the clock – it is not clear, however, if the same device can be used to generate the clock signal as well [30]. Practical concerns about the operability of this device are related to the fact that so far theoretical predictions of the possible existence of the new collective state require certain alignment of the top and bottom graphene layer (either A-A or A-B stacking) [31], little is known what will happen if these alignment conditions are disturbed. In addition it seems that any surface charges in adjacent dielectrics might destroy the delicate balance between electrons and holes that is needed to achieve the spontaneous coherence. Since graphene has an inert surface the deposition of reliable continuous and extremely thin dielectric materials with thickness <3 nm is difficult to accomplish.

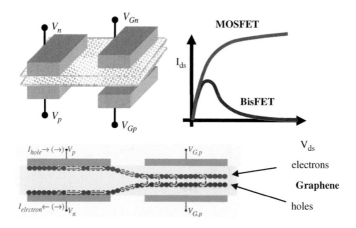

Figure 21.5   BisFET structure and IV

One consequence of the pseudo spin character of the wave function in graphene is that the dynamic of electrons in the presentence of potential barrier or potential wells is similar to that of photons in refractive media, however with a negative index of refraction. It can be shown that potential barriers in graphene can have collimating properties like optical lenses or can act like mirrors [32]. In the pn junction device the optical equivalence of a mirror is used and the barriers are created by pn junctions [33, 34]. The pn junctions are created by electro static doping by applying a different gate voltage in adjacent parts of the device. The transfer rate through this barrier turns out to be strongly dependent of the incident angle with respect to the normal of the surface that defines the barrier. Proper choice of these geometries will allow a modulation of the transfer ratio and provide a switching element. In it is not clear at the moment if structures can be build that fulfill all requirements to make this device work. In Figure 21.6 we show a technology consistent layout of this device assuming a generic ground rule feature λ with its corresponding overlay tolerance. All needed contacts are included. We also assume that the separations of the back gates can be accomplished with a spacer technique. To make the device work as proposed it would need a precise definition of the geometry to avoid "leakage" (Figure 21.6), since at normal incidence the transfer coefficient is at its maximum. For density reasons the separation of the back gates needs to be minimized. This will directly impact the scaling potential of this device because the transfer coefficient of these barriers will change with the sharpness of the potential profile. This sharpness is measured in units of the Fermi vector k_F. If the junction is "smooth", which means that its width w is $k_F w > 1$ the junction will only be transmissive for normal incident waves. In the case of a "sharp" profile the junction will be transmissive for all incident directions and junction modulation will have no significant effect. It has to be shown if these contradictive requirements can be optimized for a competitive device design.

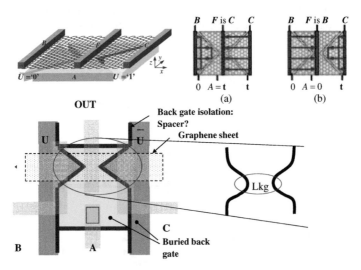

Figure 21.6 Graphene pn junction device simple logic gate: (a) device operation (b) realistic device layout and patterning problem

21.4 Closing Remarks

Conventional device solutions will allow scaling to the 8-nm node with a gate length requirement somewhat larger than 10 nm and a contact length of the same order. This will result in a contacted gate pitch of about 40 nm. The structure build in these dimensions will depend on available patterning solution and material innovation is needed to minimize the impact of parasitic effects that will degrade the intrinsic device performance. In particular the contacts will become an integrated part of the device. The device will be a fully depleted multi-gated device either in Fin or wire configuration. To meet the performance goals at a power constraint environment most likely the supply voltage will be significantly reduced at the 8-nm node timeframe at the cost of single thread performance. High mobility channel materials will be used, with a conventional FET architecture configuration, where high performance is needed. It is not clear at this moment, however, if these devices can be accommodated in the density trend or if their implementation will come at the cost of density.

The situation becomes less clear beyond the 8-nm node. For very low power and speed uncritical applications the tunnel FET might play a role in a transition phase. A predicted gate length in the sub 10 nm regime with in a contacted gate pitch of around 35 nm is the target dimension for this space that will go into manufacturing in the early 2020s. Many device ideas are considered for this time frame. There is however no candidate in place that is a sure bet. Looking at the available solutions the question has to be asked which candidate will be ready in 10 years from now for prime time? CNTs look promising since they would not change the general infrastructure of the industry and have already demonstrated a scalability potential into the sub 10-nm gate length regime. However, many of the material properties remain unsolved. Other devices that are more or less in the laboratory state have to demonstrate their compatibility with the general circuit requirements and show that they can provide an advantage in power, speed and density. In particular the spin based devices fall into this category. The impact of scaled contacts and the compatibility with existing wiring solutions is a key question that might degrade intrinsic device performance. Even further out concepts, that at present are only available as "soft copy" will most like not be ready by the 2020s time frame and have to be pushed into the next decade.

References

[1] Jeffrey Welser, Stu Wolf, Phaedon Avouris, and Tom Theis, "Applications: Nanoelectronics and Nanomagnetics", in *Nanotechnology Long-term Impacts and Research Directions*: *2000–2020*, Springer, p. 261 (2010).

[2] M. H. Na, E. J. Nowak, W. Haensch, and J. Cai, "The Effective Drive Current in CMOS Inverters", *Techn. Dig. IEEE International Electron Devices Meeting (IEDM)*, pp. 121–124 (2002).

[3] R. H. Dennard, F. H. Gaensslen, H.-N. Yu, V. L. Rideout, E. Bassous, and A. LeBlanc, "Design of Ion-ImplantedMOSFETs with Very Small Physical Dimensions", *IEEE J. Solid-State Circuits*, SC-9, 256–268 (1974).

[4] G. Baccarani, M. R. Wordeman, and R. H. Dennard, "Generalized Scaling Theory and Its Application to a $1/4$ Micrometer MOSFET Design", *IEEE Trans. Electron Devices*, ED-31, 452 (1984).

[5] W. Haensch, E. Nowak, R. H. Dennard, *et al.*, "Silicon Scaling and Beyond", *IBM J. Res. Dev.*, vol. 50, pp. 339–361, July (2006).

[6] L. Chang, David J. Frank, Robert K. Montoye, *et al.*, "Practical Strategies for Power-Efficient Computing Technologies", *Proc. IEEE*, vol. 98, no. 2, pp. 215–236 (2010).

[7]   E. J. Nowak, "Maintaining the benefits of CMOS scaling when scaling bogs down", *IBM J. Res. Dev.*, vol. 46, pp. 169–180, Mar/May (2002).

[8]   D. J. Frank, R. H. Dennard, E. Nowak, P. M. Solomon, Y. Taur, and H.-S. P. Wong, "Device Scaling Limits of Si MOSFETs and Their Application Dependencies", *Proc. IEEE* 89, 259–288 (2001).

[9]   D. Esseni, M. Mastrapasqua, G. K. Celler, C. Fiegna, L. Selmi, and E. Sangiorgi, "Low field electron and hole mobility of SOI transistors fabricated on ultrathin silicon films for deep submicrometer technology application", *IEEE Transactions on Electron Devices*, vol. 48 (no. 12) pp. 2842–2850, Dec. (2001).

[10] J. B. Chang, M. Guillorn, P. M. Solomon, C.-H. Lin, S. U. Engelmann, A. Pyzyna, J. A. Ott, and W. E. Haensch, "Scaling of SOI FinFETs down to Fin Width of 4 nm for the 10 nm technology node", *Dig. Techn. P. Symposium on VLSI Technology*, pp. 12 (2011).

[11] NSM Archive, *Physical Properties of Semiconductors*, Available at http://www.ioffe.rssi.ru/SVA/NSM/Semicond/ (accessed October 10, 2012) (n.d.).

[12] S. Bangsaruntip, A. Majumdar, G. M. Cohen, *et al.*, "Gate-all-around silicon nanowire 25-stage CMOS ring oscillators with diameter down to 3 nm", *Dig. Techn. P. Symposium on VLSI Technology, Honolulu, June 15–17*, pp. 21–22 (2010).

[13] H. Yan, H. S. Choe, S. Nam, *et al.*, "Programmable nanowire circuits for nanoprocessors", *Nature*, vol. 470 (7333), pp. 240–244, February 10 (2011).

[14] Adrian M. Ionescu and Heike Riel, "Tunnel field-effect transistors as energy-efficient electronic switches", *Nature*, vol. 479, pp. 329–337 (2011).

[15] Kerry Bernstein, Ralph K. Cavin, Wolfgang Porod, *et al.*, "Device and Architecture Outlook for Beyond CMOS Switches", *Proceedings of the IEEE*, vol. 98 (no. 12), pp. 2169–2184, (2010).

[16] A. D. Franklin and Z. Chen, "Length scaling of carbon nanotube transistors", *Nature Nanotechnology*, vol. 5, pp. 858–862, Nov. (2010).

[17] Aaron D. Franklin, Shu-Jen Han, George S. Tulevski, Mathieu Luisier, Chris M. Breslin, Lynne Gignac, Mark S. Lundstrom, and Wilfried Haensch, "Sub-10 nm Carbon Nanotube Transistor", *Techn. Dig. IEEE International Electron devices Meeting (IEDM)*, pp. 23.7.1–23.7.3 (2011).

[18] S. W. Hong, T. Banks and J. A. Rogers, "Improved Density in Aligned Arrays of Single-Walled Carbon Nanotubes by Sequential Chemical Vapor Deposition on Quartz", *Advanced Materials*, vol. 22, pp. 1826–1830 (2010).

[19] Jie Zhang, Nishant Patil, Arash Hazeghi, and Subhasish Mitra, "Carbon Nanotube Circuits in the Presence of Carbon Nanotube Density Variations", *Proceedings of the 46th Annual Design Automation Conference – DAC 2009* (2009).

[20] I. Zutic, J. Fabian, and S. Das Sarma, "Spintronics: Fundamentals and applications", *Rev. Mod. Phys.*, 76, 323 (2004).

[21] S. Datta and B. Das, "Electronic analog of the electro optic modulator", *Applied Physics Letters*, vol. 56, pp. 665–667 (1990).

[22] G. Schmidt, "Concepts for spin injection into semiconductors – a review", *J. Phys. D: Appl. Phys.*, vol. 38, pp. 107–122 (2005).

[23] J. A. Currivan, Y. Jang, M. D. Mascaro, M. A. Baldo, and C. A. Ross, "Low Energy Magnetic Domain Wall Logic in Short, Narrow, Ferromagnetic Wires", *IEEE Mag. Lett.*, vol. 3, article doi: 3000104 (2012).

[24] B. Behin-Aein, A. Sarkar, S. Srinivasan and S. Datta, "Switching energy-delay of all spin logic devices", *Appl. Phys. Lett.*, vol. 98, p.123510, (2011).

[25] A. Khitun, M. Bao, J.-Y. Lee, K. L. Wang, D. W. Lee, S. X. Wang, and I. V. Roshchin, "Inductively Coupled Circuits with Spin Wave Bus for Information Processing", *J. of Nanoelectronics and Optoelectronics*, vol. 3, p. 24 (2008).

[26] Sayeef Salahuddin and Supriyo Datta, "Use of Negative Capacitance to Provide Voltage Amplification for Low Power Nanoscale Devices", *Nano Lett.*, vol. 8 (no. 2) pp. 405–410 (2008).

[27] Frank Schwierz, "Graphene Transistors", *Nature Nanotechnology*, vol. 5, pp. 487–496 (2010).

[28] Alan C. Seabaugh and Qin Zhang, "Low-Voltage Tunnel Transistors for Beyond CMOS Logic", *Proceedings of the IEEE*, vol. 98 (no. 12), pp. 2095–2110 (2010).

[29] S. K. Banerjee, L. F. Register, E. Tutuc, *et al.*, "Bilayer PseudoSpin Field-Effect Transistor (BiSFET): A Proposed New Logic Device", *IEEE Electron Device Letters*, vol. 30, pp. 158–160 (2009).

[30] L. Tiemann, W. Dietsche, M. Hauser, and K. von Klitzing, "Critical tunneling currents in the regime of bilayer excitons", *New J. Phys.*, vol. 10, pp. 045–018 (2008).

[31] D. Basu, L. F. Register, D. Reddy, A. H. MacDonald, and S. K. Banerjee, "Tight-binding study of electron–hole pair condensation in graphene bilayers: Gate control and system-parameter dependence", *Phys. Rev. B*, vol. 82, p. 075409 (2010).

[32] P. E. Allain and J. N. Fuchs, "Klein tunneling in graphene: optics with massless electrons", *Eur. Phys. J. B*, vol. 83, pp. 301–317 (2011).

[33] T. Low and J. Appenzeller, "Electronic transport properties of a tilted graphene p-n junction", *Phys. Rev. B*, vol. 80, p. 155406 (2009).

[34] Sansiri Tanachutiwat, Ji Ung Lee, Wei Wang and C. Y Sung, "Reconfigurable Multi-Function Logic Based on Graphene P-N Junctions", *Proc. DAC'10*, June 13–18, 2010, Anaheim, California, USA (2010).

to be continued to be continued ELECTRON DEVICES SOCIETY® to be continued to be continued

Index

Guide to State-of-the-Art Electron Devices, First Edition. Edited by Joachim N. Burghartz.
© 2013 John Wiley & Sons, Ltd. Published 2013 by John Wiley & Sons, Ltd.

Printed and bound by CPI Group (UK) Ltd, Croydon, CR0 4YY

27/10/2024

14580215-0001